Y0-ARX-157

MODAL LOGIC: AN INTRODUCTION

MODAL LOGIC

AN INTRODUCTION

BRIAN F. CHELLAS

CAMBRIDGE UNIVERSITY PRESS
CAMBRIDGE
LONDON NEW YORK NEW ROCHELLE
MELBOURNE SYDNEY

To the memory of my mother

Published by the Press Syndicate of the University of Cambridge
The Pitt Building, Trumpington Street, Cambridge CB2 1RP
32 East 57th Street, New York, NY 10022, USA
296 Beaconsfield Parade, Middle Park, Melbourne 3206, Australia

First published 1980

Printed in the United States of America
Typeset by the University Press, Cambridge
Printed and bound by Halliday Lithograph Corporation
West Hanover, Mass.

Library of Congress Cataloguing in Publication Data

Chellas, Brian F.
Modal logic: an introduction

Bibliography: p.
Includes index.
1. Modality (Logic) I. Title.
BC199.M6C47 160 76-47197
ISBN 0 521 22476 4 hard covers
ISBN 0 521 29515 7 paperback

CONTENTS

PREFACE

This book is an introductory text in modal logic, the logic of necessity and possibility. It is intended for readers with the equivalent of a first course in formal logic, and it is designed to be used as a basic text in courses at the advanced undergraduate or beginning graduate level. The material in the book can easily be covered in a full-year course; with selectivity most of the material can be covered in a single term.

There are three parts to the book. Part I consists of two chapters, meant to introduce the reader to the subject of modal logic and to furnish a sufficient background for the parts that follow. Chapter 1 is a relatively informal examination of *S5*, one of the best-known systems of modal logic. Chapter 2 – 'Logical preliminaries' – contains almost everything needed for an understanding of the rest of the book. Some readers may prefer to go quickly through this chapter and then reread as necessary sections required in the context of succeeding chapters.

Part II comprises four chapters on standard models and normal systems of modal logic. The models, sometimes called 'Kripke models', are explained in chapter 3. In chapter 4 normal systems are presented from an axiomatic standpoint. Chapter 5 contains theorems on completeness and decidability, which bring together the model-theoretic and deductive-theoretic treatments of the preceding chapters. As an illustration of normal systems chapter 6 offers a discussion of deontic logic, the logic of obligation.

Part III is patterned like its predecessor, but here the topics are minimal models and classical modal logics. Thus chapter 7 is about the models (also known as 'neighborhood' or 'Scott-Montague' models), chapter 8 is an axiomatic account of the logics, and chapter 9 deals with completeness and decidability. Chapter 10 presents conditionality and (again) obligation by way of example.

An important feature of the book is the exercises that follow the sections of the chapters. These have been constructed both to consolidate understanding of the preceding material and to anticipate subsequent developments. They are an integral part of the text, and I have high hopes that the reader will attempt them as they appear.

I have appended to the text a short bibliography citing most of the works I found useful in writing this book. Many of these books and articles will take the reader farther afield to topics and results not treated here, and several contain good bibliographies.

I have a number of debts to record. First among these is to Lee Bowie, who several years ago suggested that we author a textbook in modal logic – I to write the chapters on propositional modal logic, he to write on quantification, identity, naming, and description. When it later became apparent that the material on propositional modal logic was bulky enough to warrant separate publication, Bowie graciously encouraged me to proceed alone.

In this connection I also want to express my gratitude to Richard Jeffrey and David Lewis, for their advice and support, and for recommending my project to Cambridge University Press and its distinguished editor Jeremy Mynott.

My debts to several of the works cited in the bibliography will perhaps be obvious to those already acquainted with the subject of modal logic. In particular I should mention Lemmon and Scott's *Introduction to modal logic* and Segerberg's *Essay in classical modal logic*.

The contents of chapters 6 and 10 are largely adapted from my papers 'Imperatives', 'Conditional obligation', and 'Basic conditional logic', cited in the bibliography. I wish to thank Krister Segerberg, editor of *Theoria*, Sören Stenlund, editor of *Logical theory and semantic analysis*, Richmond Thomason, editor of the *Journal of philosophical logic*, and the D. Reidel Publishing Company for permission to use this material.

Steven Kuhn and Audrey McKinney read much of my manuscript at different stages of its development, and I am grateful to them for criticism and advice.

Krister Segerberg has been a mainstay of counsel and encouragement for many years. I have learned a great deal about modal logic from Segerberg, and I have benefited enormously from conversations with him in the course of writing this book.

Among many others who have contributed in various ways to this book I would like to thank Roy Benton, Paul Golden, Deborah Mayo, and Robert Pelcovits.

Finally, I owe an enduring debt to Dana Scott, who introduced me to modal logic, who taught me how to think about it, and whose conception of the subject fundamentally influenced my own.

Woodland Valley, New York B.F.C.
July 1978

PART I

1

INTRODUCTION

In this chapter we introduce the subject of modal logic by surveying some of the main features of the system of modal logic known as *S5*. This system is but one of many we shall study. Because it is one of the simplest, we choose it to begin with.

The system *S5* is determined semantically by an account of necessity and possibility that dates to the philosopher Leibniz: a proposition is *necessary* if it holds at all possible worlds, *possible* if it holds at some. The idea is that different things may be true at different possible worlds, but whatever holds true at every possible world is necessary, while that which holds at at least one possible world is possible.

In section 1.1 we develop this semantic idea by means of a definition of truth at a possible world in a model for a language of necessity and possibility. This leads to a definition of validity, and we set out some valid sentences and principles governing validity, as well as some examples of invalidity.

The totality of valid sentences forms the modal logic *S5*. In terms of the principles set out in section 1.1 it is possible to deduce all the valid sentences. Some evidence of this appears in section 1.2, where we take the principles in section 1.1 as axioms and rules of inference, formulate *S5* as a deductive system, and derive a number of further principles.

Sections 1.1 and 1.2 exemplify in miniature our approach to the study of modal logic throughout this book: first, semantically in terms of the notion of truth; second, syntactically by means of deductive systems.

The exposition in this chapter is quite casual, and intentionally so. The purpose, in part, is to acquaint the reader with many of the notions and notations used in the rest of the book; but formality is deferred to subsequent chapters. This leads to occasional wordiness, but not, it is hoped, to loss of intelligibility.

We study modal logic in the context of a language of necessity and possibility. The sentences of the language are of the following forms.

$$\mathbb{P}_0, \mathbb{P}_1, \mathbb{P}_2, \ldots$$

$$\top, \quad \bot, \quad \neg A, \quad A \wedge B, \quad A \vee B, \quad A \rightarrow B, \quad A \leftrightarrow B, \quad \Box A, \quad \Diamond A$$

Sentences of the form $\mathbb{P}_n$ (for $n = 0, 1, 2, \ldots$) are atomic. $\top$ is a constant for truth; $\bot$ is a constant for falsity. $\neg$, $\wedge$, $\vee$, $\rightarrow$, and $\leftrightarrow$ are signs of negation, conjunction, disjunction, conditionality, and biconditionality, respectively. $\Box$ is the necessity sign; $\Diamond$ is the possibility sign.

A more detailed account of the syntax of this language appears in section 2.1, but it is not essential for an understanding of the rest of this chapter.

1.1. Truth and possible worlds

According to the leibnizian idea, necessity is what is true at every possible world and possibility is what is true at some. Linguistically: a sentence of the form $\Box A$ – *necessarily* A – is true if and only if A itself is true at every possible world; and a sentence of the form $\Diamond A$ – *possibly* A – is true just in case A is true at some possible world.

The picture is of a collection of possible worlds – including our own, the real world – at which sentences of the language are variously true and false. Our purpose is to model this, and we do so by means of an infinite sequence of sets of possible worlds,

$$P_0, P_1, P_2, \ldots.$$

The intuition behind this modeling is that, for each natural number n, the set P_n collects just those possible worlds at which the corresponding atomic sentence $\mathbb{P}_n$ is true. In other words, the sequence $P_0, P_1, P_2, \ldots$ interprets the atomic sentences by stipulating at which possible worlds they are true (and, by omission, at which they are false): $\mathbb{P}_n$ is true at a possible world α if and only if α is in the set P_n.

More precisely, a model is a pair

$$\langle W, P \rangle$$

in which W is a set of possible worlds and P abbreviates an infinite sequence $P_0, P_1, P_2, \ldots$ of subsets of W. Note that W may contain possible worlds not in any of the sets P_n; indeed, any or all of these sets may be empty. Also, we do not require that the actual world appear in every model.

In terms of a possible world in a model we state the truth conditions for sentences according to their forms. Where A is a sentence and α is a possible world in a model $\mathcal{M} = \langle W, P \rangle$, we use the symbolism

$$\models^{\mathcal{M}}_{\alpha} \mathrm{A}$$

as short for

A is true at α in $\mathcal{M}$.

The truth conditions are stated thus:

(1) $\models^{\mathcal{M}}_{\alpha} \mathbb{P}_n$ iff $\alpha \in P_n$, for $n = 0, 1, 2, \ldots$.

(2) $\models^{\mathcal{M}}_{\alpha} \top$.

(3) Not $\models^{\mathcal{M}}_{\alpha} \bot$.

(4) $\models^{\mathcal{M}}_{\alpha} \neg \mathrm{A}$ iff not $\models^{\mathcal{M}}_{\alpha} \mathrm{A}$.

(5) $\models^{\mathcal{M}}_{\alpha} \mathrm{A} \wedge \mathrm{B}$ iff both $\models^{\mathcal{M}}_{\alpha} \mathrm{A}$ and $\models^{\mathcal{M}}_{\alpha} \mathrm{B}$.

(6) $\models^{\mathcal{M}}_{\alpha} \mathrm{A} \vee \mathrm{B}$ iff either $\models^{\mathcal{M}}_{\alpha} \mathrm{A}$ or $\models^{\mathcal{M}}_{\alpha} \mathrm{B}$, or both.

(7) $\models^{\mathcal{M}}_{\alpha} \mathrm{A} \rightarrow \mathrm{B}$ iff if $\models^{\mathcal{M}}_{\alpha} \mathrm{A}$ then $\models^{\mathcal{M}}_{\alpha} \mathrm{B}$.

(8) $\models^{\mathcal{M}}_{\alpha} \mathrm{A} \leftrightarrow \mathrm{B}$ iff $\models^{\mathcal{M}}_{\alpha} \mathrm{A}$ if and only if $\models^{\mathcal{M}}_{\alpha} \mathrm{B}$.

(9) $\models^{\mathcal{M}}_{\alpha} \Box \mathrm{A}$ iff for every β in $\mathcal{M}$, $\models^{\mathcal{M}}_{\beta} \mathrm{A}$.

(10) $\models^{\mathcal{M}}_{\alpha} \Diamond \mathrm{A}$ iff for some β in $\mathcal{M}$, $\models^{\mathcal{M}}_{\beta} \mathrm{A}$.

Some discussion of this definition may be helpful.

Clause (1) reflects our remarks about the sets $P_0, P_1, P_2, \ldots$ in a model: an atomic sentence $\mathbb{P}_n$ is true at a possible world α just in case α is a member of the set P_n. According to clause (2), the truth constant $\top$ is always true at α. By (3), the falsity constant $\bot$ is always false at α. Clause (4) states that a negation $\neg$A is true at α if and only if its negate A is false at α. The content of (5) is that a conjunction $\mathrm{A} \wedge \mathrm{B}$ is true at α just in case both its conjuncts, A and B, are. According to (6), a disjunction $\mathrm{A} \vee \mathrm{B}$ is true at α just when at least one of its disjuncts, A and B, is. Our intention in clause (7) is that a conditional $\mathrm{A} \rightarrow \mathrm{B}$ is to be understood as true at α just so long as it fails to be the case that its antecedent, A, is true at α while its consequent, B, is false. And, similarly, in (8) we intend that a biconditional $\mathrm{A} \leftrightarrow \mathrm{B}$ be accounted true at α just in case its members, A and B, are either both true at α or both false. Clause (9) formulates the leibnizian interpretation of necessity: a necessitation $\Box$A is true at α if and only if its necessitate, A, is true at every possible world β in the model. Finally, according to (10), $\Diamond$A is true at α just in case there is at least one possible world β in the model at which A is true.

A sentence true at every possible world in every model is said to be valid. We use the symbol $\models$ again – this time without subscripts or superscripts – and write

$$\models A$$

to mean that the sentence A is valid. More formally, then, the definition of validity may be expressed:

$$\models A \text{ iff for every model } \mathcal{M} \text{ and every possible world } \alpha \text{ in } \mathcal{M}, \models^{\mathcal{M}}_{\alpha} A.$$

In asking after the logic of necessity and possibility we seek to know which sentences are valid – true, no matter how interpreted, at every possible world – and which are not. For example, as we shall see, every sentence of the form $\Box A \to A$ is valid, whereas not every sentence of the form $A \to \Box A$ is. In what follows we first set out some valid sentences and principles governing validity, enough to form an axiomatic basis for the derivation of all valid sentences. Then we mention some prominent cases of invalidity.

Let us begin our survey of validity with the principle just mentioned:

$$\text{T.} \quad \Box A \to A$$

According to T, whatever is necessary is so: *if necessarily* A, *then* A. To see that this schema – i.e. every sentence of this form – is valid, it is sufficient to prove that where α is any possible world in any model $\mathcal{M}$, $\models^{\mathcal{M}}_{\alpha} \Box A \to A$. And for this it will be enough to show that if $\models^{\mathcal{M}}_{\alpha} \Box A$ then $\models^{\mathcal{M}}_{\alpha} A$ (compare clause (7) in the definition of truth). So suppose that $\models^{\mathcal{M}}_{\alpha} \Box A$. By clause (9) of the truth definition, this means that $\models^{\mathcal{M}}_{\beta} A$ for every possible world β in $\mathcal{M}$. In particular, then, this holds for α, i.e. $\models^{\mathcal{M}}_{\alpha} A$.

Next let us consider the schema

$$5. \quad \Diamond A \to \Box \Diamond A.$$

The import of 5 is that what is possible is necessarily possible: *if possibly* A, *then necessarily possibly* A. To see that 5 is valid, suppose that $\models^{\mathcal{M}}_{\alpha} \Diamond A$, for possible world α in model $\mathcal{M}$. By clause (10) of the definition of truth this means that $\mathcal{M}$ has a possible world β such that $\models^{\mathcal{M}}_{\beta} A$. It follows from this (again by (10)) that no matter what possible world in the model we choose, $\Diamond A$ holds – i.e. $\models^{\mathcal{M}}_{\beta} \Diamond A$ for every possible world β in $\mathcal{M}$. But by clause (9) this means that $\models^{\mathcal{M}}_{\alpha} \Box \Diamond A$, which is what we wished to show.

The schemas T and 5 are rather special in that they do not hold in every system of modal logic we shall study. The next two principles are more widely accepted however; it is not until chapter 7 that they are called into question.

The first of these expresses a principle of distributivity of necessity with respect to the conditional:

$$\text{K.}\quad \Box(A \rightarrow B) \rightarrow (\Box A \rightarrow \Box B)$$

This means that if a conditional and its antecedent are both necessary, then so is the consequent. For the validity of K, suppose that α is a possible world in a model $\mathcal{M}$ such that both $\models_{\alpha}^{\mathcal{M}} \Box(A \rightarrow B)$ and $\models_{\alpha}^{\mathcal{M}} \Box A$. Then for every possible world β in $\mathcal{M}$, both $\models_{\beta}^{\mathcal{M}} A \rightarrow B$ and $\models_{\beta}^{\mathcal{M}} A$, from which it follows that for every possible world β in $\mathcal{M}$, $\models_{\beta}^{\mathcal{M}} B$. Thus, $\models_{\alpha}^{\mathcal{M}} \Box B$.

The second principle corresponds to a rule of inference in the next section (RN, the rule of necessitation). It states that the necessitation of a valid sentence is itself always valid. In symbols:

$$\textit{If} \models A, \textit{then} \models \Box A.$$

For suppose that $\models A$, i.e. that $\models_{\alpha}^{\mathcal{M}} A$ for every possible world α in every model $\mathcal{M}$. Then for every possible world α in every model $\mathcal{M}$, $\models_{\alpha}^{\mathcal{M}} \Box A$, which is to say that $\models \Box A$.

The last specifically modal validity we wish to mention holds in every modal logic we shall discuss in this book.

$$\text{Df}\Diamond.\quad \Diamond A \leftrightarrow \neg\Box\neg A$$

This schema embodies the idea that what is possible is just what is not-necessarily-not. Its validity means that possibility is always expressible in terms of necessity and negation, and so is theoretically superfluous. In this sense $\Diamond$ is *definable* in terms of $\Box$ and $\neg$. Df$\Diamond$ is valid because to say that for some possible world β in a model $\mathcal{M}$, $\models_{\beta}^{\mathcal{M}} A$, is just to say that it is not the case that for every possible world β in $\mathcal{M}$ it is not the case that $\models_{\beta}^{\mathcal{M}} A$. Reference to clauses (4), (9), and (10) of the truth definition reveals that the former expression means that $\models_{\alpha}^{\mathcal{M}} \Diamond A$, while the latter expression means that $\models_{\alpha}^{\mathcal{M}} \neg\Box\neg A$. Hence the biconditional $\Diamond A \leftrightarrow \neg\Box\neg A$ holds at every possible world in every model.

Let us turn now to the relationship between our modal logic and ordinary propositional, or truth-functional, logic. The relationship is simple: the modal logic includes the propositional. In part, this means that every propositionally valid sentence is modally valid, i.e.:

$$\textit{If}\ A\ \textit{is a tautology, then} \models A.$$

The explanation of this is as follows (there is a more careful account in chapter 2). By a *tautology* we mean a sentence true in every valuation of its propositionally atomic constituents. A sentence is *propositionally atomic* if it is either atomic in the ordinary sense ($\mathbb{P}_n$) or modal ($\Box$A or $\Diamond$A). A *valuation* is an assignment of truth values (truth and falsity) to the propositionally atomic sentences. The truth conditions in a valuation of the rest of the sentences in the language – those of the forms $\top$, $\bot$, $\neg$A, etc. – are determined just as they are by a possible world in a model. Thus $\top$ is true in every valuation, $\bot$ is false in every valuation, $\neg$A is true in a valuation if and only if A is false, and so on; compare clauses (2)–(8) in the definition of truth above.

In short, a valuation analyzes sentences semantically from the point of view of their truth-functional structure, counting as atomic the modal structure of sentences of the forms $\Box$A and $\Diamond$A, as well as those of the form $\mathbb{P}_n$. A sentence is a tautology, thus, if it comes out true no matter how truth values are assigned to its propositional atoms. For example, any sentence of the form $\Box A \rightarrow \Box A$ is a tautology, since $\Box$A is propositionally atomic and such a conditional is true in a valuation whether $\Box$A is assigned truth or falsity.

Now observe that in any model $\mathcal{M}$ each possible world α is a valuation in the sense just explained, since α assigns truth or falsity to each sentence of the form $\mathbb{P}_n$, $\Box$A, and $\Diamond$A, i.e. to each propositionally atomic sentence. The world α assigns truth to a propositionally atomic sentence A when $\models^{\mathcal{M}}_{\alpha}$ A, and falsity otherwise.

To prove, finally, that every tautology is valid, assume that A is a tautology – that A is true in every valuation. Then A is true at every possible world in every model, i.e. $\models^{\mathcal{M}}_{\alpha}$ A for every possible world α in every model $\mathcal{M}$. This means that $\models$ A, i.e. that A is valid.

To say that all tautologies are valid does not exhaust what is meant by saying that modal logic includes propositional logic. It means moreover that validity is preserved by propositionally correct patterns of inference. For example, the inference from $A \rightarrow B$ and A to B is propositionally correct; whenever both $A \rightarrow B$ and A are true in a valuation, so is B. Corresponding to this we have the principle that whenever a conditional and its antecedent are both valid, so is the consequent:

$$\textit{If} \models A \rightarrow B \textit{ and} \models A, \textit{ then} \models B.$$

This emerges as the rule of inference modus ponens, MP, in the next section. To prove the principle, suppose that both $\models A \rightarrow B$ and $\models$ A. This means that for every possible world α in every model $\mathcal{M}$, $\models^{\mathcal{M}}_{\alpha} A \rightarrow B$, and that for every possible world α in every model $\mathcal{M}$, $\models^{\mathcal{M}}_{\alpha}$ A. It follows at once that for every possible world α in every model $\mathcal{M}$, $\models^{\mathcal{M}}_{\alpha}$ B, i.e. that $\models$ B.

In terms of this principle and the fact that every tautology is valid we can prove that validity is preserved by propositionally correct patterns of inference generally. For suppose that it is propositionally correct to infer a sentence A from sentences $A_1, \ldots, A_n$, i.e. that A is true in every valuation in which all of $A_1, \ldots, A_n$ are. Then the sentence

$$A_1 \rightarrow (\ldots (A_n \rightarrow A) \ldots)$$

is a tautology. Thus this sentence is valid. Hence if each of $A_1, \ldots, A_n$ is valid, then by applying the modus ponens principle n times we arrive at the result that A is valid.

For example, it is propositionally correct to infer $A \rightarrow C$ from $A \rightarrow B$ and $B \rightarrow C$ (we leave it for the reader to check that $A \rightarrow C$ is true in any valuation that verifies both $A \rightarrow B$ and $B \rightarrow C$). So the sentence $(A \rightarrow B) \rightarrow ((B \rightarrow C) \rightarrow (A \rightarrow C))$ is a tautology. Hence if $\models A \rightarrow B$, then by the principle of modus ponens, $\models (B \rightarrow C) \rightarrow (A \rightarrow C)$. And so $\models A \rightarrow C$, if also $\models B \rightarrow C$.

This ends our short survey of valid sentences and principles governing validity. Let us turn now to some examples of invalidity.

To begin, the schema $A \rightarrow \Box A$ – the converse of T – is not valid. To see this, let α and β be distinct possible worlds, let $W = \{\alpha, \beta\}$, and let $P_n = \{\alpha\}$ for every natural number n (i.e. $n = 0, 1, 2, \ldots$). Then $\mathcal{M} = \langle W, P \rangle$ is a model in which $\models^{\mathcal{M}}_{\alpha} \mathbb{P}_0$ (since P_0 contains α) and not $\models^{\mathcal{M}}_{\alpha} \Box \mathbb{P}_0$ (since there is a world in $\mathcal{M}$, viz. β, not in P_0). Thus, not $\models^{\mathcal{M}}_{\alpha} \mathbb{P}_0 \rightarrow \Box \mathbb{P}_0$, which proves that the schema $A \rightarrow \Box A$ is not valid. We say in this case that $\mathcal{M}$ is a countermodel to $A \rightarrow \Box A$.

Notice that if $A \rightarrow \Box A$ were valid it would mean that whatever is the case is so necessarily. Indeed, if this schema were valid, then given the validity of T, the biconditional $A \leftrightarrow \Box A$ would be valid, so that truth and necessity would be the same. The reader should contrast the invalidity of $A \rightarrow \Box A$ with the correctness of the principle of necessitation, that if A is valid so is $\Box$A. This will help in understanding the difference between theorems and rules of inference in the next section.

Another example of invalidity is the schema $\Box(A \vee B) \rightarrow (\Box A \vee \Box B)$. The model $\mathcal{M} = \langle W, P \rangle$ in which $W = \{\alpha, \beta\}$, $P_0 = \{\alpha\}$, and $P_n = \{\beta\}$ for $n > 0$ is a countermodel to this schema. For $\models^{\mathcal{M}}_{\alpha} \mathbb{P}_0$ and $\models^{\mathcal{M}}_{\beta} \mathbb{P}_1$, which means that $\models^{\mathcal{M}}_{\alpha} \mathbb{P}_0 \vee \mathbb{P}_1$ and $\models^{\mathcal{M}}_{\beta} \mathbb{P}_0 \vee \mathbb{P}_1$. So, $\models^{\mathcal{M}}_{\alpha} \Box(\mathbb{P}_0 \vee \mathbb{P}_1)$, since the disjunction $\mathbb{P}_0 \vee \mathbb{P}_1$ is true at every possible world in $\mathcal{M}$. On the other hand, not $\models^{\mathcal{M}}_{\alpha} \mathbb{P}_1$ and also not $\models^{\mathcal{M}}_{\beta} \mathbb{P}_0$. So neither $\models^{\mathcal{M}}_{\alpha} \Box \mathbb{P}_0$ nor $\models^{\mathcal{M}}_{\alpha} \Box \mathbb{P}_1$, and hence not $\models^{\mathcal{M}}_{\alpha} \Box \mathbb{P}_0 \vee \Box \mathbb{P}_1$. Therefore, not $\models^{\mathcal{M}}_{\alpha} \Box(\mathbb{P}_0 \vee \mathbb{P}_1) \rightarrow (\Box \mathbb{P}_0 \vee \Box \mathbb{P}_1)$, i.e. $\mathcal{M}$ is a countermodel to the schema.

An even simpler way to see the invalidity of the schema $\Box(A \vee B) \rightarrow (\Box A \vee \Box B)$ is to consider the instance $\Box(\mathbb{P}_0 \vee \neg \mathbb{P}_0) \rightarrow (\Box \mathbb{P}_0 \vee \Box \neg \mathbb{P}_0)$. The disjunction $\mathbb{P}_0 \vee \neg \mathbb{P}_0$ is a tautology, so it is valid – and hence so is its necessitation, $\Box(\mathbb{P}_0 \vee \neg \mathbb{P}_0)$. Thus it is sufficient to show that the disjunction $\Box \mathbb{P}_0 \vee \Box \neg \mathbb{P}_0$ is not valid. The model described above in connection with $A \rightarrow \Box A$ does the job, as the reader may verify.

We have just shown that the necessity sign does not distribute into a disjunction; the validity of K, above, means that $\Box$ does distribute into a conditional. As a final example of invalidity, we describe a countermodel to $\Diamond(A \rightarrow B) \rightarrow (\Diamond A \rightarrow \Diamond B)$, thus showing that the possibility sign does not distribute into a conditional. The model is $\mathcal{M} = \langle W, P \rangle$, where $W = \{\alpha, \beta\}$, $P_0 = \{\alpha\}$, and $P_n = \varnothing$ (the empty set) for $n > 0$. We leave it for the reader to check that $\models^{\mathcal{M}}_{\alpha} \Diamond(\mathbb{P}_0 \rightarrow \mathbb{P}_1)$ and $\models^{\mathcal{M}}_{\alpha} \Diamond \mathbb{P}_0$, but not $\models^{\mathcal{M}}_{\alpha} \Diamond \mathbb{P}_1$. This being so, it follows that not $\models^{\mathcal{M}}_{\alpha} \Diamond(\mathbb{P}_0 \rightarrow \mathbb{P}_1) \rightarrow (\Diamond \mathbb{P}_0 \rightarrow \Diamond \mathbb{P}_1)$.

This concludes our semantical exposition of the modal logic *S5*.

EXERCISES

1.1. Prove that the following schemas are valid.

(*a*) $\Box A \rightarrow \Diamond A$
(*b*) $A \rightarrow \Box \Diamond A$
(*c*) $\Box A \rightarrow \Box \Box A$
(*d*) $\Diamond \top$
(*e*) $A \rightarrow \Diamond A$
(*f*) $\Diamond \Box A \rightarrow A$
(*g*) $\Diamond \Diamond A \rightarrow \Diamond A$
(*h*) $\Diamond \Box A \rightarrow \Box A$

1.2. Prove that the following schemas are valid.

(*a*) $\Box \top$
(*b*) $\Box(A \wedge B) \rightarrow (\Box A \wedge \Box B)$
(*c*) $(\Box A \wedge \Box B) \rightarrow \Box(A \wedge B)$
(*d*) $\neg \Diamond \bot$
(*e*) $(\Diamond A \vee \Diamond B) \rightarrow \Diamond(A \vee B)$
(*f*) $\Diamond(A \vee B) \rightarrow (\Diamond A \vee \Diamond B)$

1.3. Prove that the schema $\Box A \leftrightarrow \neg \Diamond \neg A$ is valid.

1.4. Prove each of the following.

(*a*) If $\models (A \wedge B) \to C$, then $\models (\Box A \wedge \Box B) \to \Box C$.
(*b*) If $\models A \to B$, then $\models \Box A \to \Box B$.
(*c*) If $\models A \leftrightarrow B$, then $\models \Box A \leftrightarrow \Box B$.
(*d*) If $\models A \to B$, then $\models \Diamond A \to \Diamond B$.
(*e*) If $\models A \leftrightarrow B$, then $\models \Diamond A \leftrightarrow \Diamond B$.

1.5. Prove that for any $n \geqslant 0$, if $\models (A_1 \wedge \ldots \wedge A_n) \to A$, then $\models (\Box A_1 \wedge \ldots \wedge \Box A_n) \to \Box A$. (When $n = 0$ this just means if $\models A$ then $\models \Box A$.)

1.6. Referring to the model $\mathscr{M}$ defined in connection with showing the invalidity of $\Diamond(A \to B) \to (\Diamond A \to \Diamond B)$ (see the penultimate paragraph of section 1.1), verify that $\models^{\mathscr{M}}_{\alpha} \Diamond(\mathbb{P}_0 \to \mathbb{P}_1)$ and $\models^{\mathscr{M}}_{\alpha} \Diamond \mathbb{P}_0$, but not $\models^{\mathscr{M}}_{\alpha} \Diamond \mathbb{P}_1$.

1.7. Prove that the following schemas are invalid (i.e. that each has an invalid instance).

(*a*) $\Diamond A \to A$
(*b*) $\Diamond A \to \Box A$
(*c*) $\Box \Diamond A \to A$
(*d*) $(\Diamond A \wedge \Diamond B) \to \Diamond(A \wedge B)$
(*e*) $(\Box A \to \Box B) \to \Box(A \to B)$

1.8. For each of the following, decide whether or not it is valid, and prove it.

(*a*) $\Box\Box A \to \Box A$
(*b*) $\Box \Diamond A \to \Diamond A$
(*c*) $\Diamond(A \wedge B) \to (\Diamond A \wedge \Diamond B)$
(*d*) $\Diamond \Box A \to \Box \Diamond A$
(*e*) $\Box \Diamond A \to \Diamond \Box A$
(*f*) $(\neg \Diamond A \wedge \Diamond B) \to \Diamond (\neg A \wedge B)$
(*g*) $\Diamond A \to \Diamond \Diamond A$
(*h*) $\Box A \to \Diamond \Box A$
(*i*) $(\Box A \vee \Box B) \to \Box(A \vee B)$
(*j*) $\Box \bot$

1.9. Suppose that in every model there is just one possible world and prove that under this assumption the schema $A \to \Box A$ is valid.

1.10. Consider a structure $\mathscr{M} = \langle W, R, P \rangle$ in which W and P are as they are in a model, and R is an equivalence relation on W. That is, W is a set of possible worlds, P is an infinite sequence $P_0, P_1, P_2, \ldots$ of subsets of W, and R is a binary relation on W that is reflexive (for every α in $\mathscr{M}$, $\alpha R \alpha$) and euclidean (for every α, β, and γ in $\mathscr{M}$, if $\alpha R \beta$ and $\alpha R \gamma$, then $\beta R \gamma$). Structures of this sort are models for *S5*, where the truth conditions for non-modal sentences are given as usual (i.e. (1)–(8) in the fourth paragraph of section 1.1) and those for sentences of the forms $\Box$A and $\Diamond$A are given by:

$$(9') \models^{\mathscr{M}}_{\alpha} \Box A \text{ iff for every } \beta \text{ in } \mathscr{M} \text{ such that } \alpha R \beta, \models^{\mathscr{M}}_{\beta} A.$$

$$(10') \models^{\mathscr{M}}_{\alpha} \Diamond A \text{ iff for some } \beta \text{ in } \mathscr{M} \text{ such that } \alpha R \beta, \models^{\mathscr{M}}_{\beta} A.$$

Intuitively, R is a relation that relates a world to those that are possible with respect to it; $\alpha R \beta$ means that the world β is possible with respect to the world α. Thus according to (9′) $\Box$A is true at α just in case A is true at all worlds possible with respect to α; and according to (10′) $\Diamond$A is true at α just in case A is true at some world possible with respect to α. Obviously, these models represent a generalization of the analysis of necessity and possibility in section 1.1: it is no longer assumed that every world is possible with respect to every other world.

As before, validity means truth at every possible world in every model.

Show that these models are adequate for an analysis of the system *S5* by proving that the schemas T, 5, K, and Df$\Diamond$ are all valid, that any tautology is valid, that $\Box$A is valid if A is, and that if A → B and A are valid so is B.

Hint: The validity of T depends upon the reflexivity of R, and the validity of 5 depends on the euclideanness of R. Nothing special is needed for the others.

1.11. (This exercise presupposes an acquaintance with elementary quantificational logic.) The reader may have noticed an analogy between the signs of necessity and possibility, $\Box$ and $\Diamond$, on the one hand, and the universal and existential quantifiers, $\forall$ and $\exists$, on the other. $\Box$A is true at a possible world just in case A holds at *every* world; $\Diamond$A is true at a world if and only if A holds at *some* world.

Let us specify a language of elementary quantificational logic by stipulating that its formulas are of the following forms:

$$\mathbb{P}_n(\alpha),\quad \top,\quad \bot,\quad \neg A,\quad A \wedge B,\quad A \vee B,\quad A \rightarrow B,$$

$$A \leftrightarrow B,\quad \forall \alpha A,\quad \exists \alpha A,$$

where $\mathbb{P}_n$ is one of denumerably many one-place predicates and α is a variable, so that $\mathbb{P}_n(\alpha)$ is an atomic formula.

We define the mapping τ, from the language of modal logic to the quantificational language, as follows.

(1) $\tau(\mathbb{P}_n) = \mathbb{P}_n(\alpha)$, for $n = 0, 1, 2, \ldots$.

(2) $\tau(\top) = \top$.

(3) $\tau(\bot) = \bot$.

(4) $\tau(\neg A) = \neg\tau(A)$.

(5) $\tau(A \wedge B) = \tau(A) \wedge \tau(B)$.

(6) $\tau(A \vee B) = \tau(A) \vee \tau(B)$.

(7) $\tau(A \rightarrow B) = \tau(A) \rightarrow \tau(B)$.

(8) $\tau(A \leftrightarrow B) = \tau(A) \leftrightarrow \tau(B)$.

(9) $\tau(\Box A) = \forall\alpha\tau(A)$.

(10) $\tau(\Diamond A) = \exists\alpha\tau(A)$.

Thus τ associates with each sentence A in the modal language a unique formula $\tau(A)$ in the quantificational language by replacing each atomic sentence $\mathbb{P}_n$ by $\mathbb{P}_n(\alpha)$ and putting $\forall\alpha$ and $\exists\alpha$ respectively for occurrences of $\Box$ and $\Diamond$. For example, let us calculate the results of applying τ to $\Box\mathbb{P}_0 \rightarrow \mathbb{P}_0$ and to $\Diamond\mathbb{P}_0 \rightarrow \Box\Diamond\mathbb{P}_0$.

$$\begin{aligned}\tau(\Box\mathbb{P}_0 \rightarrow \mathbb{P}_0) &= \tau(\Box\mathbb{P}_0) \rightarrow \tau(\mathbb{P}_0)\\ &= \forall\alpha\tau(\mathbb{P}_0) \rightarrow \tau(\mathbb{P}_0)\\ &= \forall\alpha\mathbb{P}_0(\alpha) \rightarrow \mathbb{P}_0(\alpha)\end{aligned}$$

$$\begin{aligned}\tau(\Diamond\mathbb{P}_0 \rightarrow \Box\Diamond\mathbb{P}_0) &= \tau(\Diamond\mathbb{P}_0) \rightarrow \tau(\Box\Diamond\mathbb{P}_0)\\ &= \tau(\Diamond\mathbb{P}_0) \rightarrow \forall\alpha\tau(\Diamond\mathbb{P}_0)\\ &= \exists\alpha\tau(\mathbb{P}_0) \rightarrow \forall\alpha\exists\alpha\tau(\mathbb{P}_0)\\ &= \exists\alpha\mathbb{P}_0(\alpha) \rightarrow \forall\alpha\exists\alpha\mathbb{P}_0(\alpha)\end{aligned}$$

It should be apparent that τ is in effect a specification of the truth conditions of modal sentences at a possible world α in a model. The transformation shows that $\Box\mathbb{P}_0 \rightarrow \mathbb{P}_0$ holds at α just in case $\forall\alpha\mathbb{P}_0(\alpha) \rightarrow \mathbb{P}_0(\alpha)$ is true, and that $\Diamond\mathbb{P}_0 \rightarrow \Box\Diamond\mathbb{P}_0$ holds at α just in case $\exists\alpha\mathbb{P}_0(\alpha) \rightarrow \forall\alpha\exists\alpha\mathbb{P}_0(\alpha)$ is true. Generally, we can see that a modal sentence A is valid just when $\tau(A)$ is; i.e.

$\models$ A iff $\tau(A)$ is a valid formula of elementary quantificational logic.

For example, the instances of T and 5 above are valid, and so are their transformations. So τ provides a way of investigating questions of validity in the modal language.

(*a*) Apply τ to K, Df$\Diamond$, and selected tautologies, to see that their transformations are valid quantificational formulas.
(*b*) Show that if $\tau(A)$ is a valid formula of elementary quantificational logic so is $\tau(\Box A)$, and that if $\tau(A \to B)$ and $\tau(A)$ are quantificationally valid so is $\tau(B)$.
(*c*) Use τ on the schemas in exercises 1.1–1.3, 1.7, and 1.8.
(*d*) Show that the principles in exercises 1.4 and 1.5 hold with respect to quantificational validity and transformations of the schemas.
(*e*) Explain how models $\mathscr{M} = \langle W, P \rangle$ for the modal language serve equally well for the quantificational language.

1.2. The system *S5*

In this section we examine necessity and possibility in *S5* from an axiomatic point of view. We begin with an axiomatization based on the principles in the preceding section. That is, we adopt as axioms, or basic theorems, all sentences of the following forms.

T. $\Box A \to A$

5. $\Diamond A \to \Box \Diamond A$

K. $\Box(A \to B) \to (\Box A \to \Box B)$

Df$\Diamond$. $\Diamond A \leftrightarrow \neg \Box \neg A$

PL. A, where A is a tautology

And we assume the following rules of inference.

RN. $\dfrac{A}{\Box A}$

MP. $\dfrac{A \to B,\ A}{B}$

By a theorem, generally, we mean any sentence that can be proved on the basis of the axioms and rules of inference. (Axioms are automatically theorems.) Where A is a sentence, we also write

$$\vdash A$$

to mean that A is a theorem.

Note that a rule of inference is properly understood as meaning that its conclusion is a theorem if each of its hypotheses is. For example, the rule RN means that $\vdash \Box A$ whenever $\vdash A$.

Before moving to proofs of genuinely modal theorems and further rules of inference, let us see again that *S5*, now formulated as a deductive system, includes propositional logic. Here this means that we can derive within the system the rule of inference

RPL. $\dfrac{A_1, \ldots, A_n}{A}$ $(n \geqslant 0)$,

where the inference from $A_1, \ldots, A_n$ to A is propositionally correct.

The proof that the rule RPL holds is like that for the analogous result in section 1.1. We show that if the inference from $A_1, \ldots, A_n$ to A is propositionally correct and each of $A_1, \ldots, A_n$ is a theorem, then A is a theorem, too. The supposition that the inference is propositionally correct means that A is true in every valuation in which each of $A_1, \ldots, A_n$ is, which in turn means that the sentence

$$A_1 \rightarrow (\ldots (A_n \rightarrow A) \ldots)$$

is a tautology (*PL*), which means that it is a theorem. If each of $A_1, \ldots, A_n$ is a theorem, then by n applications of the rule MP, so is A.

We may illustrate RPL, as in section 1.1, with the rule of inference sometimes called hypothetical syllogism:

$$\frac{A \rightarrow B, \quad B \rightarrow C}{A \rightarrow C}$$

Because $A \rightarrow C$ is true in every valuation in which both $A \rightarrow B$ and $B \rightarrow C$ are, the sentence $(A \rightarrow B) \rightarrow ((B \rightarrow C) \rightarrow (A \rightarrow C))$ is a tautology and hence a theorem. Thus if both $\vdash A \rightarrow B$ and $\vdash B \rightarrow C$, successive applications of MP yield first that $\vdash (B \rightarrow C) \rightarrow (A \rightarrow C)$ and then that $\vdash A \rightarrow C$. So this rule is covered by RPL.

The rule MP is obviously also a special case of RPL, but it should be noted that RPL covers the axioms *PL* as well. For when $n = 0$, RPL is the rule

$$\frac{\quad}{A},$$

where the inference to A is propositionally correct.

And this simply means that A is a theorem whenever A is true in every valuation, i.e. whenever A is a tautology. Thus it is a matter of indifference whether we adopt *PL* and MP, on the one hand, or simply RPL, on the other, in our axiomatization of *S5*. We choose *PL* and MP here because this is closer to the traditional approach; in the rest of the book we use RPL.

In any case, we shall hereinafter freely make use of tautologies and propositionally correct patterns of inference in deducing theorems and deriving rules of inference. Wherever we do, we signal this by *PL* (for 'propositional logic').

Turning now to specifically modal principles, let us begin by proving that the schema

$$\text{T}\Diamond.\quad A \rightarrow \Diamond A$$

– whatever is so is possibly so – is a theorem of *S5*. First we note that as a special case of the axiom T we have that $\vdash \Box\neg A \rightarrow \neg A$. By *PL*, it follows from this that $\vdash A \rightarrow \neg\Box\neg A$. In view of the axiom Df$\Diamond$, i.e. that $\vdash \Diamond A \leftrightarrow \neg\Box\neg A$, we may infer by *PL* that $\vdash A \rightarrow \Diamond A$.

We can put this discursive proof that T$\Diamond$ is a theorem more neatly as an annotated sequence of theorems:

1. $\Box\neg A \rightarrow \neg A$	T
2. $A \rightarrow \neg\Box\neg A$	1, *PL*
3. $\Diamond A \leftrightarrow \neg\Box\neg A$	Df$\Diamond$
4. $A \rightarrow \Diamond A$	2, 3, *PL*

The annotations are meant to indicate the reasoning involved as the proof proceeds. Thus line 1 is justified as an instance of T, line 2 comes from line 1 by *PL* (i.e. RPL), line 3 is a statement of Df$\Diamond$, and line 4 is inferred from lines 2 and 3 by *PL* (again, RPL). This way of setting out proofs is perspicuous and often useful, especially where the discursive mode is lengthy or tortuous. (But notice that line 2 might have been omitted, since line 4 follows from lines 1 and 3 by *PL*. We prefer the longer proof here for the sake of perspicuity.)

Next let us show that whatever is necessary is possible, i.e. that the schema

$$\text{D.}\quad \Box A \rightarrow \Diamond A$$

is a theorem of *S5*. The proof is simple: Since both $\Box A \rightarrow A$ and $A \rightarrow \Diamond A$ are theorems, by *PL* (in fact, hypothetical syllogism) so is $\Box A \rightarrow \Diamond A$.

Likewise, using T$\Diamond$, we may show that the schema

$$\text{B.}\quad A \rightarrow \Box\Diamond A$$

– whatever is so is necessarily possibly so – is a theorem of *S5*.

1. $\Diamond A \to \Box\Diamond A$ 5
2. $A \to \Diamond A$ T$\Diamond$
3. $A \to \Box\Diamond A$ 1, 2, *PL*

Before going on to prove more theorems of *S5*, it will be convenient to derive two further rules of inference.

RM. $\dfrac{A \to B}{\Box A \to \Box B}$

RE. $\dfrac{A \leftrightarrow B}{\Box A \leftrightarrow \Box B}$

The rule RM may be understood as asserting that a proposition is necessary if it is implied by a necessary proposition. To show that *S5* has this rule we argue that its conclusion is a theorem if its hypothesis is, as follows.

1. $A \to B$ hypothesis
2. $\Box(A \to B)$ 1, RN
3. $\Box(A \to B) \to (\Box A \to \Box B)$ K
4. $\Box A \to \Box B$ 2, 3, *PL*

Given RM, it is easy to derive the rule RE (which says in effect that equivalent propositions are equally necessary). We leave the derivation as an exercise for the reader.

Now let us prove that *S5* has the theorem

Df$\Box$. $\Box A \leftrightarrow \neg\Diamond\neg A$,

i.e. that necessity is definable in terms of possibility and negation. Our proof uses just *PL*, RE, and the definability of possibility in terms of necessity and negation, Df$\Diamond$.

1. $\Diamond\neg A \leftrightarrow \neg\Box\neg\neg A$ Df$\Diamond$
2. $\Box\neg\neg A \leftrightarrow \neg\Diamond\neg A$ 1, *PL*
3. $A \leftrightarrow \neg\neg A$ *PL*
4. $\Box A \leftrightarrow \Box\neg\neg A$ 3, RE
5. $\Box A \leftrightarrow \neg\Diamond\neg A$ 2, 4, *PL*

(Notice that line 2 might have been omitted, since line 5 follows from lines 1 and 4 by *PL*.)

Dual to the theorem B, *S5* has the theorem

$$\text{B}\Diamond. \quad \Diamond\Box A \to A,$$

which means that whatever is possibly necessary is simply so. By way of proof, note first that in virtue of B, $\vdash \neg A \to \Box\Diamond\neg A$, and so by *PL*, $\vdash \neg\Box\Diamond\neg A \to A$. Thus it is sufficient to show that $\vdash \Diamond\Box A \leftrightarrow \neg\Box\Diamond\neg A$. Our proof of this demonstrates the usefulness of being able to call upon Df$\Box$ and RE, as well as Df$\Diamond$:

1. $\Box A \leftrightarrow \neg\Diamond\neg A$	Df$\Box$
2. $\neg\Box A \leftrightarrow \Diamond\neg A$	1, *PL*
3. $\Box\neg\Box A \leftrightarrow \Box\Diamond\neg A$	2, RE
4. $\neg\Box\neg\Box A \leftrightarrow \neg\Box\Diamond\neg A$	3, *PL*
5. $\Diamond\Box A \leftrightarrow \neg\Box\neg\Box A$	Df$\Diamond$
6. $\Diamond\Box A \leftrightarrow \neg\Box\Diamond\neg A$	4, 5, *PL*

Here again our proof is spelled out in more detail than is necessary; line 6 follows by *PL* from lines 3 and 5.

By a similar argument we can also show that *S5* contains the following dual of the axiom 5.

$$5\Diamond. \quad \Diamond\Box A \to \Box A$$

For as a special case of 5, $\vdash \Diamond\neg A \to \Box\Diamond\neg A$, and hence $\vdash \neg\Box\Diamond\neg A \to \neg\Diamond\neg A$, by *PL*. Then 5$\Diamond$ follows by *PL*, using Df$\Box$ and the theorem on line 6 above. The import of 5$\Diamond$ is, of course, that a proposition is necessary if it is at least possibly necessary.

We come now to the schema

$$4. \quad \Box A \to \Box\Box A.$$

According to 4, whatever is necessary is necessarily necessary. We may prove that 4 is a theorem of *S5* as follows.

1. $\Diamond\Box A \to \Box A$	5$\Diamond$
2. $\Box\Diamond\Box A \to \Box\Box A$	1, RM
3. $\Box A \to \Box\Diamond\Box A$	B
4. $\Box A \to \Box\Box A$	2, 3, *PL*

Corresponding to the theorem 4 is the dual schema

$$4\Diamond. \quad \Diamond\Diamond A \to \Diamond A,$$

according to which whatever is possibly possible is possible simply. To show that 4$\Diamond$ is a theorem of *S5* we would argue that because of 4 and *PL*, $\vdash \neg\Box\Box\neg A \to \neg\Box\neg A$, and then prove that $\vdash \Diamond\Diamond A \leftrightarrow \neg\Box\Box\neg A$. For 4$\Diamond$ follows from these by *PL*. We leave the actual proof, however, as an exercise for the reader.

The system *S5* has the following noteworthy rule of inference.

$$\text{RK.} \quad \frac{(A_1 \wedge \ldots \wedge A_n) \to A}{(\Box A_1 \wedge \ldots \wedge \Box A_n) \to \Box A} \quad (n \geqslant 0)$$

RK expresses a general rule of modal consequence: a proposition is necessary if it is a consequence of a collection of propositions each of which is necessary. The condition that $n \geqslant 0$ is intentional, for we make the convention that in the absence of antecedents – when $n = 0$ – the conditionals are identified with their consequents, A and $\Box$A. Thus when $n = 0$ we have the rule RN as a special case of RK. Moreover, when $n = 1$ the rule RK becomes RM.

A proper proof that *S5* has the rule RK proceeds by induction on the number n of conjuncts in the antecedents. The basis of the induction, where $n = 0$, is trivial, since in this case RK is RN, a basic rule in the axiomatization of *S5*. For the inductive part of the proof we suppose – as an *inductive hypothesis* – that the rule holds for any number of conjuncts in the antecedents up to (but not including) some number $n > 0$, and show from this that it holds when the number of conjuncts is exactly n. The argument for this is as follows. Suppose that

$$(A_1 \wedge \ldots \wedge A_n) \to A$$

is a theorem. By *PL* this is equivalent to

$$(A_1 \wedge \ldots \wedge A_{n-1}) \to (A_n \to A).$$

By the inductive hypothesis the rule RK applies to this theorem, since the number of conjuncts in the antecedent is less than n. Thus we have the theorem

$$(\Box A_1 \wedge \ldots \wedge \Box A_{n-1}) \to \Box(A_n \to A).$$

Now from this and the axiom K, in the form

$$\Box(A_n \to A) \to (\Box A_n \to \Box A),$$

we infer by *PL* the theorem

$$(\Box A_1 \wedge \ldots \wedge \Box A_{n-1}) \to (\Box A_n \to \Box A),$$

which is equivalent by *PL* to

$$(\Box A_1 \wedge \ldots \wedge \Box A_n) \to \Box A.$$

This completes the inductive part of the proof. It follows now that the rule RK holds for any number $n \geqslant 0$ of conjuncts in the antecedents, since it holds for $n = 0$ and also for any $n > 0$ whenever it holds up to n.

Notice that only *PL*, RN, and K are used in the derivation of the rule RK. Moreover, using only *PL* and RK we can prove RN (trivially) and K (by RK on the tautology $((A \rightarrow B) \wedge A) \rightarrow B$ we get the theorem $(\Box(A \rightarrow B) \wedge \Box A) \rightarrow \Box B$, which is equivalent to K by *PL*). The moral of this is that we could equally well have chosen RK instead of RN and K in our axiomatization of *S5*.

Another special case of RK, when $n = 2$, is the rule of inference

$$\text{RR.} \quad \frac{(A \wedge B) \rightarrow C}{(\Box A \wedge \Box B) \rightarrow \Box C},$$

which expresses a limited principle of consequence (a proposition is necessary if it follows from a pair of propositions each of which is necessary). A direct proof of RR – using *PL*, RM, and K – can also be had, and it may illuminate the inductive part of the proof above for RK; we leave it as an exercise.

Three further theorems are worth mention.

N. $\Box \top$

M. $\Box(A \wedge B) \rightarrow (\Box A \wedge \Box B)$

C. $(\Box A \wedge \Box B) \rightarrow \Box(A \wedge B)$

Proofs of N, M, and C – using RN, RM, and RR, respectively – are not hard to find, so we leave them as exercises.

It is clear from our results in section 1.1 that every theorem is valid: all the axioms, T, 5, K, Df$\Diamond$, and *PL*, are valid, and the rules of inference RN and MP preserve validity. In short, the axiomatization is *sound*. It is moreover *complete*: every valid sentence is a theorem. This may not be so obvious, however, and it is not until chapter 5 that we are in a position to prove it.

We thus have two ways of characterizing the modal logic *S5* – one semantic, the other deductive. It bears emphasis, moreover, that the set of principles T, 5, K, Df$\Diamond$, *PL*, RN, and MP is not the only selection that provides an axiomatization – a deductive characterization – of *S5*. We have seen already, for example, that the rule RPL would do just as well as MP plus *PL*, and that RK could take the place of RN plus K. Such alterations result in equivalent, alternative axiomatizations of *S5* – equivalent since the axioms and rules of inference of each are derivable from the others, so that any sentence provable in one axiomatization is provable in the others.

Let us conclude this section with yet another axiomatization of *S5*, one of the best known. It is formulated on the basis of propositional logic by means of the rule RN together with the schemas T, B, 4, K, and Df$\Diamond$ as axioms. In other words, this axiomatization differs from the one with which we began only by having B and 4 as axioms in place of 5. For the sake of exposition we dub the set of theorems axiomatized in this way *S5′*.

Clearly, every theorem of *S5′* is also a theorem of *S5*, since every axiom and rule of *S5′* can be (and has been) proved in *S5*. Showing the reverse, that *S5′* includes *S5*, boils down to proving that the schema 5 is a theorem of *S5′* – i.e. that 5 can be derived on the basis of T, B, 4, K, Df$\Diamond$, *PL*, RN, and MP. To this end, observe that *S5′* has the rule RPL (because of *PL* and MP), that *S5′* has the rule RM (because of RPL, RN, and K), and that *S5′* has the theorem 4$\Diamond$ (exercise). So we may argue as follows.

1.	$\Diamond\Diamond A \rightarrow \Diamond A$	4$\Diamond$
2.	$\Box\Diamond\Diamond A \rightarrow \Box\Diamond A$	1, RM
3.	$\Diamond A \rightarrow \Box\Diamond\Diamond A$	B
4.	$\Diamond A \rightarrow \Box\Diamond A$	2, 3, *PL*

According to the last line the schema 5 is indeed a theorem of *S5′*. Therefore, the two axiomatizations are equivalent.

EXERCISES

Except where otherwise noted use any theorem or rule of inference established in section 1.2, and any theorems and rules established in previous exercises.

1.12. Derive the rule of inference RE in *S5*.

1.13. Derive the following rules of inference in *S5*.

$$(a)\ \frac{A \rightarrow B}{\Diamond A \rightarrow \Diamond B} \qquad (b)\ \frac{A \leftrightarrow B}{\Diamond A \leftrightarrow \Diamond B}$$

1.14. Derive the rule of inference RR in *S5* using only *PL*, RM, and K.

1.15. Prove that N, M, and C are theorems of *S5*.

1.16. Prove that the following schemas are theorems of *S5*.

(*a*) $\neg\Diamond\bot$

(*b*) $(\Diamond A \vee \Diamond B) \rightarrow \Diamond(A \vee B)$

(*c*) $\Diamond(A \vee B) \rightarrow (\Diamond A \vee \Diamond B)$

1.17. Prove that the following schemas are theorems of *S5*.

(*a*) $(\Box A \vee \Box B) \rightarrow \Box(A \vee B)$

(*b*) $\Diamond(A \wedge B) \rightarrow (\Diamond A \wedge \Diamond B)$

(*c*) $\Box(A \vee B) \rightarrow (\Diamond A \vee \Box B)$

(*d*) $(\Box A \wedge \Diamond B) \rightarrow \Diamond(A \wedge B)$

1.18. Derive the following rules of inference in *S5*.

(*a*) $\dfrac{A}{\Diamond A}$ (*b*) $\dfrac{\neg A}{\neg \Diamond A}$

1.19. Prove that the sentence $\Diamond\top$ is a theorem of *S5*.

1.20. Prove that the schema $4\Diamond$ is a theorem of *S5*.

1.21. Prove that the following schemas are theorems of *S5*.

(*a*) $\Box A \leftrightarrow \Box\Box A$

(*b*) $\Diamond A \leftrightarrow \Diamond\Diamond A$

(*c*) $\Diamond A \leftrightarrow \Box\Diamond A$

(*d*) $\Box A \leftrightarrow \Diamond\Box A$

1.22. Prove that the schema $\Diamond\Box A \rightarrow \Box\Diamond A$ is a theorem of *S5*.

1.23. Derive the following rules of inference in *S5*.

(*a*) $\dfrac{\Diamond A \rightarrow B}{A \rightarrow \Box B}$ (*b*) $\dfrac{A \rightarrow \Box B}{\Diamond A \rightarrow B}$

1.24. Prove that the schema $4\Diamond$ is a theorem of *S5′* (see the last paragraph of section 1.2).

1.25. Prove that *S5* is equivalently axiomatized if in the original axiomatization the axiom 5 is deleted in favor of the schema $A \rightarrow \Box\Box\Diamond A$.

1.26. Prove that *S5* is equivalently axiomatized if in the original axiomatization the axiom K is replaced by the schemas N, M, and C, and the rule of inference RN is replaced by RE.

1.27. We say that a system of modal logic is *consistent* when it does not contain $\bot$ as a theorem. It is clear that the system *S5* is consistent: the axiomatization is sound – i.e. every theorem is valid – and $\bot$ is not valid. Below we argue the consistency of *S5* in another way.

Let the mapping ϵ on the set of sentences be defined by the following clauses.

(1) $\epsilon(\mathbb{P}_n) = \mathbb{P}_n$, for $n = 0, 1, 2, \ldots$.

(2) $\epsilon(\top) = \top$.

(3) $\epsilon(\bot) = \bot$.
(4) $\epsilon(\neg A) = \neg\epsilon(A)$.
(5) $\epsilon(A \wedge B) = \epsilon(A) \wedge \epsilon(B)$.
(6) $\epsilon(A \vee B) = \epsilon(A) \vee \epsilon(B)$.
(7) $\epsilon(A \rightarrow B) = \epsilon(A) \rightarrow \epsilon(B)$.
(8) $\epsilon(A \leftrightarrow B) = \epsilon(A) \leftrightarrow \epsilon(B)$.
(9) $\epsilon(\Box A) = \epsilon(A)$.
(10) $\epsilon(\Diamond A) = \epsilon(A)$.

Thus ϵ is an 'erasure' transformation. It erases all occurrences of the modal operators $\Box$ and $\Diamond$ in a sentence A, but leaves A otherwise intact.

Now let us see that ϵ transforms axioms of *S5* into tautologies, and rules of inference of *S5* into rules of propositional logic. Clearly the erasure of a tautology (*PL*) is always a tautology. Moreover:

T. $\epsilon(\Box A \rightarrow A) = \epsilon(A) \rightarrow \epsilon(A)$.

5. $\epsilon(\Diamond A \rightarrow \Box\Diamond A) = \epsilon(A) \rightarrow \epsilon(A)$.

K. $\epsilon(\Box(A \rightarrow B) \rightarrow (\Box A \rightarrow \Box B)) =$
$(\epsilon(A) \rightarrow \epsilon(B)) \rightarrow (\epsilon(A) \rightarrow \epsilon(B))$.

Df$\Diamond$. $\epsilon(\Diamond A \leftrightarrow \neg\Box\neg A) = \epsilon(A) \leftrightarrow \neg\neg\epsilon(A)$.

The schemas on the right-hand side of these identities are all tautologies; so the erasure of any non-propositional axiom is a tautology. Finally, under erasure the rules of inference RN and MP become

$$\frac{\epsilon(A)}{\epsilon(A)} \quad \text{and} \quad \frac{\epsilon(A) \rightarrow \epsilon(B),\ \epsilon(A)}{\epsilon(B)}.$$

The first of these is merely a rule of repetition, and the second is just MP again.

It follows that under the mapping ϵ every theorem of *S5* is transformed into a tautology. Therefore, since $\epsilon(\bot)$ – i.e. $\bot$ – is not a tautology, $\bot$ is not a theorem of *S5*. So we have proved the consistency of *S5* once again.

(*a*) Apply ϵ to T, 5, K, Df$\Diamond$, and selected instances of *PL* to see that their erasures are tautologies.
(*b*) Show that if $\epsilon(A)$ is a tautology so is $\epsilon(\Box A)$, and that if $\epsilon(A \rightarrow B)$ and $\epsilon(A)$ are tautologies so is $\epsilon(B)$.
(*c*) Use ϵ on the schemas in exercises 1.15–1.17, 1.19–1.22, and 1.25.

(*d*) Show that the rules of inference in exercises 1.12–1.14, 1.18, and 1.23 hold under erasure.

(*e*) Consider the system that results when the schema $A \to \Box A$ is added as an axiom to *S5*. This system is not sound, since $A \to \Box A$ is not valid (see section 1.1), but it is consistent. Use the erasure transformation to prove this.

(*f*) Prove the consistency of the system that results when the schema $\Box\Diamond A \to \Diamond\Box A$ is added as an axiom to *S5*. (Is this system sound?)

(*g*) Referring to exercise 1.11, observe that the transformation τ can be employed like ϵ to prove the consistency of *S5*, since $\bot$ is not quantificationally valid. How do the schemas $A \to \Box A$ and $\Box\Diamond A \to \Diamond\Box A$ fare under τ?

1.28. Sometimes there is confusion about the meaning of rules of inference. For example, because of RN it might be thought that $A \to \Box A$ is a theorem. Similarly the rules RM and RE might mistakenly be regarded as evidence for the theoremhood of the schemas

$$(A \to B) \to (\Box A \to \Box B) \quad \text{and} \quad (A \leftrightarrow B) \to (\Box A \leftrightarrow \Box B).$$

Dispel this illusion by showing that neither schema is valid.

1.29. Show that $A \to \Box A$ is a theorem of the system that results when the schema $(A \to B) \to (\Box A \to \Box B)$ is added as an axiom to *S5*. Is this true if $(A \leftrightarrow B) \to (\Box A \leftrightarrow \Box B)$ is added to *S5*? What about the consistency of these systems?

1.30. Prove that $A \to \Box A$ is a theorem of the system that results when the schema $\Box\Diamond A \to \Diamond\Box A$ is added as an axiom to *S5*.

2

LOGICAL PRELIMINARIES

This chapter is an introduction to most of the concepts we shall use in studying modal logic.

In section 2.1 we set out most of the syntactic concepts. Section 2.2 introduces semantic concepts: the general idea of a model, truth conditions for non-modal sentences, and definitions of truth in a model and validity in a class of models. Filtrations of models are described in section 2.3. In section 2.4 the idea of a system of modal logic is explained, along with such relevant notions as theoremhood, deducibility, and consistency. Axiomatizability is discussed in section 2.5. Maximal sets of sentences and Lindenbaum's lemma occupy section 2.6. In section 2.7 we define determination and explain our approach, using canonical models, to proofs of determination. Finally in section 2.8 we outline our method of proving the decidability of systems of modal logic.

As the need arises the reader may wish to return to various sections of this chapter, for important definitions and theorems.

2.1. Syntax

This section is devoted to a recital of the basic syntactic concepts for the language of modal logic, many of which the reader has likely gleaned from chapter 1. The ideas are very simple. The few formal definitions we offer may be helpful, but they are not essential; we state them mainly for the sake of completeness and future reference.

Sentences. The language is founded on a denumerable set of *atomic sentences*:

$$\mathbb{P}_0, \mathbb{P}_1, \mathbb{P}_2, \ldots$$

These are the simplest sentences.

The non-atomic *molecular sentences* are formed by means of nine *syntactic operations*, or *operators*:

$$\top, \bot, \neg, \wedge, \vee, \rightarrow, \leftrightarrow, \Box, \Diamond$$

As we said in chapter 1, $\top$ and $\bot$ are zero-place operators, or constants; $\neg$, $\Box$, and $\Diamond$ are one-place operators; and $\wedge$, $\vee$, $\rightarrow$, and $\leftrightarrow$ are two-place operators.

The set of sentences may be defined formally as follows.

DEFINITION 2.1

(1) $\mathbb{P}_n$ is a *sentence*, for $n = 0, 1, 2, \ldots$.
(2) $\top$ is a *sentence*.
(3) $\bot$ is a *sentence*.
(4) $\neg$A is a *sentence* iff A is a *sentence*.
(5) A $\wedge$ B is a *sentence* iff A and B are *sentences*.
(6) A $\vee$ B is a *sentence* iff A and B are *sentences*.
(7) A $\rightarrow$ B is a *sentence* iff A and B are *sentences*.
(8) A $\leftrightarrow$ B is a *sentence* iff A and B are *sentences*.
(9) $\Box$A is a *sentence* iff A is a *sentence*.
(10) $\Diamond$A is a *sentence* iff A is a *sentence*.

We use A, B, C, ..., sometimes with superscripts or subscripts, for sentences. And we use Γ, Δ, E, ..., with occasional appurtenances, for sets of sentences. When a set of sentences Δ includes a set of sentences Γ – i.e. when $\Gamma \subseteq \Delta$ – we often say that Δ is an *extension* of Γ.

The sentence $\top$ is called the *verum* or *truth constant*, and $\bot$ is called the *falsum* or the *falsity constant*. $\neg$A is the *negation* of A, the operand A being the *negate*. A $\wedge$ B is the *conjunction* of A and B, the operands A and B being respectively the *left* and *right conjuncts*. A $\vee$ B is the *disjunction* of A and B, where A and B are the *left* and *right disjuncts*. The sentence A $\rightarrow$ B is the *conditional* of A and B; A is the *antecedent* and B is the *consequent*. A $\leftrightarrow$ B is the *biconditional* of A and B, of which A and B are the *left* and *right members*. The sentence $\Box$A is the *necessitation* of A, the operand A being the *necessitate*. There seems to be no standard terminology for a sentence of the form $\Diamond$A: let us call it the *possibilitation* of A, and call A the *possibilitate*. In $\Box$A and $\Diamond$A, A is also called the *matrix*, and $\Box$ and $\Diamond$ are called *prefixes*.

(These identifications of the operators as representing truth, falsity, negation, and so on, reflect the intended semantic analysis of the language. Properly speaking such characterizations are out of place in an account of the syntax of the language. Nevertheless, they do serve to reveal that the language has a surfeit of operators. For example, the meanings of all the operators can be given in terms of $\bot$, $\rightarrow$, and $\Box$. So we might have adopted just these operators as primitive and defined the rest. Our practice in this book is otherwise, however.)

The atomic sentences are all distinct. And of course it is intended that sentences of different forms really be distinct – so that, for example, no conditional is a necessitation. This is guaranteed by the assumption that the ranges of the syntactic operations are disjoint from one another and also from the set of atomic sentences. The unique readability of the sentences – the lack of ambiguity in their structures – is secured by the assumption that the operations are all one-to-one (so that, for example, two conditionals are identical if and only if their antecedents are and their consequents are).

It is important to remark that the set of sentences is enumerably infinite (denumerable). This means that the sentences of the language can be *enumerated*,

$$A_1, A_2, A_3, \ldots,$$

completely in an infinite list. This can be done in many ways; for our purposes it matters not *how* it can be done, but only *that* it can be done.

Conventions. There are some important conventions we observe throughout the book. One is that expressions of the forms

$$A_1 \wedge \ldots \wedge A_n$$

and

$$A_1 \vee \ldots \vee A_n$$

represent arbitrary but unspecified conjunction and disjunction of the sentences $A_1, \ldots, A_n$. The point here is that $\wedge$ and $\vee$ obey the logical (but not the syntactic) laws of associativity, so that it does not matter how such conjunctions or disjunctions are formed.

Another convention concerns sentences of the form

$$(A_1 \wedge \ldots \wedge A_n) \rightarrow A.$$

When $n = 0$, so that there are no conjuncts in the antecedent, we stipulate that the conditional is identical with its consequent, i.e. that $(A_1 \wedge \ldots \wedge A_n) \rightarrow A$ is just A when $n = 0$. Similarly, when $n = 0$ in a sentence of the form

$$A \rightarrow (A_1 \vee \ldots \vee A_n)$$

we shall say that the conditional is identical with the negation of its antecedent; i.e. $A \rightarrow (A_1 \vee \ldots \vee A_n)$ is $\neg A$ when $n = 0$. These conventions facilitate the expression of several principles.

Subsentences. A subsentence of a sentence A is any sentence that is a part of A, including A itself. This rather obvious idea is captured in the following recursive definition of the set Sn(A) of subsentences of A.

DEFINITION 2.2

(1) $\mathrm{Sn}(\mathbb{P}_n) = \{\mathbb{P}_n\}$, for $n = 0, 1, 2, \ldots$.
(2) $\mathrm{Sn}(\top) = \{\top\}$.
(3) $\mathrm{Sn}(\bot) = \{\bot\}$.
(4) $\mathrm{Sn}(\neg \mathrm{A}) = \{\neg \mathrm{A}\} \cup \mathrm{Sn}(\mathrm{A})$.
(5) $\mathrm{Sn}(\mathrm{A} \wedge \mathrm{B}) = \{\mathrm{A} \wedge \mathrm{B}\} \cup \mathrm{Sn}(\mathrm{A}) \cup \mathrm{Sn}(\mathrm{B})$.
(6) $\mathrm{Sn}(\mathrm{A} \vee \mathrm{B}) = \{\mathrm{A} \vee \mathrm{B}\} \cup \mathrm{Sn}(\mathrm{A}) \cup \mathrm{Sn}(\mathrm{B})$.
(7) $\mathrm{Sn}(\mathrm{A} \rightarrow \mathrm{B}) = \{\mathrm{A} \rightarrow \mathrm{B}\} \cup \mathrm{Sn}(\mathrm{A}) \cup \mathrm{Sn}(\mathrm{B})$.
(8) $\mathrm{Sn}(\mathrm{A} \leftrightarrow \mathrm{B}) = \{\mathrm{A} \leftrightarrow \mathrm{B}\} \cup \mathrm{Sn}(\mathrm{A}) \cup \mathrm{Sn}(\mathrm{B})$.
(9) $\mathrm{Sn}(\Box \mathrm{A}) = \{\Box \mathrm{A}\} \cup \mathrm{Sn}(\mathrm{A})$.
(10) $\mathrm{Sn}(\Diamond \mathrm{A}) = \{\Diamond \mathrm{A}\} \cup \mathrm{Sn}(\mathrm{A})$.

Thus, for example, the set of subsentences of the sentence $\Box(\mathbb{P}_0 \rightarrow \neg \Diamond(\mathbb{P}_1 \wedge \neg \mathbb{P}_0))$ – i.e. $\mathrm{Sn}(\Box(\mathbb{P}_0 \rightarrow \neg \Diamond(\mathbb{P}_1 \wedge \neg \mathbb{P}_0)))$ – is

$$\{\Box(\mathbb{P}_0 \rightarrow \neg \Diamond(\mathbb{P}_1 \wedge \neg \mathbb{P}_0)), \quad \mathbb{P}_0 \rightarrow \neg \Diamond(\mathbb{P}_1 \wedge \neg \mathbb{P}_0),$$
$$\mathbb{P}_0, \quad \neg \Diamond(\mathbb{P}_1 \wedge \neg \mathbb{P}_0), \quad \Diamond(\mathbb{P}_1 \wedge \neg \mathbb{P}_0), \quad \mathbb{P}_1 \wedge \neg \mathbb{P}_0,$$
$$\mathbb{P}_1, \quad \neg \mathbb{P}_0\},$$

as the reader should verify using the definition.

Another way of looking at the structure of a sentence is by way of a 'construction tree', as in figure 2.1 for the sentence $\Box(\mathbb{P}_0 \rightarrow \neg \Diamond(\mathbb{P}_1 \wedge \neg \mathbb{P}_0))$. The subsentences of the sentence appear at the nodes of the tree, and the branches indicate the order of application of the syntactic operations. Notice that a construction tree not only shows the subsentences of a sentence but also indicates their *occurrences* (for example, $\mathbb{P}_0$ has two occurrences in $\Box(\mathbb{P}_0 \rightarrow \neg \Diamond(\mathbb{P}_1 \wedge \neg \mathbb{P}_0))$ and so appears at two nodes on the construction tree).

We say that a set of sentences is *closed under subsentences* just in case it contains every subsentence of every sentence it contains. Thus the set of subsentences of a sentence is closed under subsentences; but so also is the set $\{\mathbb{P}_0, \neg \mathbb{P}_0, \neg\neg \mathbb{P}_0, \ldots\}$, which shows that a set of sentences closed under subsentences need not be the set of subsentences of any single sentence.

Modalities. Here it is easiest to think of the operators as symbols. By a *modality* we mean any finite sequence, possibly empty, of the operators $\neg$, $\Box$, and $\Diamond$; for example $\Box$, $\neg\neg$, $\Box\Diamond$, $\Diamond\Box\Box$, $\Box\neg\Diamond$, and $\Diamond\Box\neg$. The empty, or null, modality is signified by $\cdot$; thus $\cdot$ A is the same as A.

A modality is classified as *affirmative* just in case $\neg$ occurs in it an even number of times (including zero); it is classified as *negative* otherwise, i.e. just in case it contains an odd number of occurrences of $\neg$. The first four modalities mentioned above are affirmative, and the last two are negative.

The *dual of a modality* ϕ is the modality ϕ^* that results from interchanging $\Box$ and $\Diamond$ throughout ϕ. So, for example, the duals of the modalities above are: $\Diamond$, $\neg\neg$, $\Diamond\Box$, $\Box\Diamond\Diamond$, $\Diamond\neg\Box$, and $\Box\Diamond\neg$. Notice that the affirmative or negative quality of a modality is preserved by duality, and that the dual of the dual of ϕ is just ϕ (i.e. $\phi^{**} = \phi$).

We write

$$\phi^n \mathrm{A}$$

to indicate that the sentence A is subject to n iterations of the modality ϕ. Here n can be any number, 0, 1, 2, ... ϕ^0 A is just A, and ϕ^1 A is ϕA.

Figure 2.1. Construction tree for $\Box(\mathbb{P}_0 \to \neg\Diamond(\mathbb{P}_1 \wedge \neg\mathbb{P}_0))$.

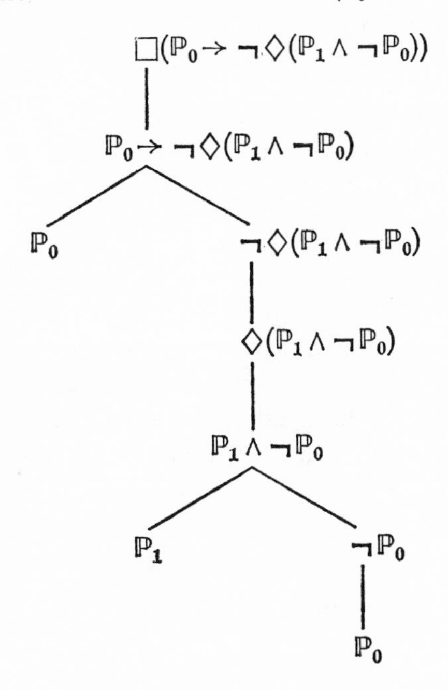

Thus our usage slightly extends the idea of iteration, but it helps to simplify the statement of many generalities. Formally, the idea may be defined as follows.

DEFINITION 2.3

(1) $\phi^0 A = A$.

(2) $\phi^n A = \phi\phi^{n-1}A$, for $n > 0$.

Finally, we say that a set of sentences is *modally closed* just in case ϕA is in the set whenever A is, for every modality ϕ. By the *modal closure* of a set of sentences we mean the result of adding to the set every modalization ϕA of any sentence A in the set.

Replacement. Given sentences A, B, and B′, we often wish to consider a sentence that results from the replacement of fixed occurrences of B in A by B′. As a matter of convenience we designate such a replacement by A[B/B′]. For example, let A be the sentence $\Box(\mathbb{P}_0 \to \neg\Diamond(\mathbb{P}_1 \wedge \neg\mathbb{P}_0))$, let B be $\mathbb{P}_0$, and let B′ be the sentence $\mathbb{P}_2 \vee \Box\neg\mathbb{P}_1$. Then A[B/B′] is any one of the following:

$$\Box(\mathbb{P}_0 \to \neg\Diamond(\mathbb{P}_1 \wedge \neg\mathbb{P}_0))$$
$$\Box((\mathbb{P}_2 \vee \Box\neg\mathbb{P}_1) \to \neg\Diamond(\mathbb{P}_1 \wedge \neg\mathbb{P}_0))$$
$$\Box(\mathbb{P}_0 \to \neg\Diamond(\mathbb{P}_1 \wedge \neg(\mathbb{P}_2 \vee \Box\neg\mathbb{P}_1)))$$
$$\Box((\mathbb{P}_2 \vee \Box\neg\mathbb{P}_1) \to \neg\Diamond(\mathbb{P}_1 \wedge \neg(\mathbb{P}_2 \vee \Box\neg\mathbb{P}_1)))$$

The point is that in any given context A[B/B′] is just one of these sentences: replacement is occurrence specific. Each of the four sentences can be obtained by making the replacement of $\mathbb{P}_0$ by $\mathbb{P}_2 \vee \Box\neg\mathbb{P}_1$ at the appropriate place (node) in the construction tree for $\Box(\mathbb{P}_0 \to \neg\Diamond(\mathbb{P}_1 \wedge \neg\mathbb{P}_0))$ (figure 2.1). Note that in the first result no occurrences of $\mathbb{P}_0$ are replaced.

Duality. If the language did not have $\to$ and $\leftrightarrow$ we could define the *dual of a sentence* A – written A* – simply as the result of replacing each atomic sentence in A by its negation and interchanging all occurrences of $\top$ and $\bot$, $\wedge$ and $\vee$, and $\Box$ and $\Diamond$ throughout. Because of the presence of $\to$ and $\leftrightarrow$, however, a more complicated definition is called for.

DEFINITION 2.4

(1) $\mathbb{P}_n{}^* = \neg\mathbb{P}_n$, for $n = 0, 1, 2, \ldots$.

(2) $\top^* = \bot$.

(3) $\bot^* = \top$.

(4) $(\neg A)^* = \neg(A^*)$.

(5) $(A \wedge B)^* = A^* \vee B^*$.

(6) $(A \vee B)^* = A^* \wedge B^*$.

(7) $(A \to B)^* = \neg(A^*) \wedge B^*$.

(8) $(A \leftrightarrow B)^* = A^* \leftrightarrow \neg(B^*)$.

(9) $(\Box A)^* = \Diamond(A^*)$.

(10) $(\Diamond A)^* = \Box(A^*)$.

According to clause (4) of this definition we may write $\neg A^*$ indifferently for $\neg(A^*)$ or $(\neg A)^*$. We do so frequently. Some examples of duality:

$$(\Diamond\top)^* = \Box\bot.$$
$$(\Box(A \vee B))^* = \Diamond(A^* \wedge B^*).$$
$$(\Box(A \to B) \to (\Box A \to \Box B))^* =$$
$$\neg\Diamond(\neg A^* \wedge B^*) \wedge \neg\Diamond(A^*) \wedge \Diamond(B^*).$$

Note that $(\phi A)^{**} = \phi^*(A^*)$, for any modality ϕ and sentence A.

Schemas. By a *schema* we mean a set of sentences, usually of a particular form. For example, when we refer to the schema

5. $\Diamond A \to \Box\Diamond A$

we mean the set of sentences of this form – a conditional with a possibilitation as antecedent and the necessitation of the possibilitation as consequent. An *instance* of a schema is thus a member of the set of sentences that constitutes the schema.

It is natural to think of schemas as linguistic items of a special kind, akin to but more abstract than the sentences that are their instances. Though we often speak this way, we do so mainly as a matter of convenience.

Occasionally we speak of the *dual of a schema*. This usage is easier to illustrate than to explain. So, for example, we regard the schemas in the following pairs as dual to each other.

T.	$\Box A \to A$	T$\Diamond$.	$A \to \Diamond A$
5.	$\Diamond A \to \Box\Diamond A$	5$\Diamond$.	$\Diamond\Box A \to \Box A$
M.	$\Box(A \wedge B) \to (\Box A \wedge \Box B)$	M$\Diamond$.	$(\Diamond A \vee \Diamond B) \to \Diamond(A \vee B)$

Decidability. Finally, we need to say something about decidability, at least as it applies to sets of sentences.

A set of sentences Γ is *decidable* just in case there is a decision procedure – an effective finitary method – for determining of any sentence in the language whether or not it is in Γ. For example, the set of sentences containing the negation sign is decidable, since there is a routine way of discovering whether or not a sentence has $\neg$ in its construction.

A related but weaker notion is that of effective enumerability. We say that a set of sentences Γ is *effectively enumerable* just when there is an effective method for telling of any sentence in Γ that it is in Γ. When a set has this property we also say that there is a *positive test* for membership in it. (Similarly, a *negative test* for Γ is a positive test for its complement $-\Gamma$, the set of sentences not in Γ.)

Clearly, Γ is decidable if and only if both it and its complement are effectively enumerable. For if Γ has a positive test there is an effective way of enumerating its members:

$$A_1, A_2, A_3, \ldots;$$

and if Γ has a negative test there is an effective way of enumerating the members of its complement:

$$\bar{A}_1, \bar{A}_2, \bar{A}_3, \ldots.$$

So to produce a decision procedure for Γ it is enough to combine the tests, for example by defining the effective enumeration

$$A_1, \bar{A}_1, A_2, \bar{A}_2, A_3, \bar{A}_3, \ldots.$$

Then to discover whether or not a sentence A is in Γ it is sufficient to check A against this series. Sooner or later (i.e. some finite distance into the series) A must appear, either as A_n or as $\bar{A}_n$, for some n.

EXERCISES

2.1. The schema $\neg A \leftrightarrow (A \rightarrow \bot)$ is a tautology and is always valid in this book. Hence negation might as well have been introduced definitionally in terms of the falsum and the conditional, by the stipulation that $\neg A = A \rightarrow \bot$. Similarly the verum could have been defined in terms of the falsum and negation ($\top \leftrightarrow \neg\bot$ is a tautology), which means that $\top$ is also definable in terms of $\bot$ and $\rightarrow$. Formulate tautological biconditionals for the operators $\wedge$, $\vee$, and $\leftrightarrow$, to show that they might have been defined in terms of $\perp$ and $\rightarrow$. (Refer to section 1.1 for the meanings.) Since the schema Df$\Diamond$, $\Diamond A \leftrightarrow \neg\Box\neg A$, is always valid in this book we see that all the operators are definable in terms of $\perp$, $\rightarrow$, and $\Box$.

2.2. Explain informally why the set of sentences is enumerable.

2.3. Describe construction trees for the following sentences.

(*a*) $\Diamond\top$
(*b*) $(\Box P_0 \vee \Box P_1) \rightarrow \Box(P_0 \vee P_1)$
(*c*) $P_0 \rightarrow \Diamond P_0$
(*d*) $\Diamond(P_0 \wedge P_1) \rightarrow (\Diamond P_0 \wedge \Diamond P_1)$
(*e*) $\Diamond\Box P_0 \rightarrow P_0$
(*f*) $\Box(P_0 \vee P_1) \rightarrow (\Diamond P_0 \vee \Box P_1)$
(*g*) $\Diamond\Diamond P_0 \rightarrow \Diamond P_0$
(*h*) $\Diamond(P_0 \rightarrow P_1) \leftrightarrow (\Box P_0 \rightarrow \Diamond P_1)$
(*i*) $\Diamond\Box P_0 \rightarrow \Box P_0$
(*j*) $\Diamond\top \leftrightarrow (\Box P_0 \rightarrow \Diamond P_0)$

2.4. Describe the sets of subsentences of the sentences in the preceding exercise.

2.5. Classify the following modalities as affirmative or negative, and then describe their duals.

(*a*) $\Box\Box\Diamond$
(*b*) $\Box\neg\Diamond\neg\Box$
(*c*) $\neg\neg\Box\neg\Diamond$
(*d*) $\Diamond\Diamond\neg\Diamond\neg$
(*e*) $\Box\neg\Diamond\neg\Diamond\Box\neg\neg\Diamond\Box\Diamond\Box\neg\Diamond\Box\Box$

2.6. Prove that for every $n > 0$, $\phi^n = \phi^{n-1}\phi$. (The proof is by induction on n.)

2.7. Using results from chapter 1 argue that in *S5* every modality is equivalent to one of the following six.

$$\cdot,\ \Box,\ \Diamond,\ \neg,\ \neg\Box,\ \neg\Diamond$$

That is, show that where ϕ is any modality one of the following schemas is a theorem of *S5*.

$$A \leftrightarrow \phi A \qquad \Box A \leftrightarrow \phi A \qquad \Diamond A \leftrightarrow \phi A$$
$$\neg A \leftrightarrow \phi A \qquad \neg\Box A \leftrightarrow \phi A \qquad \neg\Diamond A \leftrightarrow \phi A$$

Then argue that there are no further reductions of modalities in *S5*, i.e. that the list of six modalities is minimal for this system.

2.8. Describe the various results of replacing $\Diamond \mathbb{P}_0$ by $\neg\Box\neg\mathbb{P}_0$ in the following sentence.

$$(\Diamond \mathbb{P}_0 \wedge \Diamond \mathbb{P}_1) \rightarrow (\Diamond(\mathbb{P}_0 \wedge \mathbb{P}_1) \vee \Diamond(\Diamond \mathbb{P}_0 \wedge \mathbb{P}_1) \vee \Diamond(\mathbb{P}_0 \wedge \Diamond \mathbb{P}_1))$$

2.9. Describe the duals of the following sentences.

(*a*) $\Box\bot$
(*b*) $\Diamond \mathbb{P}_0 \leftrightarrow \neg\Box\neg\mathbb{P}_0$
(*c*) $\Diamond(\mathbb{P}_0 \wedge \mathbb{P}_1) \rightarrow (\Diamond \mathbb{P}_0 \wedge \Diamond \mathbb{P}_1)$
(*d*) $\Box(\mathbb{P}_0 \vee \mathbb{P}_1) \rightarrow (\Diamond \mathbb{P}_0 \vee \Box \mathbb{P}_1)$
(*e*) $\Diamond(\mathbb{P}_0 \rightarrow \mathbb{P}_1) \leftrightarrow (\Box \mathbb{P}_0 \rightarrow \Diamond \mathbb{P}_1)$

2.10. Describe the duals of the following schemas.

D. $\Box A \rightarrow \Diamond A$
B. $A \rightarrow \Box\Diamond A$
4. $\Box A \rightarrow \Box\Box A$

2.2. Models, truth, and validity

In chapter 1 a model is a structure $\langle W, P\rangle$ in which W is a set of possible worlds and P is an assignment of truth values to atomic sentences at possible worlds. The truth values of non-atomic sentences at possible worlds are determined by this structure, from which a definition of validity – as truth at all worlds in all models – emerges. In chapters 3 and 7 we successively generalize this notion of model to provide semantic analyses of ever more general systems of modal logic. Although it is impossible to say once for all what a model is, it will help to avoid repetition and redundancy later on if we state here some of the common features of the kinds of model introduced in chapters 3 and 7.

The set W of possible worlds and the assignment P of truth values are elements of the models in both chapter 3 and chapter 7. So we may describe a model in these senses as a structure

$$\mathcal{M} = \langle W, \ldots, P\rangle,$$

where:

(i) W is a set;

(ii) P is a function on the set $\{0, 1, 2, \ldots\}$ of natural numbers such that for each such number n, P_n is a subset of W (i.e. P: $\{0, 1, 2, \ldots\} \to \mathscr{P}(W)$);

and the ellipsis indicates the possibility of additional elements. (In (ii) we abandon talk of P as an abbreviation of an infinite sequence of subsets of W, in favor of describing it as a mapping from natural numbers to sets of possible worlds.)

A model is said to be *finite* if its set of worlds has only finitely many elements; otherwise the model is *infinite*. Finite models are important in connection with questions of decidability for modal logics.

Our description of a model is of sufficient detail to permit a statement of the truth conditions, at a possible world, of non-modal sentences, and to permit definitions of various degrees of validity.

Again we write $\models^{\mathscr{M}}_{\alpha}$ A to mean that A is *true at the possible world* α *in the model* $\mathscr{M}$. For non-modal sentences – those not of the forms $\Box$A and $\Diamond$A – this notion is defined as follows.

DEFINITION 2.5. Let α be a world in a model $\mathscr{M} = \langle W, \ldots, P \rangle$.

(1) $\models^{\mathscr{M}}_{\alpha} \mathbb{P}_n$ iff $\alpha \in P_n$, for $n = 0, 1, 2, \ldots$.

(2) $\models^{\mathscr{M}}_{\alpha} \top$.

(3) Not $\models^{\mathscr{M}}_{\alpha} \bot$.

(4) $\models^{\mathscr{M}}_{\alpha} \neg A$ iff not $\models^{\mathscr{M}}_{\alpha} A$.

(5) $\models^{\mathscr{M}}_{\alpha} A \wedge B$ iff both $\models^{\mathscr{M}}_{\alpha} A$ and $\models^{\mathscr{M}}_{\alpha} B$.

(6) $\models^{\mathscr{M}}_{\alpha} A \vee B$ iff either $\models^{\mathscr{M}}_{\alpha} A$ or $\models^{\mathscr{M}}_{\alpha} B$, or both.

(7) $\models^{\mathscr{M}}_{\alpha} A \to B$ iff if $\models^{\mathscr{M}}_{\alpha} A$ then $\models^{\mathscr{M}}_{\alpha} B$.

(8) $\models^{\mathscr{M}}_{\alpha} A \leftrightarrow B$ iff $\models^{\mathscr{M}}_{\alpha} A$ if and only if $\models^{\mathscr{M}}_{\alpha} B$.

Statements of the truth conditions of sentences of the forms $\Box$A and $\Diamond$A – as in section 1.1 – appear in chapter 3, in connection with standard models, and again in chapter 7, in connection with minimal models.

Definition 2.5 may be regarded as an account of the lowest degree of validity – truth at a possible world in a model. In chapter 1 we distinguished only one other degree of validity, the highest – truth at all worlds in all models. This was appropriate, since we were interested there only in an introductory exposition of one of the simplest systems of modal logic, *S5*. Because we seek both more subtlety and more generality in the chapters that follow, we need to define some intermediate degrees of validity, of which the sort in chapter 1 is but a limiting case.

Let us say that a sentence is *true in a model* just in case it is true at every world in the model. This is a level of truth, or validity, a degree above that of definition 2.5. Next, let us say that a sentence is *valid in a class of models* if and only if it is true in every model in the class, i.e. true at every world in every model in the class. Thus a sentence valid in the class of *all* models is valid simpliciter, as in chapter 1.

We write $\vDash^{\mathcal{M}} A$ to mean that A is true in the model $\mathcal{M}$, and $\vDash_{\mathsf{C}} A$ to mean that A is valid in the class C of models. We record these definitions formally.

DEFINITION 2.6. $\vDash^{\mathcal{M}} A$ iff for every world α in $\mathcal{M}$, $\vDash^{\mathcal{M}}_{\alpha} A$.

DEFINITION 2.7. $\vDash_{\mathsf{C}} A$ iff for every model $\mathcal{M}$ in C, $\vDash^{\mathcal{M}}_{\alpha} A$.

When $\vDash^{\mathcal{M}} A$ we also say that $\mathcal{M}$ is a *model of* (or *for*) A; and that $\mathcal{M}$ is a model of (for) a set of sentences Γ when $\mathcal{M}$ is a model of every sentence in Γ.

Falsity always means non-truth, so that to say A is false at α in $\mathcal{M}$ just means that not $\vDash^{\mathcal{M}}_{\alpha} A$, and to say A is false in $\mathcal{M}$ means that A is false at some world in $\mathcal{M}$. When A is false in $\mathcal{M}$ we also say that $\mathcal{M}$ is a *counter-model to* (or *for*) A. Similarly, $\mathcal{M}$ is a countermodel to (or for) a set of sentences Γ just in case some sentence in Γ is false at some world in $\mathcal{M}$.

Let us say that a sentence A *$\mathcal{M}$-implies* a sentence B if and only if B is true at every world in $\mathcal{M}$ at which A is, and that A and B are *$\mathcal{M}$-equivalent* just when they are true (and hence false) at exactly the same worlds in $\mathcal{M}$. Evidently, A and B are $\mathcal{M}$-equivalent just in case A $\mathcal{M}$-implies B and B $\mathcal{M}$-implies A. By saying that a set of sentences Γ is *logically finite relative to the model $\mathcal{M}$* we mean that every sentence in Γ is $\mathcal{M}$-equivalent to one or another of a finite number of sentences in Γ.

Two further concepts are important. We say that two possible worlds agree on a sentence when they both verify or both falsify it, and that two models agree on a sentence when they are both models or both counter-models for it. Two models are said to be *pointwise equivalent* if and only if their world sets can be put in one-to-one correspondence in such a way that corresponding worlds agree on all sentences. Two models are (simply) *equivalent* just in case they agree on all sentences. It follows that pointwise equivalent models are always equivalent, though not vice versa.

The notion of model under discussion here is of course too insubstantial to produce any results about modal logic proper. But an important result about propositional logic does emerge from the account so far given, to wit, that this logic is a part of every modal logic we shall study. By this we mean, here, that whenever a sentence A is a tautological consequence of sentences $A_1, \ldots, A_n$, then A is valid in any class of models in which all of $A_1, \ldots, A_n$ are. We state this formally.

THEOREM 2.8. *Let* A *be a tautological consequence of* $A_1, \ldots, A_n$ $(n \geqslant 0)$. *Then if* C *is any class of models such that* $\models_{\mathsf{C}} A_1, \ldots, \models_{\mathsf{C}} A_n$, *then* $\models_{\mathsf{C}} A$.

Proof. We prove the theorem by showing first that where C is any class of models:

(1) If A is a tautology, then $\models_{\mathsf{C}} A$.

(2) If $\models_{\mathsf{C}} A \to B$ and $\models_{\mathsf{C}} A$, then $\models_{\mathsf{C}} B$.

The reasoning for (1) and (2) repeats that for the corresponding results in section 1.1. Nevertheless, we go through it again here in detail.

For (1). Recall that a propositionally atomic sentence is one of the form $\mathbb{P}_n$, $\Box A$, or $\Diamond A$; that a valuation is an assignment of truth values (truth, falsity) to each of the propositionally atomic sentences; that sentences of other kinds are assigned truth values in the usual way (compare definition 2.5); and that a tautology is a sentence true in every valuation. In any model $\mathcal{M}$ each possible world α determines a valuation V_α, in the sense that for each propositionally atomic sentence A,

$$V_\alpha(A) = \text{truth iff } \models^{\mathcal{M}}_{\alpha} A.$$

Clearly, this definition yields the result that for *any* sentence A,

$$A \text{ is true in } V_\alpha \text{ iff } \models^{\mathcal{M}}_{\alpha} A.$$

(Clearly. But the proof is given as an exercise.)

Now to see that every tautology is valid in any class C of models. Suppose that A is a tautology, so that it is true in every valuation. Then for every world α in any model in C, A is true in the corresponding valuation V_α. From this it follows by the result above that for every world α in any model $\mathcal{M}$ in C, $\models^{\mathcal{M}}_{\alpha} A$, which means that $\models_{\mathsf{C}} A$.

For (2). Let C be a class of models, and suppose that both $\models_{\mathsf{C}} A \to B$ and $\models_{\mathsf{C}} A$. Then for every world α in every model $\mathcal{M}$ in C, both $\models^{\mathcal{M}}_{\alpha} A \to B$

and $\models_\alpha^{\mathcal{M}} A$. From this it follows that for every world α in every model $\mathcal{M}$ in C, $\models_\alpha^{\mathcal{M}} B$, which means that $\models_{\mathsf{C}} B$.

Now we can prove the theorem. Suppose A is a tautological consequence of $A_1, \ldots, A_n$ ($n \geqslant 0$). Then

$$A_1 \to (\ldots(A_n \to A)\ldots)$$

is a tautology, and hence by (1) is valid. If we suppose further that each of $A_1, \ldots, A_n$ is valid in a class C of models, then by reference to (2) – n times – we find that A is also valid in C. (Again, compare section 1.1.) This concludes the proof.

In practical terms, theorem 2.8 means that all propositionally correct modes of inference will be available when it comes to theorem proving later on; more precisely, that the rule of inference RPL introduced in chapter 1 is always correct.

We close this section with the idea of *truth sets*. The truth set, $\|A\|^{\mathcal{M}}$, of the sentence A in the model $\mathcal{M}$ is the set of worlds in $\mathcal{M}$ at which A is true. Formally:

DEFINITION 2.9. $\|A\|^{\mathcal{M}} = \{\alpha \text{ in } \mathcal{M} : \models_\alpha^{\mathcal{M}} A\}$.

The following theorem reveals the structures of truth sets of the several sorts of sentences.

THEOREM 2.10. *Let $\mathcal{M} = \langle W, \ldots, P\rangle$ be a model. Then:*

$$\begin{aligned}
&(1)\ \|\mathbb{P}_n\|^{\mathcal{M}} = P_n, \text{ for } n = 0, 1, 2, \ldots.\\
&(2)\ \|\top\|^{\mathcal{M}} = W.\\
&(3)\ \|\bot\|^{\mathcal{M}} = \varnothing.\\
&(4)\ \|\neg A\|^{\mathcal{M}} = W - \|A\|^{\mathcal{M}}.\\
&(5)\ \|A \wedge B\|^{\mathcal{M}} = \|A\|^{\mathcal{M}} \cap \|B\|^{\mathcal{M}}.\\
&(6)\ \|A \vee B\|^{\mathcal{M}} = \|A\|^{\mathcal{M}} \cup \|B\|^{\mathcal{M}}.\\
&(7)\ \|A \to B\|^{\mathcal{M}} = (W - \|A\|^{\mathcal{M}}) \cup \|B\|^{\mathcal{M}}.\\
&(8)\ \|A \leftrightarrow B\|^{\mathcal{M}} = ((W - \|A\|^{\mathcal{M}}) \cup \|B\|^{\mathcal{M}})\\
&\qquad\qquad\qquad\qquad \cap((W - \|B\|^{\mathcal{M}}) \cup \|A\|^{\mathcal{M}}).
\end{aligned}$$

The proof of this theorem, using definitions 2.5 and 2.9, is easy and is left as an exercise.

In a way, $\|A\|^{\mathcal{M}}$ can be regarded as the *proposition* expressed by the sentence A in the model $\mathcal{M}$. See exercise 2.22.

EXERCISES

2.11. Using definitions 2.6 and 2.7 prove that

$\models_{C} A$ iff $\models^{\mathcal{M}}_{\alpha} A$ for every world α in every model $\mathcal{M}$ in C.

2.12. Prove:

(*a*) A and B are $\mathcal{M}$-equivalent iff A $\mathcal{M}$-implies B and B $\mathcal{M}$-implies A.

(*b*) A $\mathcal{M}$-implies B iff $\models^{\mathcal{M}} A \rightarrow B$.

(*c*) A and B are $\mathcal{M}$-equivalent iff $\models^{\mathcal{M}} A \leftrightarrow B$.

2.13. Prove that a finite set of sentences is logically finite with respect to any model.

2.14. Prove that the set $\{\mathbb{P}_0, \neg\mathbb{P}_0, \neg\neg\mathbb{P}_0, \ldots\}$ is logically finite (with respect to any model).

2.15. Prove that two models are equivalent if they are pointwise equivalent.

2.16. Prove theorem 2.10.

2.17. Prove:

(*a*) $\|A\|^{\mathcal{M}} \subseteq \|B\|^{\mathcal{M}}$ iff $\models^{\mathcal{M}} A \rightarrow B$.

(*b*) $\|A\|^{\mathcal{M}} = \|B\|^{\mathcal{M}}$ iff $\models^{\mathcal{M}} A \leftrightarrow B$.

2.18. Prove:

(*a*) A $\mathcal{M}$-implies B iff $\|A\|^{\mathcal{M}} \subseteq \|B\|^{\mathcal{M}}$.

(*b*) A and B are $\mathcal{M}$-equivalent iff $\|A\|^{\mathcal{M}} = \|B\|^{\mathcal{M}}$.

2.19. In the proof of theorem 2.8 we say that the definition of the valuation V_α yields the result that for any sentence A,

A is true in V_α iff $\models^{\mathcal{M}}_{\alpha} A$.

The proof of this is by induction on the complexity of A. That is, we show that it holds if A is propositionally atomic, that it holds if A is $\top$ or $\bot$, and that it holds when A is a conjunction, disjunction, conditional, or biconditional given that it holds for all sentences of less complexity than A. We give the proof for the cases in which A is (*a*) propositionally atomic, (*b*) a negation, $\neg B$, and (*c*) a conditional, $B \rightarrow C$.

For (*a*). If A is propositionally atomic, then by the definition of truth in a valuation we have that A is true in V_α if and only if $V_\alpha(A)$ = truth. But by definition, $V_\alpha(\mathrm{A})$ = truth if and only if $\models_\alpha^{\mathcal{M}} \mathrm{A}$. So the result follows: A is true in V_α if and only if $\models_\alpha^{\mathcal{M}} \mathrm{A}$.

For the inductive cases (*b*) and (*c*) we make the hypothesis that the result holds for all sentences shorter than A.

For (*b*). $\neg\mathrm{B}$ is true in V_α if and only if B is not true in V_α (by the definition of truth in a valuation) if and only if not $\models_\alpha^{\mathcal{M}} \mathrm{B}$ (by the inductive hypothesis) if and only if $\models_\alpha^{\mathcal{M}} \neg\mathrm{B}$ (definition 2.5(4)). So the result obtains when A is a negation.

For (*c*):

$\mathrm{B} \to \mathrm{C}$ is true in V_α iff if B is true in V_α then C is true in V_α
– definition of truth in a valuation;

iff if $\models_\alpha^{\mathcal{M}} \mathrm{B}$ then $\models_\alpha^{\mathcal{M}} \mathrm{C}$
– inductive hypothesis;

iff $\models_\alpha^{\mathcal{M}} \mathrm{B} \to \mathrm{C}$
– definition 2.5(7).

So the result holds when A is a conditional.

The remaining cases are left for the reader. (Note that the inductive hypothesis does not apply to the cases in which A is $\top$ or $\bot$.)

2.20. Let us consider the models $\mathcal{M} = \langle W, R, P \rangle$ of exercise 1.10 and the associated truth conditions (9′) and (10′) for modal sentences. If we drop the assumption that the relation R is reflexive in such models, then the schema T is no longer valid. Similarly, if we drop the assumption that R is euclidean, then the schema 5 is not valid. For example, consider the model $\mathcal{M} = \langle W, R, P \rangle$ in which $W = \{\alpha, \beta\}$ (where $\alpha \neq \beta$), $R = \{\langle\alpha, \beta\rangle, \langle\beta, \beta\rangle\}$, and $P_n = \{\beta\}$ for every natural number n. In this model the relation R is euclidean, but not reflexive (it lacks the pair $\langle\alpha, \alpha\rangle$). Clearly, $\models_\beta^{\mathcal{M}} \mathbb{P}_0$, and since β is the only world to which α is related by R, $\models_\alpha^{\mathcal{M}} \Box\mathbb{P}_0$. But not $\models_\alpha^{\mathcal{M}} \mathbb{P}_0$. So the instance $\Box\mathbb{P}_0 \to \mathbb{P}_0$ of T fails at α in $\mathcal{M}$ – i.e. not $\models_\alpha^{\mathcal{M}} \Box\mathbb{P}_0 \to \mathbb{P}_0$ – which means that the schema T is not valid in the class of models of this sort in which the relation R is not reflexive.

Describe a model of this sort in which R is reflexive but not euclidean, such that an instance of the schema 5 is false at some possible world.

2.21. Consider models like those of the preceding exercise and exercise 1.10 except that no assumption whatever is made about the structure of the binary relation R. Show that instances of the following schemas can be falsified at possible worlds in such models.

D. $\Box\mathrm{A} \to \Diamond\mathrm{A}$

T. $\Box\mathrm{A} \to \mathrm{A}$

B. $A \to \Box\Diamond A$

4. $\Box A \to \Box\Box A$

5. $\Diamond A \to \Box\Diamond A$

Prove that K, Df$\Diamond$, and all tautologies are valid in any class of such models, and that validity in a class of such models is preserved by RN and MP.

2.22. If we think of the proposition expressed by a sentence as something that has within it the information about how the sentence comes true or false in various possible situations, then it is natural to regard a truth set $\|A\|^{\mathcal{M}}$ as the proposition expressed by the sentence A in the model $\mathcal{M}$. For such sets behave just like propositions in that for any possible world α in $\mathcal{M}$ $\|A\|^{\mathcal{M}}$ determines whether A is true or false at α (according as α is or is not a member of $\|A\|^{\mathcal{M}}$).

Notice that on this construal the truth set $\|\top\|^{\mathcal{M}}$ is a (perhaps *the*) necessary proposition in a model $\mathcal{M}$ (since it determines $\top$ as true everywhere in $\mathcal{M}$), and the truth conditions in chapter 1 of a necessitation can be stated:

$$\models_{\alpha}^{\mathcal{M}} \Box A \text{ iff } \|\top\|^{\mathcal{M}} = \|A\|^{\mathcal{M}}$$

– or, what comes to the same thing:

$$\models_{\alpha}^{\mathcal{M}} \Box A \text{ iff } \|\top\|^{\mathcal{M}} \subseteq \|A\|^{\mathcal{M}}.$$

In this latter formulation we have that $\Box A$ is true at α just in case the proposition expressed by A in $\mathcal{M}$ is implied by the necessary proposition in $\mathcal{M}$. (It is appropriate to regard inclusion between truth sets as implication between propositions, since $\|A\|^{\mathcal{M}} \subseteq \|B\|^{\mathcal{M}}$ if and only if $\models^{\mathcal{M}} A \to B$; see exercise 2.17.)

Using the truth sets $\|\top\|^{\mathcal{M}}$ and $\|A\|^{\mathcal{M}}$, give a succinct formulation of the truth conditions of $\Diamond A$ at the world α in $\mathcal{M}$.

2.3. Filtrations

Let Γ be a set of sentences closed under subsentences. For any model $\mathcal{M} = \langle W, \ldots, P \rangle$ we define the equivalence relation $\equiv$ on the worlds in $\mathcal{M}$ by the stipulation that, for α and β in $\mathcal{M}$,

$$\alpha \equiv \beta \text{ iff for every } A \in \Gamma, \models_{\alpha}^{\mathcal{M}} A \text{ if and only if } \models_{\alpha}^{\mathcal{M}} A.$$

That is to say, worlds in $\mathcal{M}$ are equivalent under $\equiv$ just in case they agree on every sentence in Γ. The relation $\equiv$ is indeed an equivalence, and so it divides the set of worlds in $\mathcal{M}$ into mutually exclusive, non-empty equivalence classes $[\alpha]$ for each α in $\mathcal{M}$; i.e. for each α in $\mathcal{M}$ we define:

$$[\alpha] = \{\beta \text{ in } \mathcal{M} : \alpha \equiv \beta\}.$$

Let us also define $\equiv$-equivalence classes of sets of worlds in $\mathcal{M}$, by saying

$$[X] = \{[\alpha] : \alpha \in X\},$$

whenever X is a set of worlds in $\mathcal{M}$; i.e. $[X]$ is the set of equivalence classes of worlds in X.

Notice that the definitions of $\equiv$, $[\alpha]$, and $[X]$ depend essentially on Γ, so that we should properly write $\equiv_\Gamma$, $[\alpha]_\Gamma$, and $[X]_\Gamma$. But we do not, since the omission is almost always harmless.

In terms of these constructions we can say in part what a filtration is. To wit, a *filtration of the model* $\mathcal{M} = \langle W, \ldots, P \rangle$ *through* Γ is a model $\mathcal{M}^* = \langle W^*, \ldots, P^* \rangle$ in which:

(i) $W^* = [W]$.

(ii) $P^*_n = [P_n]$, for each n such that $\mathbb{P}_n \in \Gamma$.

In other words, each world in $\mathcal{M}^*$ is an equivalence class of worlds in $\mathcal{M}$ (and each such class is a world in $\mathcal{M}^*$), and for each such world $[\alpha]$ and each atomic sentence $\mathbb{P}_n$ in Γ,

$$[\alpha] \in P^*_n \text{ iff } \alpha \in P_n.$$

We also call $\mathcal{M}^*$ a Γ-*filtration of* $\mathcal{M}$. A model is a *filtration*, simply, if and only if it is a filtration through some set of sentences (closed under subsentences).

The important thing about a filtration $\mathcal{M}^*$ of $\mathcal{M}$ through Γ is that a world α in $\mathcal{M}$ and its equivalence class $[\alpha]$ in $\mathcal{M}^*$ agree on every sentence in Γ; i.e. for every α in $\mathcal{M}$ and every A in Γ,

$$\models^{\mathcal{M}}_{\alpha} \text{A iff } \models^{\mathcal{M}^*}_{[\alpha]} \text{A}.$$

Equivalently: $[\|\text{A}\|^{\mathcal{M}}] = \|\text{A}\|^{\mathcal{M}^*}$, for every A in Γ. This fundamental filtration theorem is proved in chapter 3 for standard models, and again in chapter 7 for minimal models. (Indeed, it can be proved already for the cases in which A is non-modal. See exercise 2.27.)

It follows that a model $\mathcal{M}$ and a Γ-filtration $\mathcal{M}^*$ of it are always equivalent modulo Γ, which is to say that for every A in Γ,

$$\models^{\mathcal{M}} \text{A iff } \models^{\mathcal{M}^*} \text{A}.$$

For the theorem implies that, where A is a sentence in Γ

$$\models^{\mathcal{M}}_{\alpha} \text{A, for every } \alpha \text{ in } \mathcal{M}, \text{ iff, for every } \alpha \text{ in } \mathcal{M}, \models^{\mathcal{M}^*}_{[\alpha]} \text{A}.$$

The left side of this means that $\models^{\mathcal{M}} \mathrm{A}$ (definition 2.6), and the right side means that for every $[\alpha]$ in $\mathcal{M}^*$, $\models^{\mathcal{M}^*}_{[\alpha]} \mathrm{A}$, and hence that $\models^{\mathcal{M}^*} \mathrm{A}$.

And, quite generally, it follows from this that a sentence in Γ is valid in a class of models just in case it is valid in the class of Γ-filtrations of those models. That is, for every A in Γ,

$$\models_{\mathsf{C}} \mathrm{A} \text{ iff } \models_{\Gamma(\mathsf{C})} \mathrm{A},$$

where C is any class of models and $\Gamma(\mathsf{C})$ is the class of Γ-filtrations of them. The argument here is left as an exercise.

We explain in section 2.8 how filtrations are used to demonstrate the decidability of modal logics. In this connection it is important to observe that filtration through a finite set of sentences always yields a finite model. For if n is the number of sentences in Γ, then a filtration through Γ is a model having at most 2^n worlds (equivalence classes of worlds in the original model), that being the maximum number of ways that worlds can agree on sentences in Γ. And 2^n is finite if n is. In particular, if Γ is the set of subsentences of a sentence A, then Γ is finite and closed under subsentences, and every Γ-filtration is a finite model.

More generally, if Γ is logically finite relative to a model $\mathcal{M}$, then every Γ-filtration of $\mathcal{M}$ is finite. For where n is the smallest number of finitely many sentences in Γ to which all others in Γ are $\mathcal{M}$-equivalent, 2^n is finite and is the maximum number of ways that worlds in $\mathcal{M}$ can agree on the sentences in Γ.

EXERCISES

Let Γ be a set of sentences closed under subsentences, and let $\mathcal{M}^* = \langle W^*, \ldots, P^* \rangle$ be a Γ-filtration of a model $\mathcal{M} = \langle W, \ldots, P \rangle$.

2.23. Prove that $\equiv$ is an equivalence relation, i.e. that for every α, β, and γ in $\mathcal{M}$:

(*a*) $\alpha \equiv \alpha$;

(*b*) if $\alpha \equiv \beta$ and $\alpha \equiv \gamma$, then $\beta \equiv \gamma$.

2.24. Prove that for every world α and β and every set X of worlds in $\mathcal{M}$:

(*a*) $\beta \in [\alpha]$ iff $\alpha \equiv \beta$.

(*b*) If $\alpha \in X$, then $[\alpha] \in [X]$.

Give an example to show that (*b*) does not generally hold in reverse.

2.25. Let $\mathbb{P}_n$ be an atomic sentence in Γ. Prove that for every α and β in $\mathcal{M}$:

(*a*) If $\alpha \equiv \beta$, then $[\alpha] \in P_n^*$ iff $[\beta] \in P_n^*$.

(*b*) $[\alpha] \in P_n^*$ iff $\alpha \in P_n$.

2.26. Prove the equivalence of (*a*) and (*b*).

(*a*) For every A in Γ and every α in $\mathcal{M}$, $\models_\alpha^{\mathcal{M}}$ A if and only if $\models_{[\alpha]}^{\mathcal{M}^*}$ A.

(*b*) For every A in Γ, $[\|A\|^{\mathcal{M}}] = \|A\|^{\mathcal{M}^*}$.

2.27. We can prove (*a*) (equivalently, (*b*)) in the preceding exercise for non-modal sentences A in Γ. The proof is inductive with respect to the complexity of A in Γ. We give it for the cases in which A in Γ is (*a*) atomic, $\mathbb{P}_n$, (*b*) a negation, $\neg$B, and (*c*) a conditional, B$\rightarrow$C. The remaining cases are left for the reader.

For (*a*):

$\models_\alpha^{\mathcal{M}} \mathbb{P}_n$ iff $\alpha \in P_n$
– definition 2.5 (1);
iff $[\alpha] \in P_n^*$
– since $\mathbb{P}_n \in \Gamma$;
iff $\models_{[\alpha]}^{\mathcal{M}^*} \mathbb{P}_n$
– definition 2.5 (1).

So the result holds when A in Γ is atomic.

For the inductive cases we assume as an inductive hypothesis that the result holds for all sentences in Γ of less complexity than A; in particular, then, we assume that the result holds for every subsentence of A.

For (*b*):

$\models_\alpha^{\mathcal{M}} \neg$B iff not $\models_\alpha^{\mathcal{M}}$ B
– definition 2.5 (4);
iff not $\models_{[\alpha]}^{\mathcal{M}^*}$ B
– inductive hypothesis;
iff $\models_{[\alpha]}^{\mathcal{M}^*} \neg$B
– definition 2.5 (4).

So the result holds when A in Γ is a negation.

For (*c*):

$\models_\alpha^{\mathcal{M}}$ B$\rightarrow$C iff if $\models_\alpha^{\mathcal{M}}$ B then $\models_\alpha^{\mathcal{M}}$ C
– definition 2.5 (7);

iff if $\models_{[\alpha]}^{\mathcal{M}^*} \mathrm{B}$ then $\models_{[\alpha]}^{\mathcal{M}^*} \mathrm{C}$
– inductive hypothesis;

iff $\models_{[\alpha]}^{\mathcal{M}^*} \mathrm{B} \rightarrow \mathrm{C}$
– definition 2.5 (7).

So the result holds when A in Γ is a conditional.

2.28. Argue in detail that the result in the preceding exercise implies that for every A in Γ,

(*a*) $\models^{\mathcal{M}} \mathrm{A}$ iff $\models^{\mathcal{M}^*} \mathrm{A}$.

Then show that this in turn implies that for every A in Γ,

(*b*) $\models_{\mathsf{C}} \mathrm{A}$ iff $\models_{\Gamma(\mathsf{C})} \mathrm{A}$,

where $\Gamma(\mathsf{C})$ is the class of Γ-filtrations of models in C.

2.29. Let Γ be the set of subsentences of the sentence $\neg \mathbb{P}_0 \rightarrow \mathbb{P}_1$, and suppose that $W = \{\alpha, \beta, \gamma\}$, $P_0 = \{\alpha, \gamma\}$, $P_1 = \{\beta\}$, and $P_n = \{\alpha\}$ for $n > 1$. Describe W^*, P_0^*, and P_1^*. How many worlds are there in $\mathcal{M}^*$?

2.30. Suppose Γ is the set $\{\mathbb{P}_0, \neg \mathbb{P}_0, \neg\neg \mathbb{P}_0, \ldots\}$, which is logically finite relative to $\mathcal{M}$ (exercise 2.14) and closed under subsentences. What is the maximum number of possible worlds in $\mathcal{M}^*$?

2.31. Notice that the idea of a Γ-filtration does not depend upon Γ being closed under subsentences. But give a simple example of the failure of the filtration theorem in exercise 2.27 with respect to a filtration through a set of sentences *not* closed under subsentences.

In general, when speaking of a Γ-filtration we presuppose that Γ is closed under subsentences.

2.4. Systems of modal logic

We encountered a number of *rules of inference* in connection with *S5*, the system of modal logic presented in chapter 1. For example, modus ponens:

$$\text{MP.} \quad \frac{\mathrm{A} \rightarrow \mathrm{B},\ \mathrm{A}}{\mathrm{B}}$$

In general, a rule of inference has the form

$$\frac{\mathrm{A}_1, \ldots, \mathrm{A}_n}{\mathrm{A}},$$

where $n \geqslant 0$. The sentences $A_1, \ldots, A_n$ are the *hypotheses* of the rule; A is the *conclusion*. A set of sentences is said to be *closed under* – or, sometimes, simply to *have* – a rule of inference just in case the set contains the conclusion of the rule whenever it contains the hypotheses (or just contains the conclusion if, when $n = 0$, there are no hypotheses). Thus a set of sentences Γ is closed under the rule MP if and only if whenever $A \to B$ and A are in Γ, so is B.

We define systems of modal logic in terms of closure under the rule of inference

$$\text{RPL.} \quad \frac{A_1, \ldots, A_n}{A} \quad (n \geqslant 0),$$

where A is a tautological consequence of $A_1, \ldots, A_n$.

DEFINITION 2.11. A set of sentences is a *system of modal logic* iff it is closed under RPL.

Thus a system of modal logic is any set of sentences closed with respect to all propositionally correct modes of inference. We shall reserve Σ as a variable for sets of sentences that are systems of modal logic, and for brevity and variety we shall often call them systems or modal logics, or even, simply, logics. Examples of systems abound already: *S5*; the sets of sentences true at a world in a model, true in a model, valid in a class of models (theorem 2.8); the set of sentences itself (it is the largest system).

The *theorems* of a system are just the sentences in it. We usually write $\vdash_\Sigma A$ to mean that A is a theorem of Σ:

DEFINITION 2.12. $\vdash_\Sigma A$ iff $A \in \Sigma$.

Because systems are simply sets of sentences, relative strength is measured in terms of inclusion: a system is at least as strong as a system Σ – is a Σ-*system* – just in case it contains every theorem of Σ. (So Σ is always itself a Σ-system.)

We make official our usage in chapter 1 and denote the set of tautologies by *PL*.

THEOREM 2.13

(1) *PL is a system of modal logic.*

(2) *Every system of modal logic is a PL-system.*

(3) *PL is the smallest system of modal logic.*

Proof. For (1), suppose that A is a tautological consequence of tautologies $A_1, \ldots, A_n$. Then A, too, is a tautology. Thus *PL* is closed under RPL, and hence is a system of modal logic. For (2), note that when $n = 0$ the rule RPL means that A is a tautology. So any system of modal logic must contain every tautology, i.e. $PL \subseteq \Sigma$, for every system Σ. So every system of modal logic is a *PL*-system. (3) follows from (1) and (2).

It does very little violence to the conception of what constitutes a system of modal logic to count *PL* as one, and it simplifies matters enormously to do so.

In terms of theoremhood we characterize notions of *deducibility* and *consistency*. A sentence A is deducible from a set of sentences Γ in a system Σ – written $\Gamma \vdash_\Sigma A$ – if and only if Σ contains a theorem of the form

$$(A_1 \wedge \ldots \wedge A_n) \rightarrow A,$$

where the conjuncts A_i $(i = 1, \ldots, n)$ of the antecedent are sentences in Γ. A set of sentences Γ is consistent in Σ – written $\mathrm{Con}_\Sigma \Gamma$ – just in case the sentence $\perp$ is not Σ-deducible from Γ. Thus Γ is *in*consistent in Σ – $\mathrm{C\o n}_\Sigma \Gamma$ – just when $\Gamma \vdash_\Sigma \perp$. We record these definitions formally.

DEFINITION 2.14. $\Gamma \vdash_\Sigma A$ iff there are $A_1, \ldots, A_n \in \Gamma$ $(n \geqslant 0)$ such that $\vdash_\Sigma (A_1 \wedge \ldots \wedge A_n) \rightarrow A$.

DEFINITION 2.15. $\mathrm{Con}_\Sigma \Gamma$ iff not $\Gamma \vdash_\Sigma \perp$.

Theoremhood, deducibility, and consistency, so defined, have all the expected properties, many of which are enumerated in the following theorem.

THEOREM 2.16

(1) $\vdash_\Sigma A$ *iff* $\varnothing \vdash_\Sigma A$.
(2) $\vdash_\Sigma A$ *iff for every* Γ, $\Gamma \vdash_\Sigma A$.
(3) *If* $\Gamma \vdash_{PL} A$, *then* $\Gamma \vdash_\Sigma A$.
(4) *If* $A \in \Gamma$, *then* $\Gamma \vdash_\Sigma A$.
(5) *If* $\Gamma \vdash_\Sigma B$ *and* $\{B\} \vdash_\Sigma A$, *then* $\Gamma \vdash_\Sigma A$.
(6) *If* $\Gamma \vdash_\Sigma A$ *and* $\Gamma \subseteq \Delta$, *then* $\Delta \vdash_\Sigma A$.
(7) $\Gamma \vdash_\Sigma A$ *iff there is a finite subset* Δ *of* Γ *such that* $\Delta \vdash_\Sigma A$.
(8) $\Gamma \vdash_\Sigma A \rightarrow B$ *iff* $\Gamma \cup \{A\} \vdash_\Sigma B$.
(9) $\mathrm{Con}_\Sigma \Gamma$ *iff there is an* A *such that not* $\Gamma \vdash_\Sigma A$.

(10) $\text{Con}_\Sigma \Gamma$ *iff there is no* A *such that both* $\Gamma \vdash_\Sigma A$ *and* $\Gamma \vdash_\Sigma \neg A$.

(11) *If* $\text{Con}_\Sigma \Gamma$, *then* $\text{Con}_{PL} \Gamma$.

(12) *If* $\text{Con}_\Sigma \Gamma$ *and* $\Delta \subseteq \Gamma$, *then* $\text{Con}_\Sigma \Delta$.

(13) $\text{Con}_\Sigma \Gamma$ *iff for every finite subset* Δ *of* Γ, $\text{Con}_\Sigma \Delta$.

(14) $\Gamma \vdash_\Sigma A$ *iff* $\text{Cøn}\, \Gamma \cup \{\neg A\}$.

(15) $\text{Con}_\Sigma \Gamma \cup \{A\}$ *iff not* $\Gamma \vdash_\Sigma \neg A$.

Proof. We prove only some of the parts of the theorem and leave the rest as exercises.

For (1). If $\vdash_\Sigma A$, then there is a Σ-theorem of the form $(A_1 \wedge \ldots \wedge A_n) \rightarrow A$ – to wit, where $n = 0$ and the conditional is just A. Since the non-existent A_is of the antecedent are all in ø, $ø \vdash_\Sigma A$. Conversely, if $ø \vdash_\Sigma A$, it must be that $\vdash_\Sigma (A_1 \wedge \ldots \wedge A_n) \rightarrow A$, where $n = 0$, etc. That is, $\vdash_\Sigma A$.

For (2). The reasoning for left-to-right is like that for (1) and is left to the reader. For the reverse, suppose that $\Gamma \vdash_\Sigma A$, for every set of sentences Γ. Then in particular, $ø \vdash_\Sigma A$, which by (1) means that $\vdash_\Sigma A$.

For (3). This simply says that deducibility within the means of propositional logic is acceptable within any system of modal logic. For if $\Gamma \vdash_{PL} A$, then there is a *PL*-theorem of the form $(A_1 \wedge \ldots \wedge A_n) \rightarrow A$, where the antecedent's conjuncts are members of Γ. By theorem 2.13 (2), it is also a Σ-theorem, for any system Σ. So $\Gamma \vdash_\Sigma A$.

For (4). This expresses the reflexivity of the deducibility relation. Suppose that $A \in \Gamma$. The sentence $A \rightarrow A$ is a tautology, hence a *PL*-theorem, hence a Σ-theorem for any system Σ. It is also a conditional with consequent A, the antecedent of which is in Γ. Therefore, $\Gamma \vdash_\Sigma A$.

For (5). This is a statement of the transitivity of the deducibility relation. The proof is somewhat complicated, and we leave it as an exercise.

For (6). According to this, deducibility obeys a principle of augmentation, or strengthening. The proof is easy and is left for the reader.

For (7). Deducibility is compact, in the important sense that deducibility from a set of sentences always implies deducibility from a finite portion of the set. This follows at once from the fact that the number of conjuncts in the antecedent of the requisite conditional $(A_1 \wedge \ldots \wedge A_n) \rightarrow A$, is always finite. The right-to-left part follows from (6).

For (8). This states the so-called deduction theorem for systems of modal logic – that a conditional is deducible from a set of sentences just in case its consequent is deducible from the set enlarged by the addition of the antecedent. The proof is rather easy and is left as an exercise.

For (9). This is an alternative way of characterizing consistency. Suppose that $\mathrm{Con}_\Sigma \Gamma$, i.e. that not $\Gamma \vdash_\Sigma \bot$. Clearly, then, there is a sentence A such that not $\Gamma \vdash_\Sigma A$. For the reverse, suppose that $\mathrm{C\emptyset n}_\Sigma \Gamma$, i.e. that $\Gamma \vdash_\Sigma \bot$. By *PL* and part (3), $\{\bot\} \vdash_\Sigma A$, for every sentence A. So by (5), $\Gamma \vdash_\Sigma A$, for every sentence A.

For (10). Another definition of consistency, the proof is left to the reader. Note that it says that a set of sentences is inconsistent just in case there is a sentence such that both it and its negation are deducible from the set.

For (11). This is the same as saying that every *PL*-inconsistent set of sentences is also Σ-inconsistent, for any system Σ. For if $\mathrm{Con}_\Sigma \Gamma$, then not $\Gamma \vdash_\Sigma \bot$. Hence by (3), not $\Gamma \vdash_{PL} \bot$, i.e. $\mathrm{Con}_{PL} \Gamma$.

For (12). This expresses a principle of diminution for consistency – equivalently, augmentation for inconsistency. The proof uses (6) and is left as an exercise.

For (13). This uses consistency to express compactness. Note the contrapositive formulation: a set of sentences is inconsistent if and only if it has a finite inconsistent subset. The proof, which uses (7), is left for the reader.

For (14). Here is a characterization of deducibility in terms of consistency (or inconsistency). Suppose that $\Gamma \vdash_\Sigma A$. By (6) we see that $\Gamma \cup \{\neg A\} \vdash_\Sigma A$, and by (4) that $\Gamma \cup \{\neg A\} \vdash_\Sigma \neg A$. Hence by (10), $\mathrm{C\emptyset n}_\Sigma \Gamma \cup \{\neg A\}$. Conversely, suppose that $\mathrm{C\emptyset n}_\Sigma \Gamma \cup \{\neg A\}$, i.e. that $\Gamma \cup \{\neg A\} \vdash_\Sigma \bot$. Then by (8), $\Gamma \vdash_\Sigma \neg A \to \bot$. By *PL* and (3), $\{\neg A \to \bot\} \vdash_\Sigma A$. So by (5), $\Gamma \vdash_\Sigma A$.

For (15). The proof uses (14) and is left as an exercise.

EXERCISES

Use the definitions and theorems in section 2.4 and any results established in previous exercises.

2.32. With reference to definition 2.11, explain why the following sets of sentences are systems of modal logic.

(*a*) the set of theorems of *S5*

(*b*) the set of sentences true at a possible world

(*c*) the set of sentences true in a model

(*d*) the set of sentences valid in a class of models

(*e*) the set of (all) sentences

2.33. Explain why a system Σ of modal logic is always a Σ-system.

2.34. Prove:

(*a*) $\Gamma \vdash_{\Sigma} \top$ (for all Γ).
(*b*) $\{\bot\} \vdash_{\Sigma} A$ (for all A).
(*c*) $\{A, \neg A\} \vdash_{\Sigma} \bot$.
(*d*) $\{\neg A \rightarrow \bot\} \vdash_{\Sigma} A$.
(*e*) $\Gamma \vdash_{\Sigma} A \wedge B$ iff $\Gamma \vdash_{\Sigma} A$ and $\Gamma \vdash_{\Sigma} B$.
(*f*) If $\Gamma \vdash_{\Sigma} A$ or $\Gamma \vdash_{\Sigma} B$, then $\Gamma \vdash_{\Sigma} A \vee B$.
(*g*) If $\Gamma \vdash_{\Sigma} A \vee B$ and $\Gamma \vdash_{\Sigma} \neg A$, then $\Gamma \vdash_{\Sigma} B$.
(*h*) If $\Gamma \vdash_{\Sigma} A \rightarrow B$ and $\Gamma \vdash_{\Sigma} A$, then $\Gamma \vdash_{\Sigma} B$.
(*i*) If $\Gamma \vdash_{\Sigma} \neg(A \rightarrow B)$, then $\Gamma \vdash_{\Sigma} A$ and $\Gamma \vdash_{\Sigma} \neg B$.
(*j*) If $\Gamma \vdash_{\Sigma} B$, then $\Gamma \vdash_{\Sigma} A \rightarrow B$.
(*k*) If $\Gamma \vdash_{\Sigma} \neg A$, then $\Gamma \vdash_{\Sigma} A \rightarrow B$.
(*l*) If $\Gamma \vdash_{\Sigma} A \rightarrow B$ and $\Gamma \vdash_{\Sigma} B \rightarrow C$, then $\Gamma \vdash_{\Sigma} A \rightarrow C$.
(*m*) $\Gamma \vdash_{\Sigma} (A \wedge B) \rightarrow C$ iff $\Gamma \vdash_{\Sigma} A \rightarrow (B \rightarrow C)$.
(*n*) $\Gamma \vdash_{\Sigma} A \leftrightarrow B$ iff $\Gamma \vdash_{\Sigma} A \rightarrow B$ and $\Gamma \vdash_{\Sigma} B \rightarrow A$.

2.35. Complete the proof of theorem 2.16 (parts (2), (5), (6), (7), (8), (10), (12), (13), and (15)).

2.36. Prove the following generalization of part (5) of theorem 2.16.

If $\Gamma \vdash_{\Sigma} B$, for every sentence B in Δ, and $\Delta \vdash_{\Sigma} A$, then $\Gamma \vdash_{\Sigma} A$.

2.37. We may define a concept of provable equivalence, with respect to a system Σ, by saying that two sentences A and B are Σ-*equivalent* (written $A \sim_{\Sigma} B$) just in case their biconditional is a theorem of Σ, i.e. just in case $\vdash_{\Sigma} A \leftrightarrow B$. Using this definition, prove:

(*a*) $A \sim_{\Sigma} A$.
(*b*) If $A \sim_{\Sigma} B$, then $B \sim_{\Sigma} A$.
(*c*) If $A \sim_{\Sigma} B$ and $B \sim_{\Sigma} C$, then $A \sim_{\Sigma} C$.
(*d*) If $A \sim_{\Sigma} B$ and $A \sim_{\Sigma} C$, then $B \sim_{\Sigma} C$.
(*e*) For every sentence A there is a sentence B such that $A \sim_{\Sigma} B$.
(*f*) If $A \sim_{PL} B$, then $A \sim_{\Sigma} B$.

2.38. With respect to a system Σ we say that a set of sentences Γ is *deductively closed* – *Σ-closed*, for short – just in case Γ contains every sentence that is Σ-deducible from it; i.e. Γ is Σ-closed if and only if $A \in \Gamma$ for every sentence A such that $\Gamma \vdash_{\Sigma} A$.

Prove that for each system Σ the Σ-closed sets of sentences are exactly the Σ-systems, i.e. that Γ is a Σ-closed set of sentences just in case Γ is a Σ-system.

2.5. Axiomatizability

Let us say that a rule of inference is *reasonable* if there is an effective way of telling when sentences are related by it as hypotheses and conclusion. For example, the rule MP is reasonable, since it is a decidable matter whether three sentences are of the forms $A \to B$, A, and B (if so, the first two are hypotheses of MP and the last is a conclusion).

Now every system of modal logic Σ can be regarded as the set of sentences generated from some subset Γ of its theorems by a set of rules of inference. This is trivial, since Σ is always generated from the subset Σ by the rule

$$\frac{A}{A}.$$

But when Γ is a decidable set of sentences and the rules of inference are reasonable and finite in number, Σ is said to be *axiomatizable*, and Γ is said to be a set of *axioms* for Σ. Together the axioms and rules constitute an *axiomatization* of the logic. For example, the system *S5* was axiomatized in chapter 1 by means of the axioms T, 5, K, Df◇, and *PL* and the rules RN and MP. These axioms form a decidable set, and the rules are reasonable.

Though most of the systems of modal logic treated in this book are axiomatizable, this is by no means true of every system. For example, the set of sentences true at a world in a model is a system, but it is rarely axiomatizable.

Axiomatizable systems are important because they admit a notion of proof and hence a positive test for theoremhood. By a *proof* in an axiomatizable system we mean a finite sequence of sentences each of which either is an axiom or follows from previous sentences in the sequence by one of the rules of inference. Thus, for example, the following sequence of sentences is a proof in *S5* relative to the axiomatization just mentioned.

$$\Box\neg A \to \neg A$$

$$(\Box\neg A \to \neg A) \to (A \to \neg\Box\neg A)$$

$$A \to \neg\Box\neg A$$
$$(A \to \neg\Box\neg A) \to ((\Diamond A \leftrightarrow \neg\Box\neg A) \to (A \to \Diamond A))$$
$$(\Diamond A \leftrightarrow \neg\Box\neg A) \to (A \to \Diamond A)$$
$$\Diamond A \leftrightarrow \neg\Box\neg A$$
$$A \to \Diamond A$$

The first, second, fourth, and sixth sentences are instances of T, *PL*, *PL*, and Df$\Diamond$, respectively; each of the others follows from its immediate two predecessors by MP. A proof is a proof *of* its last sentence. The theorems of an axiomatizable system are thus just the sentences for which there are proofs, i.e. just the sentences that terminate proofs in an axiomatization of the system. So the sequence above is a proof of $A \to \Diamond A$, which shows that this is a theorem of *S5*.

(One must distinguish between a proof *in* a system, like the sequence above, and proofs *about* the system, like that for $A \to \Diamond A$ in section 1.2. Though the latter is set out in a sequence of lines, there is no claim that the sequence of sentences there is a proof, relative to some axiomatization, in *S5*. It is, rather, a graphic way of stating a proof – *about S5* – that $A \to \Diamond A$ is a theorem.)

Note that it is a decidable question whether a sequence of sentences is a proof relative to an axiomatization of a system. For a sentence can appear in a proof if and only if it is a member of a decidable set of axioms or is inferred from earlier sentences by a reasonable rule of inference. This means that there is a positive test for theoremhood in an axiomatizable system. For the proofs – being decidable finite sequences – can be enumerated effectively in an infinite series,

$$p_1, p_2, p_3, \ldots.$$

So if a sentence is a theorem it lies at the end of some proof p_n and hence will be discovered after inspecting at most the first n proofs. This may not be a practical test, but it is none the less foolproof: if a sentence is a theorem there is a proof of it some finite distance into the enumeration.

Of course if a sentence is not a theorem it will never be discovered at the end of a proof, and there is no guarantee that this will become known after inspecting any finite number of proofs. But this is only to say that we have a positive and not a negative test for theoremhood – that the test is not a decision procedure for theoremhood. In general, axiomatizability only implies the existence of a positive test for theoremhood; something more is required to show the existence of a negative test. One kind of negative test is described in section 2.8.

EXERCISES

2.39. Explain why the rules of inference in section 1.2 are all reasonable. Give an example of an unreasonable rule.

2.40. Explain, accurately but informally, why the proofs in an axiomatization can be effectively enumerated.

2.41. The system *PL* (the set of tautologies) is axiomatizable. Why?

2.6. Maximality and Lindenbaum's lemma

A set of sentences is *maximal* in a system Σ just in case it is Σ-consistent and has only Σ-inconsistent proper extensions. Intuitively, a set is maximal if it is consistent and contains as many sentences as it can without becoming inconsistent. We write $\text{Max}_\Sigma \Gamma$ to mean that Γ is Σ-maximal, and we state the definition as follows.

DEFINITION 2.17. $\text{Max}_\Sigma \Gamma$ iff (i) $\text{Con}_\Sigma \Gamma$, and (ii) for every A, if $\text{Con}_\Sigma \Gamma \cup \{A\}$, then $A \in \Gamma$.

Note that clause (ii) says that, where Γ is maximal, the addition of a sentence not already in Γ yields an inconsistent set of sentences.

THEOREM 2.18. *Let Γ be a Σ-maximal set of sentences. Then:*

(1) $A \in \Gamma$ *iff* $\Gamma \vdash_\Sigma A$.
(2) $\Sigma \subseteq \Gamma$.
(3) $\top \in \Gamma$.
(4) $\bot \notin \Gamma$.
(5) $\neg A \in \Gamma$ *iff* $A \notin \Gamma$.
(6) $A \wedge B \in \Gamma$ *iff both* $A \in \Gamma$ *and* $B \in \Gamma$.
(7) $A \vee B \in \Gamma$ *iff either* $A \in \Gamma$ *or* $B \in \Gamma$.
(8) $A \rightarrow B \in \Gamma$ *iff if* $A \in \Gamma$ *then* $B \in \Gamma$.
(9) $A \leftrightarrow B \in \Gamma$ *iff* $A \in \Gamma$ *if and only if* $B \in \Gamma$.
(10) Γ *is a Σ-system.*

Proof. We assume throughout that Γ is a Σ-maximal set of sentences.

For (1). According to this, deducibility from a maximal set of sentences coincides with membership in it. Left-to-right is just theorem 2.16(4).

For the reverse, suppose – to reach a contradiction – that $\Gamma \vdash_\Sigma A$, but $A \notin \Gamma$. Then by the maximality of Γ, $\text{Cøn}_\Sigma\, \Gamma \cup \{A\}$. From this by theorem 2.16(15), $\Gamma \vdash_\Sigma \neg A$. So by theorem 2.16(10), $\text{Cøn}_\Sigma\, \Gamma$, which contradicts the maximality of Γ.

For (2). To show that a Σ-maximal set of sentences always contains the theorems of Σ, suppose that $A \in \Sigma$, i.e. $\vdash_\Sigma A$. Then by theorem 2.16(2), A is Σ-deducible from every set of sentences. In particular, $\Gamma \vdash_\Sigma A$, which by (1) means that $A \in \Gamma$.

For (3). Observe that $\vdash_\Sigma \top$, since $\vdash_{PL} \top$, and use (2).

For (4). Suppose, per absurdum, that $\bot \in \Gamma$. Then $\Gamma \vdash_\Sigma \bot$, which means that $\text{Cøn}_\Sigma\, \Gamma$. This contradicts the maximality of Γ.

For (5). This is best divided into two: (i) Not both $A \in \Gamma$ and $\neg A \in \Gamma$. (ii) Either $A \in \Gamma$ or $\neg A \in \Gamma$. For (i), assume to the contrary that Γ contains both A and $\neg A$. Then by (1) both are deducible from Γ, which means that Γ is inconsistent. This contradicts the maximality of Γ. For (ii), suppose to the contrary that Γ contains neither A nor $\neg A$. Then by (1) neither is deducible from Γ. This means (theorem 2.16(14, 15)) that $\text{Con}_\Sigma\, \Gamma \cup \{A\}$ and $\text{Con}_\Sigma\, \Gamma \cup \{\neg A\}$. By definition, then, $A \in \Gamma$ and $\neg A \in \Gamma$, which we have just shown to be impossible.

For (6). Suppose that $A \wedge B \in \Gamma$, so that by (1), $\Gamma \vdash_\Sigma A \wedge B$. By *PL* each of the conjuncts is deducible from the conjunction, so – via theorem 2.16(5) – $\Gamma \vdash_\Sigma A$ and $\Gamma \vdash_\Sigma B$. By (1), $A \in \Gamma$ and $B \in \Gamma$. The reasoning for right-to-left is similar and is left for the reader.

For (7). Suppose that $A \vee B \in \Gamma$. By (1), $\Gamma \vdash_\Sigma A \vee B$. Now suppose neither disjunct is in Γ, so that by (5), $\neg A \in \Gamma$ and $\neg B \in \Gamma$. Then by (1), $\Gamma \vdash_\Sigma \neg A$ and $\Gamma \vdash_\Sigma \neg B$. We leave it to the reader to reach a contradiction from here. Conversely, suppose that either $A \in \Gamma$ or $B \in \Gamma$. By (1), either $\Gamma \vdash_\Sigma A$ or $\Gamma \vdash_\Sigma B$. By *PL* $A \vee B$ is deducible from each of its disjuncts. So by theorem 2.16(5), $\Gamma \vdash_\Sigma A \vee B$. Hence by (1), $A \vee B \in \Gamma$.

For (8). Note that left-to-right means that maximal sets of sentences are closed under MP. To show it, suppose that $A \rightarrow B \in \Gamma$ and $A \in \Gamma$. By (1), $\Gamma \vdash_\Sigma A \rightarrow B$ and $\Gamma \vdash_\Sigma A$. By *PL* B is deducible from these sentences. So by theorem 2.16(5), $\Gamma \vdash_\Sigma B$, and hence by (1), $B \in \Gamma$. For right-to-left, suppose that $A \rightarrow B \notin \Gamma$, to show that $A \in \Gamma$ and $B \notin \Gamma$. By (5) and then (1), $\Gamma \vdash_\Sigma \neg(A \rightarrow B)$. By *PL* each of A and $\neg B$ is deducible from $\neg(A \rightarrow B)$. So by theorem 2.16(5) each is deducible from Γ. By (1), $A \in \Gamma$ and $\neg B \in \Gamma$, and by (5) the latter means that $B \notin \Gamma$.

For (9). The proof is left as an exercise, with the remark that $A \leftrightarrow B$ is *PL*-interdeducible with $(A \rightarrow B) \wedge (B \rightarrow A)$.

For (10). By exercise 2.38 it is sufficient to show that Γ is Σ-closed, i.e. deductively closed with respect to Σ. But it is, by part (1).

The next theorem is known as Lindenbaum's lemma. It is the proposition that every consistent set of sentences has a maximal extension.

THEOREM 2.19. *If* $\mathrm{Con}_\Sigma \Gamma$, *then there exists a* Δ *such that* (i) $\Gamma \subseteq \Delta$, *and* (ii) $\mathrm{Max}_\Sigma \Delta$.

Proof. The proof is long and somewhat complicated. It involves several definitions and lemmas, each of which we set out separately. We begin with a set of sentences Γ which we suppose to be consistent; that is, we assume throughout that $\mathrm{Con}_\Sigma \Gamma$.

The plan now is to define a set of sentences Δ in terms of Γ which we can prove to be a Σ-maximal extension of Γ.

For the duration of the proof we assume we have a fixed enumeration of the set of sentences,

$$A_1, A_2, A_3, \ldots.$$

That is, we suppose that each sentence in the language occurs at least once in this list (it does not matter if a sentence shows up more than once).

In terms of the set Γ and this enumeration of the sentences we define an infinite sequence of sets of sentences,

$$\Delta_0, \Delta_1, \Delta_2, \ldots.$$

The definition is inductive. First we define the initial set in the sequence, Δ_0; then we specify how any other set in the sequence, Δ_n, is to be defined in terms of its immediate predecessor, Δ_{n-1}.

Definition 1

$$(1)\ \Delta_0 = \Gamma.$$

$$(2)\ \Delta_n = \left.\begin{cases} \Delta_{n-1} \cup \{A_n\}, \text{ if } \mathrm{Con}_\Sigma \Delta_{n-1} \cup \{A_n\}; \\ \Delta_{n-1}, \text{ otherwise}; \end{cases}\right\} n > 0.$$

In other words: Δ_0 is the set Γ; and, for $n > 0$, Δ_n is formed by adding the nth sentence in the enumeration, A_n, to Δ_{n-1} if that addition is Σ-consistent (if not, Δ_n is the same as Δ_{n-1}).

It is obvious from their construction that each of the sets in the sequence is Σ-consistent. For the first set in the sequence is consistent by hypothesis, and the rest are formed only by making consistent additions to their immediate predecessors. Thus:

Lemma 2. $\mathrm{Con}_{\Sigma}\Delta_n$, for $n \geqslant 0$.

A proper inductive proof of this is left as an exercise.

Now we define the set Δ to be the infinite collection of all the sentences in any of the sets in the sequence.

Definition 3. $\Delta = \bigcup_{n=0}^{\infty} \Delta_n$.

Note that Δ includes each of the sets in the sequence:

Lemma 4. $\Delta_n \subseteq \Delta$, for $n \geqslant 0$.

So in particular Δ includes Γ $(=\Delta_0)$:

Lemma 5. $\Gamma \subseteq \Delta$.

Thus Δ is an extension of Γ. It remains only to be shown that Δ is Σ-maximal. For this we need three more lemmas.

Lemma 6. $\Delta_k \subseteq \Delta_n$, for $k \leqslant n \geqslant 0$.

Lemma 7. $A_k \in \Delta_k$, whenever $A_k \in \Delta$, for $k > 0$.

Lemma 8. For every finite subset Δ' of Δ, $\Delta' \subseteq \Delta_n$ for some $n \geqslant 0$.

According to lemma 6, each set in the sequence includes all of its predecessors. This is obvious from the construction of the sequence. Lemma 7 states that a sentence in Δ with index k in the enumeration of the sentences is always in the set in the sequence having index k. This may not be so obvious, but we leave the proof for the reader. The argument for lemma 8 – that every finite subset of Δ is included in one of the sets in the sequence – is this. Suppose Δ' to be a finite subset of Δ, and let A_n be the sentence in Δ' with largest integer index n (there must be such a sentence since Δ' is finite). Now we show that $\Delta' \subseteq \Delta_n$. Let A be a sentence in Δ'. Because A occurs somewhere in the enumeration of the sentences, $A = A_k$, where $k \leqslant n$. Since A $(= A_k)$ is in Δ', it is in Δ. So by lemma 7, A $(= A_k)$ is in the set Δ_k. By lemma 6, $\Delta_k \subseteq \Delta_n$. Hence A is in Δ_n.

Lemma 9. $\mathrm{Max}_{\Sigma}\Delta$. That is, (*a*) $\mathrm{Con}_{\Sigma}\Delta$, and (*b*) for every A, if $\mathrm{Con}_{\Sigma}\, \Delta \cup \{A\}$, then $A \in \Delta$.

For (*a*) it is enough by theorem 2.16(13) to prove that every finite subset of Δ is Σ-consistent. Suppose to the contrary that Δ has a Σ-

inconsistent finite subset, Δ'. Let A_n be the sentence in Δ' with largest index n. Then Δ' is a subset of the set Δ_n in the sequence, and so – by theorem 2.16 (12) – Δ_n is also Σ-inconsistent. This contradicts lemma 2.

For (*b*), suppose that $\text{Con}_\Sigma \Delta \cup \{A\}$, for some A. Because, for some n, A appears as the nth sentence in the enumeration of the sentences, our assumption can be equivalently stated: $\text{Con}_\Sigma \Delta \cup \{A_n\}$. By theorem 2.16 (12), every subset of this set is Σ-consistent. In particular, Con_Σ $\Delta_{n-1} \cup \{A_n\}$ (since $\Delta_{n-1} \subseteq \Delta$, by lemma 4). By definition 1, then, $\Delta_n = \Delta_{n-1} \cup \{A_n\}$. Since A_n is thus in Δ_n, it is in Δ itself, i.e. $A \in \Delta$.

This completes the proof.

From Lindenbaum's lemma it follows that a sentence is deducible from a set of sentences if and only if it belongs to every maximal extension of the set, and also that a sentence is a theorem just in case it is a member of every maximal set of sentences. We state these corollaries formally.

THEOREM 2.20

(1) $\Gamma \vdash_\Sigma A$ *iff* $A \in \Delta$, *for every* $\text{Max}_\Sigma \Delta$ *such that* $\Gamma \subseteq \Delta$.

(2) $\vdash_\Sigma A$ *iff* $A \in \Delta$, *for every* $\text{Max}_\Sigma \Delta$.

Proof

For (1). Suppose that $\Gamma \vdash_\Sigma A$, $\text{Max}_\Sigma \Delta$, and $\Gamma \subseteq \Delta$. By theorem 2.16 (6), $\Delta \vdash_\Sigma A$, and so by theorem 2.18 (1), $A \in \Delta$. For the reverse, suppose that not $\Gamma \vdash_\Sigma A$, to show that there is a Σ-maximal extension of Γ not containing A. It follows by theorem 2.16 (14) that the set $\Gamma \cup \{\neg A\}$ is Σ-consistent. By Lindenbaum's lemma this set has a Σ-maximal extension Δ, which is also an extension of Γ. Because $\neg A \in \Delta$, it follows by theorem 2.18 (5) that $A \notin \Delta$.

For (2). Take $\varnothing$ for Γ in (1), and use theorem 2.16 (1).

In terms of maximality we can define what we shall call the *proof set* of a sentence. Relative to a system Σ, the proof set of a sentence A – denoted by $|A|_\Sigma$ – is the set of Σ-maximal sets of sentences containing A:

DEFINITION 2.21. $|A|_\Sigma = \{\text{Max}_\Sigma \Gamma : A \in \Gamma\}$.

In other words, where Γ is Σ-maximal,

$$\Gamma \in |A|_\Sigma \text{ iff } A \in \Gamma.$$

We conclude this section with a theorem about proof sets.

THEOREM 2.22

(1) $|A|_\Sigma = \{\mathrm{Max}_\Sigma \Gamma: \Gamma \vdash_\Sigma A\}$.
(2) $|A|_\Sigma \subseteq |B|_\Sigma$ *iff* $\vdash_\Sigma A \to B$.
(3) $|A|_\Sigma = |B|_\Sigma$ *iff* $\vdash_\Sigma A \leftrightarrow B$.
(4) $|\top|_\Sigma = \{\Gamma: \mathrm{Max}_\Sigma \Gamma\}$.
(5) $|\bot|_\Sigma = \emptyset$.
(6) $|\neg A|_\Sigma = |\top|_\Sigma - |A|_\Sigma$.
(7) $|A \wedge B|_\Sigma = |A|_\Sigma \cap |B|_\Sigma$.
(8) $|A \vee B|_\Sigma = |A|_\Sigma \cup |B|_\Sigma$.
(9) $|A \to B|_\Sigma = (|\top|_\Sigma - |A|_\Sigma) \cup |B|_\Sigma$.
(10) $|A \leftrightarrow B|_\Sigma = |A \to B|_\Sigma \cap |B \to A|_\Sigma$.

Proof

For (1). The proof here rests essentially on theorem 2.18(1) and is left to the reader. According to this the proof set of a sentence consists of just those maximal sets of sentences that prove the sentence – whence the appellation.

For (2). Let us prove this by a chain of equivalences.

$|A|_\Sigma \subseteq |B|_\Sigma$ iff for every Σ-maximal set of sentences Γ, if $\Gamma \in |A|_\Sigma$, then $\Gamma \in |B|_\Sigma$;
iff for every Σ-maximal set of sentences Γ, if $A \in \Gamma$, then $B \in \Gamma$;
iff for every Σ-maximal set of sentences Γ, $A \to B \in \Gamma$
– theorem 2.18(8);
iff $\vdash_\Sigma A \to B$
– theorem 2.20(2).

For (3). That the proof sets of two sentences are the same just when their biconditional is a theorem may be proved by reference to (2) and the fact that $\vdash_\Sigma A \leftrightarrow B$ if and only if both $\vdash_\Sigma A \to B$ and $\vdash_\Sigma B \to A$. Exercise.

For (4)–(10). These correspond to parts (3)–(9) of theorem 2.18. Proofs left as an exercise.

EXERCISES

Use definitions, theorems, and exercises from section 2.4, definitions and theorems from section 2.6, and results from previous exercises.

2.42. Complete the proof of theorem 2.18 (parts (3), (6), (7), and (9)).

2.43. Prove lemmas 2 and 7 in the proof of theorem 2.19 (Lindenbaum's lemma).

2.44. Prove part (2) of theorem 2.20.

2.45. Complete the proof of theorem 2.22 (parts (1) and (3)–(10)).

2.46. Let Γ and Δ be Σ-maximal sets of sentences. Prove that $\Gamma = \Delta$ if and only if $\Gamma \subseteq \Delta$.

2.47. Prove that Lindenbaum's lemma (theorem 2.19) follows from its corollary theorem 2.20(1) (and hence that the two are equivalent).

2.48. Let Γ be a Σ-consistent set of sentences satisfying the condition that for every sentence A, either $A \in \Gamma$ or $\neg A \in \Gamma$. Prove that Γ is Σ-maximal.

2.49. Let Γ be a Σ-system satisfying the condition that for every sentence A, $\neg A \in \Gamma$ if and only if $A \notin \Gamma$. Prove that Γ is Σ-maximal.

2.50. Let $A_1, A_2, A_3, \ldots$ be an enumeration of the sentences, and let Γ be a consistent set of sentences. Define the sequence $\Delta_0, \Delta_1, \Delta_2, \ldots$ of sets of sentences thus:

$$(1)\ \Delta_0 = \Gamma.$$

$$(2)\ \Delta_n = \begin{Bmatrix} \Delta_{n-1} \cup \{A_n\}, \text{ if } \Delta_{n-1} \vdash_\Sigma A_n; \\ \Delta_{n-1}, \text{ otherwise}; \end{Bmatrix} n > 0.$$

Show by an inductive proof that $\text{Con}_\Sigma \Delta_n$ for each $n \geqslant 0$. Could the definition above replace definition 1 in the proof of theorem 2.19 (Lindenbaum's lemma)?

2.51. Prove:

(*a*) $|A|_\Sigma \subseteq |B|_\Sigma$ iff $A \vdash_\Sigma B$ (i.e. $\{A\} \vdash_\Sigma B$).

(*b*) $|A|_\Sigma = |B|_\Sigma$ iff $A \sim_\Sigma B$ (see exercise 2.37).

2.7. Soundness, completeness, and canonical models

A system of modal logic Σ is said to be *sound* with respect to a class of models C just in case every theorem of Σ is valid in C; i.e. just in case for every sentence A,

$$\text{if } \vdash_{\Sigma} \text{A, then } \models_{\mathsf{C}} \text{A.}$$

Σ is *complete* with respect to C if and only if every sentence valid in C is a theorem of Σ; i.e. if and only if for every A,

$$\text{if } \models_{\mathsf{C}} \text{A, then } \vdash_{\Sigma} \text{A.}$$

And Σ is said to be *determined* by C just when it is both sound and complete with respect to C; i.e. just when for every A,

$$\vdash_{\Sigma} \text{A iff } \models_{\mathsf{C}} \text{A.}$$

Note that it is possible that a system of modal logic be determined by more than one class of models.

Much of this book is concerned with characterizing classes of models that determine various well-known systems of modal logic, and with proving that these systems are indeed so determined. Proof of completeness is the more difficult part; soundness is usually relatively easy and straightforward. In this section we lay the ground for the completeness theorems in chapters 5 and 9 by introducing some of the principal ideas and methods used in their proofs.

Perhaps the most important idea is that of a *canonical model* for a system of modal logic. A canonical model for a system Σ is a model $\mathcal{M} = \langle W, \ldots, P \rangle$ in which:

(i) W is the set of Σ-maximal sets of sentences.

(ii) P_n is the proof set of the atomic sentence $\mathbb{P}_n$, i.e. $P_n = |\mathbb{P}_n|_{\Sigma}$, for $n = 0, 1, 2, \ldots$.

Thus in a canonical model for a system Σ, each world is a Σ-maximal set of sentences (and each such set is a world), and for each such world α and each natural number n,

$$\alpha \in P_n \text{ iff } \mathbb{P}_n \in \alpha.$$

(Note that we use world-variables, α, β, etc., for the maximal sets of sentences that are the possible worlds in a canonical model.)

The chief feature of a canonical model $\mathcal{M}$ for a system of modal logic Σ is this: in $\mathcal{M}$ just those sentences are true at a world (Σ-maximal set of sentences) as are contained by it; i.e. for every α in $\mathcal{M}$,

$$\models_{\alpha}^{\mathcal{M}} \text{A iff A} \in \alpha.$$

Put another way, $\|\text{A}\|^{\mathcal{M}} = |\text{A}|_{\Sigma}$. We prove this in chapter 5 with respect to canonical standard models, and again in chapter 9 with respect to canonical minimal models. (Indeed, we can prove it already for the cases in which A is non-modal. See exercise 2.53.)

Because the worlds in a canonical model for a system of modal logic will always verify just those sentences they contain, it follows that the sentences true in such a model are precisely the theorems of the system. That is to say, if $\mathcal{M}$ is a canonical model for a system Σ, then

$$\models^{\mathcal{M}} \mathrm{A} \text{ iff } \vdash_{\Sigma} \mathrm{A}.$$

For it follows from the preceding result that

$$\models^{\mathcal{M}}_{\alpha} \mathrm{A}, \text{ for every } \alpha \text{ in } \mathcal{M}, \text{ iff, for every } \alpha \text{ in } \mathcal{M}, \mathrm{A} \in \alpha.$$

By definition 2.6 the left side of this means that $\models^{\mathcal{M}} \mathrm{A}$, and by theorem 2.20 (2) the right side means $\vdash_{\Sigma} \mathrm{A}$ (since the worlds are just the Σ-maximal sets of sentences). Equivalently, then, we may say that every system of modal logic is determined by each, and all, of its canonical models.

Our strategy in proving completeness should thus be apparent. In order to show that a system of modal logic Σ is complete with respect to a class C of models, it is sufficient to prove that C contains a canonical model $\mathcal{M}$ for Σ. For then we can argue that if a sentence A is valid in C, then A is true in $\mathcal{M}$, and hence that A is a theorem of Σ. So the main problem in proving the completeness of a system of modal logic becomes that of finding, or defining, a canonical model for the system that can be shown to be in the class of models in question. This is not always a trifling matter.

EXERCISES

2.52. Let $\mathcal{M}$ be a canonical model for a system of modal logic Σ. Prove the equivalence of (*a*) and (*b*).

(*a*) For every possible world α in $\mathcal{M}$ (i.e. for every Σ-maximal set of sentences α), $\models^{\mathcal{M}}_{\alpha} \mathrm{A}$ if and only if $\mathrm{A} \in \alpha$.

(*b*) $\|\mathrm{A}\|^{\mathcal{M}} = |\mathrm{A}|_{\Sigma}$.

2.53. We can prove (*a*) (equivalently, (*b*)) above, for every non-modal sentence A. The proof is by induction on the complexity of A. We give it for the cases in which A is (*a*) atomic, $\mathbb{P}_n$, (*b*) a negation, $\neg \mathrm{B}$, and (*c*) a conditional, $\mathrm{B} \rightarrow \mathrm{C}$. The remaining cases are left for the reader.

For (*a*). By the definition of truth (2.5 (1)), $\models^{\mathcal{M}}_{\alpha} \mathbb{P}_n$ if and only if $\alpha \in P_n$, i.e. $\alpha \in |\mathbb{P}_n|_{\Sigma}$. But by definition 2.21 this holds if and only if $\mathbb{P}_n \in \alpha$. So the result holds when A is atomic.

For the inductive cases we make the hypothesis that the result holds for all sentences shorter than A.

For (*b*):

$\models^{\mathcal{M}}_{\alpha} \neg B$ iff not $\models^{\mathcal{M}}_{\alpha} B$
– definition 2.5 (4);
iff not $B \in \alpha$
– inductive hypothesis;
iff $\neg B \in \alpha$
– theorem 2.18 (5).

So the result holds when A is a negation.

For (*c*):

$\models^{\mathcal{M}}_{\alpha} B \rightarrow C$ iff if $\models^{\mathcal{M}}_{\alpha} B$ then $\models^{\mathcal{M}}_{\alpha} C$
– definition 2.5 (7);
iff if $B \in \alpha$ then $C \in \alpha$
– inductive hypothesis;
iff $B \rightarrow C \in \alpha$
– theorem 2.18 (8).

So the result holds when A is a conditional.

2.8. Decidability and the finite model property

A system of modal logic is decidable just in case its set of theorems is, i.e. just in case there is an effective finitary method for deciding of any sentence whether or not it is a theorem of the system. To understand our approach to proving the decidability of modal logics we need the ideas of axiomatizability (section 2.5) and the finite model property.

We say that a modal logic Σ has the *finite model property* – the *f.m.p.* – if and only if each non-theorem of Σ is false in some finite model of Σ.

Recall that if Σ is axiomatizable then there is a positive test for theoremhood in Σ. If Σ both has the f.m.p. and is axiomatizable by means of a finite number of schemas then there is also a negative test for theoremhood in Σ. For let

$$\mathcal{M}_1, \mathcal{M}_2, \mathcal{M}_3, \ldots$$

be a complete enumeration of the finite models (this collection *is* enumerable since each model is finite). Then if A is a non-theorem of Σ it is false in $\mathcal{M}_n$ for some *n*. To find such a falsifying model it is enough to proceed through the enumeration examining each model in turn – first to see whether the model is a model of Σ (a finite task since the model is finite and Σ is axiomatized by finitely many schemas and rules of inference) and then, if it is, to see whether A is false (again a finite task since the model is finite). Sooner or later, i.e. after examining a finite number of models, a model of Σ will appear that is a countermodel to A.

So a modal logic is decidable if it has the f.m.p. and is axiomatizable by a finite number of schemas: there is both a positive and a negative test for theoremhood in the logic.

In chapters 5 and 9 we use finite determination theorems to show the f.m.p. for systems of modal logic. In outline, our strategy is as follows. Suppose we have proved that Σ is determined by a class of models C, i.e. that for every A,

$$\vdash_{\Sigma} A \text{ iff } \models_{\mathsf{C}} A.$$

From this result by means of filtrations we show that Σ is also determined by the class $\mathsf{C}_{\mathrm{FIN}}$ of finite models in C, i.e. that for every A,

$$\vdash_{\Sigma} A \text{ iff } \models_{\mathsf{C}_{\mathrm{FIN}}} A.$$

Then we know that Σ has the f.m.p.: if not $\vdash_{\Sigma} A$ then there is a finite model $\mathcal{M}$ (in $\mathsf{C}_{\mathrm{FIN}}$) such that not $\models^{\mathcal{M}} A$.

The argument for the finite determination theorem is of course trivial from left to right: if $\vdash_{\Sigma} A$ then $\models_{\mathsf{C}} A$, and so $\models_{\mathsf{C}_{\mathrm{FIN}}} A$ since $\mathsf{C}_{\mathrm{FIN}} \subseteq \mathsf{C}$. From right to left we argue, contrapositively, in the following way. Suppose that not $\vdash_{\Sigma} A$, so that not $\models^{\mathcal{M}} A$ for some $\mathcal{M}$ in C. Then where Γ is finite (or at least logically finite relative to $\mathcal{M}$) and contains the subsentences of A we define a Γ-filtration $\mathcal{M}^*$ of $\mathcal{M}$ in such a way that $\mathcal{M}^*$ is in C, and hence in $\mathsf{C}_{\mathrm{FIN}}$ since it is finite. By a general filtration theorem (compare section 2.3) we then conclude that not $\models^{\mathcal{M}^*} A$, and so not $\models_{\mathsf{C}_{\mathrm{FIN}}} A$.

The interesting and often difficult part of proofs of this kind is defining $\mathcal{M}^*$ so that it is in C.

Note that when Γ is just the set of subsentences of A it is possible to compute an upper bound on the size of $\mathcal{M}^*$; it has at most 2^n worlds, where n is the number of subsentences of A. Thus a search for a model of Σ that rejects A can be limited to models of at most this size.

In conclusion we should emphasize that the decision procedures afforded by filtrations and the f.m.p. are seldom practical, even when the number of subsentences of a sentence is relatively small. But this is not to gainsay their validity, nor their theoretical interest. Indeed, discovery of decidability by these methods may stimulate the quest for more realistic decision procedures.

EXERCISES

2.54. Explain informally why any collection of finite models is enumerable.

2.55. Suppose a system Σ has the f.m.p. and is axiomatizable. Why is this not in general sufficient for a negative test for theoremhood (of the kind described)? That is, why is it stipulated that Σ be axiomatizable by a finite number of schemas?

(This is perhaps the place to mention that in specifying the conditions for a negative test for theoremhood we could as well have stipulated that the system be axiomatizable by means of a finite number of non-tautological schemas. This is because *PL* holds in every model (theorem 2.8), and so only non-tautological axiom schemas need to be verified. And, in any case, *PL* is itself axiomatizable by a finite number of schemas and reasonable rules, though we forbear giving an example.)

2.56. Explain why *PL* is decidable by describing a negative test for theoremhood in *PL* (see exercise 2.41).

2.57. A system of modal logic is inconsistent just in case it contains $\perp$ (compare exercise 1.27). So there is just one inconsistent system, to wit, the set of all sentences (since by *PL* every sentence is deducible from $\perp$). Of course the set of sentences is decidable; so the inconsistent system is decidable.

Explain how to modify the decision procedure for theoremhood using axiomatizability and the f.m.p. in case it is not known whether the system in question is consistent.

PART II

3

STANDARD MODELS FOR MODAL LOGICS

According to the account of necessity and possibility in chapter 1, a sentence of the form $\Box A$ is true at a possible world just in case A itself is true at all possible worlds, and a sentence of the form $\Diamond A$ holds at a possible world if and only if A holds at some possible world. This idea was modeled very simply in terms of a collection of possible worlds together with an assignment of truth values, at each world, to the atomic sentences. We saw that the ensuing notion of validity is quite strict, encompassing as it does a large assortment of principles.

In the present chapter we modify this leibnizian conception of necessity and possibility by introducing an element of relative possibility. The result is a much more supple notion of validity, one that greatly reduces the stock of principles that are bound to hold.

In section 3.1 we define the idea of a standard model, state the truth conditions for modal sentences at worlds in models of this sort, and prove a theorem about validity in classes of standard models. In section 3.2 we single out the schemas D, T, B, 4, and 5 for special attention, both because of their historical prominence (recall that they are all theorems of *S5*) and because the techniques required for their treatment are illuminating and instructive. Section 3.3 contains a generalization of the truth conditions of modal sentences, and we examine in these terms a certain schema that has D, T, B, 4, and 5 as special cases. In section 3.4 we explain generated models and prove a theorem relating standard models to those in chapter 1. In section 3.5 we specify the notion of filtration for standard models, prove the basic theorem therefor, and give an example of its application. Section 3.6 continues this with some theorems that are useful in the context of decidability proofs in chapter 5.

3.1. Standard models

A *standard model* is a structure

$$\mathcal{M} = \langle W, R, P \rangle$$

in which, as usual, W is a set of possible worlds and P represents an assignment of sets of possible worlds to atomic sentences. The new element, R, is a relation between possible worlds. Formally:

DEFINITION 3.1. $\mathcal{M} = \langle W, R, P \rangle$ is a *standard model* iff:

(1) W is a set.

(2) R is a binary relation on W (i.e. $R \subseteq W \times W$).

(3) P is a mapping from natural numbers to subsets of W (i.e. $P_n \subseteq W$, for each natural number n).

The interpretation of the relation R in a standard model will vary significantly, but in general it may be thought of as relative possibility or, perhaps better, relevance. We shall write

$$\alpha R \beta$$

to mean that the world β is possible relative to – or is relevant to – the world α. (Other variants are: β is an alternative to α; β is accessible from α; β is reachable from α; β is (once) removed from α.) It bears emphasis here that R may be any sort of binary relation on W; no assumption whatsoever is made concerning its content or structure. The effect of this becomes apparent in the next section.

The truth conditions of non-modal sentences are given already in definition 2.5. The truth conditions of modal sentences use the relation R of relative possibility as follows: $\Box$A is true at a world α if and only if A is true at every world β that is possible relative to α, i.e. at every world β such that $\alpha R \beta$. And $\Diamond$A is true at α if and only if A is true at some world β possible relative to α, i.e. at some world β such that $\alpha R \beta$. These truth conditions are a simple generalization of the account in chapter 1. Instead of truth at every or some possible world, we now have truth at every or some world *possible relative to* the given one. Formally:

DEFINITION 3.2. Let α be a world in a standard model $\mathcal{M} = \langle W, R, P \rangle$.

(1) $\models_\alpha^\mathcal{M} \Box\mathrm{A}$ iff for every β in $\mathcal{M}$ such that $\alpha R \beta$, $\models_\beta^\mathcal{M} \mathrm{A}$.

(2) $\models_\alpha^\mathcal{M} \Diamond\mathrm{A}$ iff for some β in $\mathcal{M}$ such that $\alpha R \beta$, $\models_\beta^\mathcal{M} \mathrm{A}$.

To illustrate the role of the relation R in standard models, suppose that we wish the operator $\Box$ to express a concept of moral necessity or obligation like that expressed by 'ought'. Clearly this notion of necessity cannot be expected to obey all the laws of the system *S5*; $\Box\mathrm{A} \rightarrow \mathrm{A}$, for example, is wrong, since there are unfulfilled obligations. But let us understand R as a relation of moral relevance – so that $\alpha R \beta$ means that β is morally better than α, at least in the sense that whatever ought to be the case at α is in fact the case at β. Then $\Box$A (*it ought to be the case that* A) is true at α just in case A is true at every world β such that $\alpha R \beta$, i.e. such that β is morally better than α.

For another illustration, suppose $\Box$ to be a future tense modality, with a reading like 'it always will be the case that'. Here we have a notion of temporal necessity, and so it is appropriate to think of the possible worlds as points in time. The relation R then provides an ordering of the set of times, so that $\alpha R \beta$ means that the moment α precedes the moment β, or that β is later than α. Thus $\Box$A (*it will always be the case that* A) is true at a moment α if and only if A is true at every moment β such that $\alpha R \beta$, i.e. such that β is later than α.

We return in chapter 6 to these ways of understanding the modal operators and the relation of relative possibility in standard models. We mention them here in order to provide some motivation for the move from the models of chapter 1 to these.

The ideas of truth in a model and validity in a class of models are familiar from definitions 2.6 and 2.7. By theorem 2.8, validity in any class of models is preserved by the rule RPL; so this holds for standard models as well (in particular, then, tautologies are true in all standard models). We prove now that the schema Df$\Diamond$ is true in every standard model, and that validity in a class of such models is preserved by the rule RK.

THEOREM 3.3. *Let* C *be a class of standard models. Then*:

(1) $\models_{\mathsf{C}} \Diamond \mathrm{A} \leftrightarrow \neg \Box \neg \mathrm{A}$.

(2) *For* $n \geqslant 0$, *if* $\models_{\mathsf{C}} (\mathrm{A}_1 \wedge \ldots \wedge \mathrm{A}_n) \rightarrow \mathrm{A}$, *then* $\models_{\mathsf{C}} (\Box \mathrm{A}_1 \wedge \ldots \wedge \Box \mathrm{A}_n) \rightarrow \Box \mathrm{A}$.

Proof

For (1). Let α be a world in a standard model $\mathcal{M} = \langle W, R, P \rangle$. Then:

$\models_\alpha^{\mathcal{M}} \Diamond \mathrm{A}$ iff for some β in $\mathcal{M}$ such that $\alpha R \beta$, $\models_\beta^{\mathcal{M}} \mathrm{A}$
– definition 3.2(2);

iff not every β in $\mathcal{M}$ such that $\alpha R \beta$ is such that not $\models_\beta^{\mathcal{M}} \mathrm{A}$;

iff not every β in $\mathcal{M}$ such that $\alpha R \beta$ is such that $\models_\beta^{\mathcal{M}} \neg \mathrm{A}$
– definition 2.5(4);

iff not $\models^{\mathcal{M}}_{\alpha} \Box\neg A$
– definition 3.2(1);

iff $\models^{\mathcal{M}}_{\alpha} \neg\Box\neg A$
– definition 2.5(4).

Therefore, $\models^{\mathcal{M}}_{\alpha} \Diamond A \leftrightarrow \neg\Box\neg A$, for every world α in every standard model $\mathcal{M}$.

For (2). The proof here is by induction on n, and it rests on two lemmas – to wit, where C is a class of standard models:

Lemma 1. If $\models_{\mathsf{C}} A$, then $\models_{\mathsf{C}} \Box A$.

Lemma 2. $\models_{\mathsf{C}} \Box(A \to B) \to (\Box A \to \Box B)$.

That is, the rule of inference RN preserves validity in any class of standard models (lemma 1), and the schema K is valid in every class of standard models (lemma 2). The proofs of these lemmas are not difficult; we leave them for the reader.

For the inductive proof of (2), now, recall that when $n = 0$ the conditionals in question are identified with their consequents. So for the basis of the induction, we need to show that for any class C of standard models, if $\models_{\mathsf{C}} A$, then $\models_{\mathsf{C}} \Box A$. This is just lemma 1.

For the inductive part of the proof, we assume as an inductive hypothesis that the theorem holds for all natural numbers $k < n$, and show that it holds, too, at n. We put the inductive hypothesis as follows.

For $k < n$, if $\models_{\mathsf{C}} (B_1 \wedge \ldots \wedge B_k) \to B$,
then $\models_{\mathsf{C}} (\Box B_1 \wedge \ldots \wedge \Box B_k) \to \Box B$.

Now suppose that

$$\models_{\mathsf{C}} (A_1 \wedge \ldots \wedge A_n) \to A.$$

It follows by *PL* that

$$\models_{\mathsf{C}} (A_1 \wedge \ldots \wedge A_{n-1}) \to (A_n \to A).$$

Hence by the inductive hypothesis,

$$\models_{\mathsf{C}} (\Box A_1 \wedge \ldots \wedge \Box A_{n-1}) \to \Box(A_n \to A).$$

From this and lemma 2 it follows by *PL* that

$$\models_{\mathsf{C}} (\Box A_1 \wedge \ldots \wedge \Box A_{n-1}) \to (\Box A_n \to \Box A).$$

And from this by *PL*,

$$\models_{\mathsf{C}} (\Box A_1 \wedge \ldots \wedge \Box A_n) \to \Box A.$$

This completes the proof.

There is of course much more to validity in the class of standard models than the schema and rule covered in theorem 3.3. But, together with theorem 2.8, theorem 3.3 provides for the soundness of the systems of mcdal logic introduced in chapter 4. A number of further principles appear in the exercises that follow.

EXERCISES

3.1. Prove lemmas 1 and 2 in the proof of part (2) of theorem 3.3. (It may be helpful to compare the proof in section 1.2 that the system *S5* has the rule RK.)

3.2. Identify some schemas (for example from the system *S5*) that will be valid, intuitively, if $\Box$ is taken to mean:

(*a*) it ought to be the case that

(*b*) it will always be the case that

(*c*) John Doe believes that

(*d*) Jane Doe knows that

(*e*) God knows that

(Be sure to understand the meaning of $\Diamond$ – via Df$\Diamond$ – in each case.)

3.3. Show that instances of each of the following schemas can be falsified at possible worlds in standard models.

D.	$\Box A \to \Diamond A$	D_c.	$\Diamond A \to \Box A$
T.	$\Box A \to A$	T_c.	$A \to \Box A$
B.	$A \to \Box\Diamond A$	B_c.	$\Box\Diamond A \to A$
4.	$\Box A \to \Box\Box A$	4_c.	$\Box\Box A \to \Box A$
5.	$\Diamond A \to \Box\Diamond A$	5_c.	$\Box\Diamond A \to \Diamond A$

(Compare exercise 2.21.)

3.4. Let us say that standard models $\mathcal{M} = \langle W, R, P\rangle$ and $\mathcal{M}' = \langle W', R', P'\rangle$ *agree on the atoms* of a sentence A just in case (i) $W = W'$, (ii) $R = R'$, and (iii) $P_n = P'_n$ for every n such that $\mathbb{P}_n$ is an atomic subsentence of A. In other words, two standard models agree on the atoms of a sentence if and only if they have the same set of possible worlds and the same relation on them and they agree on the truth values assigned, per world, to each atomic constituent of the sentence.

Prove the following theorem.

> *If standard models $\mathcal{M}$ and $\mathcal{M}'$ agree on the atoms of* A, *then they agree on* A *itself (in the sense that for any α in $\mathcal{M}$ (equally, in $\mathcal{M}'$), $\models^{\mathcal{M}}_{\alpha}$* A *if and only if $\models^{\mathcal{M}'}_{\alpha}$* A).

The proof is by induction on the complexity of A. Give it at least for the cases in which (*a*) A is atomic, $\mathbb{P}_n$, (*b*) A is the falsum, $\bot$, (*c*) A is a conditional, $B \to C$, and (*d*) A is a necessitation, $\Box B$. N.B. for the inductive cases the inductive hypothesis must be stated: for every sentence X of complexity less than A, $\models^{\mathcal{M}}_{\alpha}$ X if and only if $\models^{\mathcal{M}'}_{\alpha}$ X, *for every α in $\mathcal{M}$.*

In virtue of this theorem it is possible to ignore the values P_n (in a standard model $\mathcal{M}$) of atomic sentences $\mathbb{P}_n$ not in a sentence A, for example when constructing a countermodel to A.

3.5. Prove that the following schemas are valid in any class of standard models.

N. $\Box\top$

M. $\Box(A \wedge B) \to (\Box A \wedge \Box B)$

C. $(\Box A \wedge \Box B) \to \Box(A \wedge B)$

3.6. Prove that the schema Df$\Box$, $\Box A \leftrightarrow \neg\Diamond\neg A$, is valid in any class of standard models.

3.7. Prove the following, where **C** is any class of standard models.

(*a*) If $\models_{\mathbf{C}} (A \wedge B) \to C$, then $\models_{\mathbf{C}} (\Box A \wedge \Box B) \to \Box C$.

(*b*) If $\models_{\mathbf{C}} A \to B$, then $\models_{\mathbf{C}} \Box A \to \Box B$.

(*c*) If $\models_{\mathbf{C}} A \leftrightarrow B$, then $\models_{\mathbf{C}} \Box A \leftrightarrow \Box B$.

(*d*) If $\models_{\mathbf{C}} A \to B$, then $\models_{\mathbf{C}} \Diamond A \to \Diamond B$.

(*e*) If $\models_{\mathbf{C}} A \leftrightarrow B$, then $\models_{\mathbf{C}} \Diamond A \leftrightarrow \Diamond B$.

3.8 Prove that the following schemas are valid in any class of standard models.

N$\Diamond$. $\neg\Diamond\bot$

M$\Diamond$. $\Diamond(A \vee B) \to (\Diamond A \vee \Diamond B)$

C$\Diamond$. $(\Diamond A \vee \Diamond B) \to \Diamond(A \vee B)$

K$\Diamond$. $(\neg\Diamond A \wedge \Diamond B) \to \Diamond(\neg A \wedge B)$

3.9. Describe a single standard model that simultaneously falsifies instances of $\Box(A \vee B) \to (\Box A \vee \Box B)$ and $(\Diamond A \wedge \Diamond B) \to \Diamond(A \wedge B)$.

3.10. For each of the following, decide whether or not it is valid in the class of all standard models, and prove it.

(*a*) $\Diamond\top$

(*b*) $(\Box A \vee \Box B) \to \Box(A \vee B)$

(*c*) $A \to \Diamond A$

(*d*) $\Diamond(A \wedge B) \to (\Diamond A \wedge \Diamond B)$

(*e*) $\Diamond\Box A \to A$

(*f*) $\Box(A \vee B) \to (\Diamond A \vee \Box B)$

(*g*) $\Diamond\Diamond A \to \Diamond A$

(*h*) $\Diamond(A \to B) \leftrightarrow (\Box A \to \Diamond B)$

(*i*) $\Diamond\Box A \to \Box A$

(*j*) $\Diamond\top \leftrightarrow (\Box A \to \Diamond A)$

3.11. Let $\mathcal{M} = \langle W, R, P\rangle$ be a standard model. Prove:

(*a*) $\Box A \to \Diamond A$ is true in $\mathcal{M}$ if R is serial, i.e. if R satisfies the condition that for every α in $\mathcal{M}$ there is a β in $\mathcal{M}$ such that $\alpha R\beta$.

(*b*) $\Box A \to A$ is true in $\mathcal{M}$ if R is reflexive (see exercise 1.10).

(*c*) $A \to \Box\Diamond A$ is true in $\mathcal{M}$ if R is symmetric, i.e. if R satisfies the condition that for every α and β in $\mathcal{M}$, if $\alpha R\beta$ then $\beta R\alpha$.

(*d*) $\Box A \to \Box\Box A$ is true in $\mathcal{M}$ if R is transitive, i.e. if R satisfies the condition that for every α, β, and γ in $\mathcal{M}$, if $\alpha R\beta$ and $\beta R\gamma$, then $\alpha R\gamma$.

(*e*) $\Diamond A \to \Box\Diamond A$ is true in $\mathcal{M}$ if R is euclidean (see exercise 1.10).

Compare the countermodels to D, T, B, 4, and 5 in exercise 3.3. (But do not peek at the proof of theorem 3.5 in the next section.)

3.12. Notice that if in a standard model $\mathcal{M} = \langle W, R, P\rangle$, $R = W \times W$ – i.e. if R is the universal relation on W, relating each world to itself and to every other – then we are once again in the situation in chapter 1:

$\models^{\mathcal{M}}_{\alpha} \Box A$ iff $\models^{\mathcal{M}}_{\beta} A$ for every world β in $\mathcal{M}$ (i.e. for every β in $\mathcal{M}$ such that $\alpha R\beta$);

$\models^{\mathcal{M}}_{\alpha} \Diamond A$ iff $\models^{\mathcal{M}}_{\beta} A$ for some world β in $\mathcal{M}$ (i.e. for some β in $\mathcal{M}$ such that $\alpha R\beta$).

Thus in this sense the models in chapter 1 are special cases of those in the present chapter.

Confirm this further by proving that the schemas T and 5 are valid in the class of standard models in which the relation is universal.

Notice that a relation is reflexive and euclidean if it is universal (prove this if necessary), but not vice versa (give an example of a relation that is reflexive and euclidean, but not universal).

3.13. The relation in a standard model can be replaced equivalently by a function that associates with each possible world in the model a set of possible worlds. Let $\mathcal{M} = \langle W, R, P \rangle$ be a standard model, and define the function f from worlds to sets of worlds (formally, $f\colon W \to \mathcal{P}(W)$) as follows, for each α in $\mathcal{M}$.

$$f(\alpha) = \{\beta \text{ in } \mathcal{M} : \alpha R \beta\}.$$

That is, $f(\alpha)$ is the set of worlds in $\mathcal{M}$ that are R-related to α; intuitively, $f(\alpha)$ is the set of worlds relevant to α.

Thus with f in place of R the structure $\mathcal{M} = \langle W, f, P \rangle$ is (as good as) a standard model, and the truth conditions of a necessitation at a world α in $\mathcal{M}$ may be given by the clause:

$$\models_{\alpha}^{\mathcal{M}} \Box \mathrm{A} \text{ iff for every } \beta \in f(\alpha), \models_{\beta}^{\mathcal{M}} \mathrm{A};$$

or, more simply, by using the notation of truth sets (definition 2.9):

$$\models_{\alpha}^{\mathcal{M}} \Box \mathrm{A} \text{ iff } f(\alpha) \subseteq \|\mathrm{A}\|^{\mathcal{M}}.$$

This latter formulation suggests another way of looking at standard models and the interpretation of necessity. To wit, we regard the set $f(\alpha)$, for each world α in $\mathcal{M}$, as a certain proposition – the *necessary proposition* with respect to the world α. (N.B. $f(\alpha)$ need not be expressed by, i.e. need not be the truth set of, any sentence.) Then we understand the formulation above as stating that a necessitation $\Box$A is true at α in $\mathcal{M}$ if and only if the proposition expressed by A in $\mathcal{M}$, $\|\mathrm{A}\|^{\mathcal{M}}$, is implied by the necessary proposition, $f(\alpha)$, for α in $\mathcal{M}$. (Recall exercise 2.22 regarding the construal of truth sets as propositions and of inclusion as implication between propositions.)

Using the propositions $f(\alpha)$ and $\|\mathrm{A}\|^{\mathcal{M}}$, give a succinct formulation of the truth conditions of $\Diamond$A at the world α in $\mathcal{M}$. (Compare, again, exercise 2.22 in which $f(\alpha)$ is always the set W of all worlds in $\mathcal{M}$.)

Given a model $\mathcal{M} = \langle W, f, P \rangle$, define the relation R of relative possibility in terms of f.

3.14. Understand by a model, for the purposes of this exercise, a structure $\mathcal{M} = \langle W, \mathcal{R}, P \rangle$ in which W and P are as usual and $\mathcal{R}$ is a class of binary relations on W. That is, in such a model $\mathscr{R} \subseteq \{R : R \subseteq W \times W\}$. Relative to a world α in $\mathscr{M}$ the truth conditions for necessitations are defined by:

$\models_\alpha^{\mathscr{M}} \Box A$ iff there is an $R \in \mathscr{R}$ such that for every β in $\mathscr{M}$ such that $\alpha R \beta$, $\models_\beta^{\mathscr{M}} A$.

(*a*) State truth conditions for possibilitations so that Df$\Diamond$ is valid in the class of such models.

(*b*) Which of the schemas N, M, C, K, D, T, B, 4, and 5 (see chapter 1) are valid in the class of such models?

(*c*) Which of the rules RN, RM, RE, RK, and RR (see chapter 1) preserve validity in the class of such models?

3.15. Understand by a model, for the purposes of this exercise, a structure $\mathscr{M} = \langle W, R, Q, P \rangle$ in which W, R, and P are as in a standard model and Q is a subset of W (the worlds in Q are 'queer'). Relative to a world α in $\mathscr{M}$ the truth conditions for necessitations are defined by:

$\models_\alpha^{\mathscr{M}} \Box A$ iff $\alpha \notin Q$ and for every β in $\mathscr{M}$ such that $\alpha R \beta$, $\models_\beta^{\mathscr{M}} A$.

(*a*) State truth conditions for possibilitations so that Df$\Diamond$ is valid in the class of such models.

(*b*) Which of the schemas N, M, C, K, D, T, B, 4, and 5 are valid in the class of such models?

(*c*) Which of the rules RN, RM, RE, RK, and RR preserve validity in the class of such models?

3.16. Recall exercise 1.11 and the mapping τ from the modal language to a language of elementary quantificational logic. We saw that τ had the effect of giving the truth conditions of modal sentences at a world α in a model of the kind described in chapter 1. Here we redefine τ for sentences of the forms $\Box A$ and $\Diamond A$. To do so we add to the quantificational language a single two-place predicate, $\mathbb{R}$, and a second variable, β. Writing A' for the result of interchanging all occurrences of α and β in the quantificational formula A, we define τ for necessitations and possibilitations as follows.

(9) $\tau(\Box A) = \forall\beta(\mathbb{R}(\alpha, \beta) \rightarrow (\tau(A))')$.

(10) $\tau(\Diamond A) = \exists\beta(\mathbb{R}(\alpha, \beta) \wedge (\tau(A))')$.

Now τ specifies the truth conditions of modal sentences at the world α in a standard model. For example, the transformation shows that $\Box \mathbb{P}_0 \rightarrow \mathbb{P}_0$ holds at α just in case the formula

$\forall\beta(\mathbb{R}(\alpha, \beta) \rightarrow \mathbb{P}_0(\beta)) \rightarrow \mathbb{P}_0(\alpha)$

is true, and $\Diamond \mathbb{P}_0 \rightarrow \Box \Diamond \mathbb{P}_0$ holds at α just in case the formula

$$\exists\beta(\mathbb{R}(\alpha,\beta) \wedge \mathbb{P}_0(\beta)) \rightarrow \forall\beta(\mathbb{R}(\alpha,\beta) \rightarrow \exists\alpha(\mathbb{R}(\beta,\alpha) \wedge \mathbb{P}_0(\alpha)))$$

is true. Writing $\models$ A now to mean that A is valid in the class of all standard models, it can be seen that in general, for every A,

$\models$ A iff τ(A) is a valid formula of elementary quantificational logic.

Neither $\Box \mathbb{P}_0 \rightarrow \mathbb{P}_0$ nor $\Diamond \mathbb{P}_0 \rightarrow \Box \Diamond \mathbb{P}_0$ is valid in the class of all standard models, since neither of their transformations is quantificationally valid.

(*a*) Check to see that τ produces the stated results when applied to $\Box \mathbb{P}_0 \rightarrow \mathbb{P}_0$ and $\Diamond \mathbb{P}_0 \rightarrow \Box \Diamond \mathbb{P}_0$.
(*b*) Apply τ to Df$\Diamond$, to see that its transformation is a valid quantificational formula.
(*c*) Show that if $\tau((A_1 \wedge \ldots \wedge A_n) \rightarrow A)$ is a valid formula of elementary quantificational logic so is $\tau((\Box A_1 \wedge \ldots \wedge \Box A_n) \rightarrow \Box A)$.
(*d*) Use τ on the schemas in exercises 3.3, 3.5, 3.6, and 3.8–3.10.
(*e*) Show that the principles in exercise 3.7 hold with respect to quantificational validity and transformations of the schemas.
(*f*) Explain how standard models for the modal language serve equally well for the quantificational language.

3.2. The schemas D, T, B, 4, and 5

Let us consider the following schemas.

D. $\Box A \rightarrow \Diamond A$

T. $\Box A \rightarrow A$

B. $A \rightarrow \Box \Diamond A$

4. $\Box A \rightarrow \Box\Box A$

5. $\Diamond A \rightarrow \Box \Diamond A$

THEOREM 3.4. *None of the schemas* D, T, B, 4, *and* 5 *is valid in the class of all standard models.*

Proof. It is enough in each case to exhibit an instance of the schema in question and describe a standard model that falsifies it.

For D. Consider the instance $\Box \mathbb{P}_0 \rightarrow \Diamond \mathbb{P}_0$, and let $\mathcal{M} = \langle W, R, P \rangle$ be a standard model in which $W = \{\alpha\}$, $R = \varnothing$, and $P_n = \varnothing$, for $n \geqslant 0$. Thus $\mathcal{M}$ contains but one world, to which no world is related, and at which every atomic sentence is false. The unaccustomed reader may find this a rather extraordinary standard model, but it is readily verified that it is a standard model: W is a set, R is a binary relation (the empty relation) on W, and P is a mapping from natural numbers to (empty) subsets of W.

It is easy to see that:

(*a*) every β in $\mathcal{M}$ such that $\alpha R \beta$ is such that $\models_{\beta}^{\mathcal{M}} \mathbb{P}_0$;
(*b*) there is not some β in $\mathcal{M}$ such that $\alpha R \beta$ and $\models_{\beta}^{\mathcal{M}} \mathbb{P}_0$.

By definition 3.2, (*a*) means that $\models_{\alpha}^{\mathcal{M}} \Box \mathbb{P}_0$, and (*b*) means that not $\models_{\alpha}^{\mathcal{M}} \Diamond \mathbb{P}_0$. So by definition 2.5 (7), not $\models_{\alpha}^{\mathcal{M}} \Box \mathbb{P}_0 \rightarrow \Diamond \mathbb{P}_0$. Thus the schema D is false in a standard model.

This countermodel to D can be pictured as in figure 3.1. The box represents the set W, and the circle represents the world α in W. Inside the world are listed the essential facts about which sentences are true there. Thus both $\Box \mathbb{P}_0$ and $\neg \Diamond \mathbb{P}_0$ appear inside α, to indicate that $\Box \mathbb{P}_0$ is true at α and that $\Diamond \mathbb{P}_0$ is false at α. In such a picture the content of the relation R – the relative possibility of worlds – is indicated by means of arrows from circle to circle; see figures 3.2–3.5, for example. In figure 3.1 there are no arrows, however, since there are no worlds relevant to α in the model. Notice that if there were an arrow leading from α to another circle, β, then α could not contain the sentences it does. For then both $\mathbb{P}_0$ and $\neg \mathbb{P}_0$ would show up inside β, which is impossible. The point is that since a picture can coherently be drawn as in figure 3.1, the conditional $\Box \mathbb{P}_0 \rightarrow \Diamond \mathbb{P}_0$ can be falsified, and hence the schema D is not valid.

For T. Consider the instance $\Box \mathbb{P}_0 \rightarrow \mathbb{P}_0$, and let $\mathcal{M} = \langle W, R, P \rangle$ be a standard model in which $W = \{\alpha, \beta\}$ (where $\alpha \neq \beta$), $R = \{\langle \alpha, \beta \rangle\}$, and $P_n = \{\beta\}$, for $n \geqslant 0$. Since P_0 contains β, the only world relevant to α, the

Figure 3.1

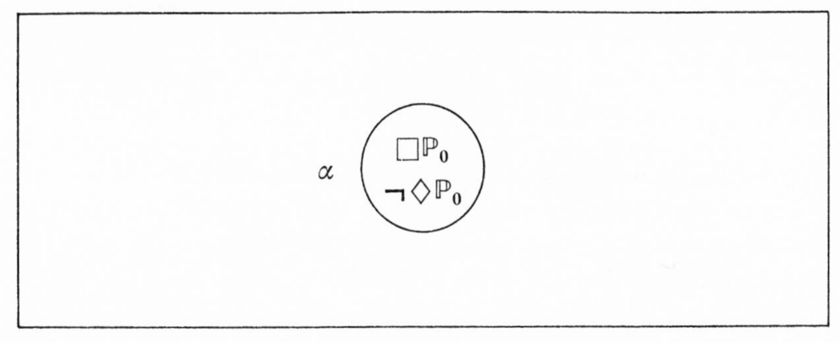

atomic sentence $\mathbb{P}_0$ is true at every world possible relative to α; i.e.

every β in $\mathcal{M}$ such that $\alpha R\beta$ is such that $\models^{\mathcal{M}}_{\beta} \mathbb{P}_0$.

So by definition 3.2(1), $\models^{\mathcal{M}}_{\alpha} \Box\mathbb{P}_0$. But since P_0 does not contain α, not $\models^{\mathcal{M}}_{\alpha} \mathbb{P}_0$. Hence, not $\models^{\mathcal{M}}_{\alpha} \Box\mathbb{P}_0 \rightarrow \mathbb{P}_0$. Thus the schema T is false in a standard model.

We have a picture of this countermodel to T in figure 3.2. Both $\Box\mathbb{P}_0$ and $\neg\mathbb{P}_0$ appear inside α, indicating that $\Box\mathbb{P}_0$ is true and $\mathbb{P}_0$ is false at this world; and $\mathbb{P}_0$ appears inside β to show that it is true there. The arrow from α to β indicates that β is relevant to α, and the absence of any other arrows in the picture shows that this is the only case of relative possibility in the model. Again, because this picture can coherently be drawn, we see that T is not in general valid. Notice that the picture would be incoherent if there were an arrow leading from α back to α.

For B. Consider the instance $\mathbb{P}_0 \rightarrow \Box\Diamond\mathbb{P}_0$, and the standard model $\mathcal{M} = \langle W, R, P\rangle$ in which $W = \{\alpha, \beta\}$ (distinct), $R = \{\langle\alpha, \beta\rangle\}$, and $P_n = \{\alpha\}$, for $n \geqslant 0$. Figure 3.3 pictures this model. Thus $\mathbb{P}_0$ is true at α,

Figure 3.2

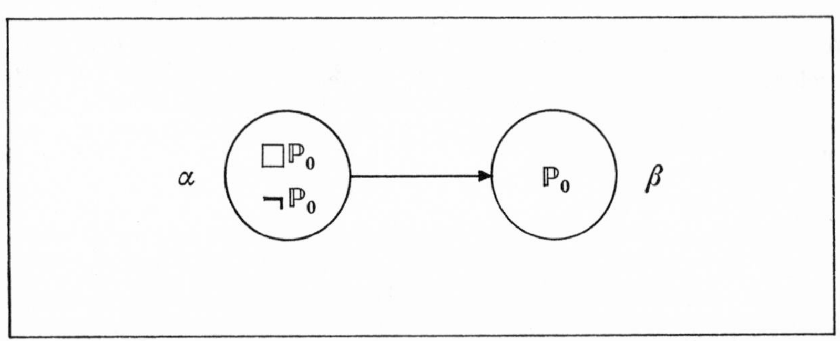

Figure 3.3

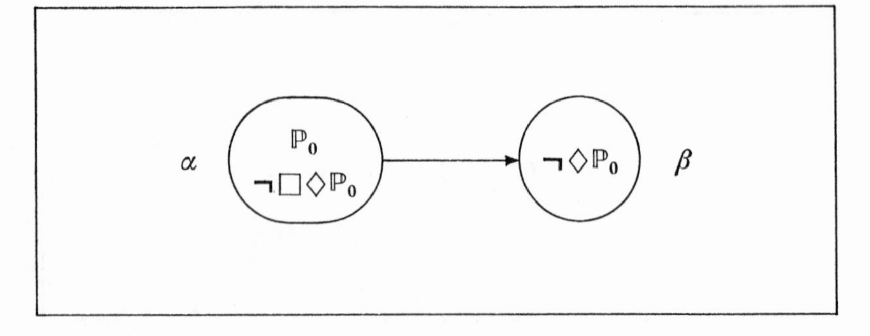

and $\Diamond \mathbb{P}_0$ is false at β (since there are no alternative worlds to β, as the lack of arrows leading away from β shows). Hence $\Box \Diamond \mathbb{P}_0$ is false at α, since β is accessible from α. So the instance of B is false at α, which means that the schema B is invalid. Note the effect on the picture if an arrow is drawn from β to α.

For 4. Consider the instance $\Box \mathbb{P}_0 \rightarrow \Box\Box \mathbb{P}_0$, and the standard model $\mathcal{M} = \langle W, R, P \rangle$ in which $W = \{\alpha, \beta, \gamma\}$ (all distinct), $R = \{\langle \alpha, \beta \rangle, \langle \beta, \gamma \rangle\}$, and $P_n = \{\beta\}$, for $n \geqslant 0$. We see, then, that $\models_\beta^\mathcal{M} \mathbb{P}_0$ and not $\models_\gamma^\mathcal{M} \mathbb{P}_0$. Thus in figure 3.4 β contains $\mathbb{P}_0$ and γ contains $\neg \mathbb{P}_0$. Because γ is the only world related to β, β also contains $\neg \Box \mathbb{P}_0$. But β is the only world related to α. So α contains both $\Box \mathbb{P}_0$ and $\neg \Box\Box \mathbb{P}_0$, which means that $\models_\alpha^\mathcal{M} \Box \mathbb{P}_0$ and not $\models_\alpha^\mathcal{M} \Box\Box \mathbb{P}_0$. Hence, not $\models_\alpha^\mathcal{M} \Box \mathbb{P}_0 \rightarrow \Box\Box \mathbb{P}_0$. Thus the schema 4 is false in a standard model. (What would happen if there were an arrow from α to γ in figure 3.4?)

For 5. We leave it to the reader to describe a countermodel to the instance $\Diamond \mathbb{P}_0 \rightarrow \Box \Diamond \mathbb{P}_0$ of the schema 5. Figure 3.5 provides a clue.

This concludes the proof of theorem 2.1. Figures 3.1–3.5 and the

Figure 3.4

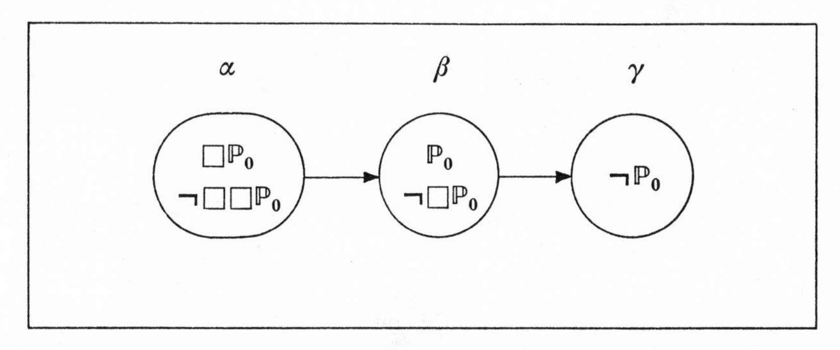

Figure 3.5

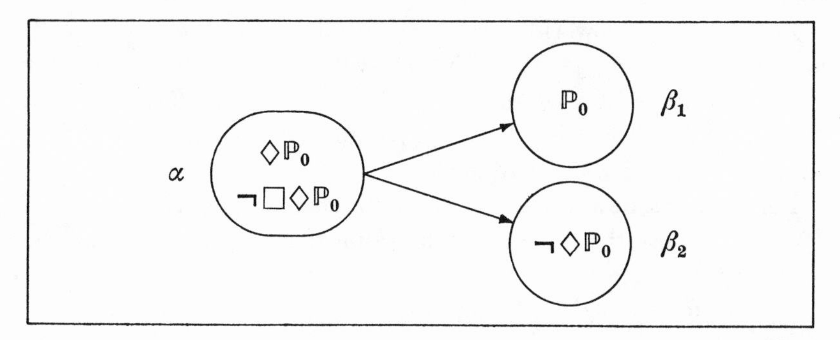

descriptions of the countermodels should enhance the reader's understanding of standard models.

In a standard model $\mathcal{M} = \langle W, R, P \rangle$ the relation R is:

> *serial* iff for every α in $\mathcal{M}$ there is a β in $\mathcal{M}$ such that $\alpha R \beta$;
>
> *reflexive* iff for every α in $\mathcal{M}$, $\alpha R \alpha$;
>
> *symmetric* iff for every α and β in $\mathcal{M}$, if $\alpha R \beta$, then $\beta R \alpha$;
>
> *transitive* iff for every α, β, and γ in $\mathcal{M}$, if $\alpha R \beta$ and $\beta R \gamma$, then $\alpha R \gamma$;
>
> *euclidean* iff for every α, β, and γ in $\mathcal{M}$, if $\alpha R \beta$ and $\alpha R \gamma$, then $\beta R \gamma$.

Let us call the model $\mathcal{M}$ itself serial, reflexive, symmetric, transitive, or euclidean, according as the relation in it has these properties.

THEOREM 3.5. *The following schemas are valid respectively in the indicated classes of standard models.*

(1) D: *serial*

(2) T: *reflexive*

(3) B: *symmetric*

(4) 4: *transitive*

(5) 5: *euclidean*

Proof

For (1). Let α be a world in a serial standard model $\mathcal{M} = \langle W, R, P \rangle$, and suppose that $\models_\alpha^\mathcal{M} \Box A$. It is enough to show that $\models_\alpha^\mathcal{M} \Diamond A$. The assumption means that $\models_\beta^\mathcal{M} A$, for every β in $\mathcal{M}$ such that $\alpha R \beta$. By the seriality of R, such a β exists. So for some β in $\mathcal{M}$ such that $\alpha R \beta$, $\models_\beta^\mathcal{M} A$; i.e. $\models_\alpha^\mathcal{M} \Diamond A$. Therefore, the schema D is valid in the class of serial standard models.

For (2). Let α be a world in a reflexive standard model $\mathcal{M} = \langle W, R, P \rangle$, and assume that $\models_\alpha^\mathcal{M} \Box A$. We wish to show that $\models_\alpha^\mathcal{M} A$. By the assumption, for every β in $\mathcal{M}$, if $\alpha R \beta$, then $\models_\beta^\mathcal{M} A$. In particular, if $\alpha R \alpha$, then $\models_\alpha^\mathcal{M} A$. Hence, $\models_\alpha^\mathcal{M} A$, since R is reflexive. Therefore, the schema T is valid in the class of reflexive standard models.

For (3). Let α be a world in a symmetric standard model $\mathcal{M} = \langle W, R, P \rangle$, and suppose that $\models_\alpha^\mathcal{M} A$. For the result that $\models_\alpha^\mathcal{M} \Box\Diamond A$, we need to show that

$$\text{for every } \beta \text{ in } \mathcal{M} \text{ such that } \alpha R \beta,\ \models_\beta^\mathcal{M} \Diamond A,$$

i.e. that

> for every β in $\mathcal{M}$ such that $\alpha R \beta$ there is a γ in $\mathcal{M}$ such that $\beta R \gamma$ and $\models_{\gamma}^{\mathcal{M}} \mathrm{A}$.

So let β be a world in $\mathcal{M}$ such that $\alpha R \beta$. By the symmetry of R, $\beta R \alpha$. So there is a γ in $\mathcal{M}$ – viz., α – such that $\beta R \gamma$ and $\models_{\gamma}^{\mathcal{M}} \mathrm{A}$. Therefore, the schema B is valid in the class of symmetric standard models.

For (4). Let α be a world in a transitive standard model $\mathcal{M} = \langle W, R, P \rangle$, and suppose that $\models_{\alpha}^{\mathcal{M}} \Box \mathrm{A}$. We wish to show that $\models_{\alpha}^{\mathcal{M}} \Box\Box \mathrm{A}$, which means that

> for every β in $\mathcal{M}$ such that $\alpha R \beta$, and for every γ in $\mathcal{M}$ such that $\beta R \gamma$, $\models_{\gamma}^{\mathcal{M}} \mathrm{A}$.

So let β and γ be worlds in $\mathcal{M}$ such that $\alpha R \beta$ and $\beta R \gamma$. It is left as an exercise to show that from this together with the assumption and the transitivity of R it follows that $\models_{\gamma}^{\mathcal{M}} \mathrm{A}$. This suffices to establish that the schema 4 is valid in the class of transitive standard models.

For (5). Let α be a world in a euclidean standard model $\mathcal{M} = \langle W, R, P \rangle$, and suppose that $\models_{\alpha}^{\mathcal{M}} \Diamond \mathrm{A}$, i.e. that there is a β in $\mathcal{M}$ such that $\alpha R \beta$ and $\models_{\beta}^{\mathcal{M}} \mathrm{A}$. To show from this that $\models_{\alpha}^{\mathcal{M}} \Box \Diamond \mathrm{A}$ is to show that

> for every β in $\mathcal{M}$ such that $\alpha R \beta$ there is a γ in $\mathcal{M}$ such that $\beta R \gamma$ and $\models_{\gamma}^{\mathcal{M}} \mathrm{A}$.

It is left to the reader to show that this follows from the assumption together with the euclideanness of R.

Theorem 3.5 is the basis of a number of soundness theorems in chapter 5. It may enhance the reader's understanding of the discursive proofs above to refer to figures 3.1–3.5, for the models pictured there violate, respectively, the conditions on the models in parts (1)–(5) of the theorem.

Let us close this section by considering briefly the schema

$$\text{G.} \quad \Diamond\Box \mathrm{A} \rightarrow \Box\Diamond \mathrm{A}.$$

That G is not valid in the class of standard models is evident from the model pictured in figure 3.6. Note that if there were arrows leading from β and γ to a circle δ – if there were a single world δ relevant to both β and γ – the diagram would be inconsistent. From this we see that G is true in every standard model $\mathcal{M}$ in which the relation R satisfies the condition that for every α, β, and γ in $\mathcal{M}$,

> if $\alpha R \beta$ and $\alpha R \gamma$, then for some δ in $\mathcal{M}$, $\beta R \delta$ and $\gamma R \delta$.

When the relation R has this property, let us call it – and the model $\mathcal{M}$ itself – *incestual* (since it means that offspring β and γ of a common parent α have themselves an offspring δ in common).

In the next section we see how each clause in theorem 3.5 is a special case of a simple theorem that relates certain generalizations of the schema G and the property of incestuality.

EXERCISES

3.17. Complete the proof of theorem 3.4 by describing a standard countermodel to the instance $\Diamond \mathbb{P}_0 \rightarrow \Box \Diamond \mathbb{P}_0$ of the schema 5. (See figure 3.5.)

3.18. Complete the proof of theorem 3.5 (parts (4) and (5)).

3.19. Prove that the schema G is true in every incestual standard model.

3.20. Prove that the schema D is true in every reflexive standard model.

3.21. Consider the duals of T, B, 4, and 5:

T$\Diamond$. $A \rightarrow \Diamond A$

B$\Diamond$. $\Diamond \Box A \rightarrow A$

4$\Diamond$. $\Diamond \Diamond A \rightarrow \Diamond A$

5$\Diamond$. $\Diamond \Box A \rightarrow \Box A$

Show that these schemas are valid respectively in the classes of reflexive, symmetric, transitive, and euclidean standard models.

3.22. The results of this exercise aid in the proof of the distinctness of the fifteen systems of modal logic registered on the diagram in figure 4.1 (see also figure 5.1).

Figure 3.6

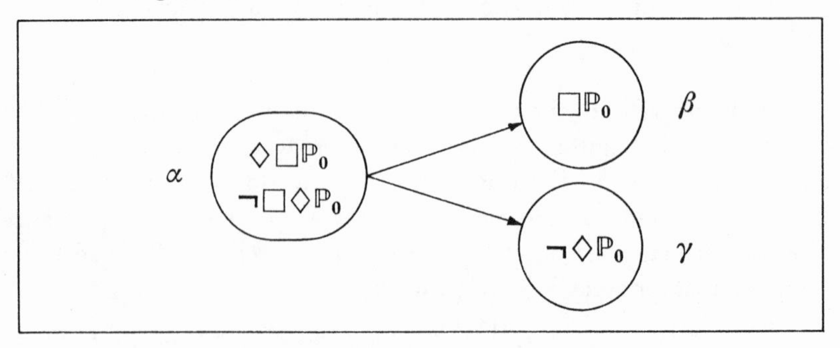

(*a*) Describe reflexive symmetric standard models that falsify instances of the schemas 4 and 5.
(*b*) Describe reflexive transitive standard models that falsify instances of the schemas B and 5.
(*c*) Describe serial transitive euclidean standard models that falsify instances of the schemas T and B.
(*d*) Describe a symmetric transitive standard model that falsifies an instance of the schema D.
(*e*) Describe a serial euclidean standard model that falsifies an instance of the schema 4.
(*f*) Describe a serial symmetric standard model that falsifies an instance of the schema T.

3.23. A binary relation on a set is said to be:

a *similarity* iff it is reflexive and symmetric;

a *quasi-ordering* iff it is reflexive and transitive;

an *equivalence* iff it is reflexive and euclidean.

Referring to theorem 3.5, prove:

(*a*) The schemas T and B are true in every standard model in which the relation is a similarity.
(*b*) The schemas T and 4 are true in every standard model in which the relation is a quasi-ordering.
(*c*) The schemas T and 5 are true in every standard model in which the relation is an equivalence.

3.24. Identify a condition on standard models to validate the sentence

P. $\Diamond\top$.

(Hint: See exercise 3.10, parts (*a*) and (*j*).)

3.25. Identify a condition on standard models to validate the sentence

$\bar{\text{P}}$. $\neg\Diamond\top$.

3.26. Identify conditions on standard models to validate the converses of D, T, B, 4, and 5:

D_c. $\Diamond A \to \Box A$

T_c. $A \to \Box A$

B_c. $\Box\Diamond A \to A$

4_c. $\Box\Box A \to \Box A$

5_c. $\Box\Diamond A \to \Diamond A$

3.27. Show that the schema G is true in any standard model that is symmetric or euclidean.

3.28. Falsify an instance of the schema G in a standard model in which the relation is a quasi-ordering (i.e. is reflexive and transitive; see exercise 3.23).

3.29. Recall the models $\mathcal{M} = \langle W, f, P \rangle$ described in exercise 3.13 and the explanation there of why they are equivalent to standard models. In a model of this sort truth conditions for possibilitations are given by:

$$\models^{\mathcal{M}}_{\alpha} \Diamond A \text{ iff } f(\alpha) \cap \|A\|^{\mathcal{M}} \neq \emptyset.$$

The following conditions are equivalent to the properties of seriality, reflexivity, symmetry, transitivity, and euclideanness. Which are which?

(*a*) if $\beta \in f(\alpha)$, then $\alpha \in f(\beta)$

(*b*) $f(\alpha) \neq \emptyset$

(*c*) if $\beta \in f(\alpha)$, then $f(\alpha) \subseteq f(\beta)$

(*d*) $\alpha \in f(\alpha)$

(*e*) if $\beta \in f(\alpha)$, then $f(\beta) \subseteq f(\alpha)$

3.30. Let X be a set. By the *identity* (or diagonal) relation, I, on X we mean the binary relation represented by the set $\{\langle x, x \rangle : x \in X\}$. In other words, for any x and y in X, xIy if and only if $x = y$.

Let R be a binary relation on the set X. By the *converse*, $\breve{R}$, of the relation R we mean the binary relation represented by the set $\{\langle x, y \rangle : yRx\}$. In other words, for any x and y in X, $x\breve{R}y$ if and only if yRx.

Let R and S be binary relations on the set X. By the *relative product* (or composition), $R|S$, of R with S we mean the binary relation represented by the set $\{\langle x, y \rangle$: for some $z \in X$, xRz and $zSy\}$. In other words, for any x and y in X, $x(R|S)y$ if and only if there is a z in X such that xRz and zSy.

Properties of binary relations can be neatly expressed using these notions and notations. For example, the reflexivity, symmetry, and transitivity of the relation R are expressed respectively as follows.

(*a*) $I \subseteq R$. (*b*) $R \subseteq \breve{R}$. (*c*) $R|R \subseteq R$.

Prove this. Then give succinct expression in these terms to the properties of seriality, euclideanness, and incestuality. Finally, prove:

(*d*) $\breve{I} = I$. (*e*) $I|R = R$. (*f*) $R|I = R$.

3.31. Equivalence relations have been characterized in terms of reflexivity and euclideanness (see exercise 3.23). Prove that a relation is an equivalence if and only if it is:

(*a*) reflexive, symmetric, and transitive
(*b*) serial, symmetric, and transitive
(*c*) serial, symmetric, and euclidean

3.32. The *field* of a binary relation is the set of things related by the relation. Let R be an equivalence relation on a set X.

(*a*) Prove that X is the field of R.

For each $x \in X$, define $[x]$ to be $\{y \in X: xRy\}$; i.e. for each $x, y \in X$, $[x] = [y]$ if and only if xRy. The set $[x]$ is called the *equivalence class* (under R) generated by x.

(*b*) Prove that any two R-equivalence classes are either identical or disjoint, i.e. that for any $x, y \in X$, either $[x] = [y]$ or $[x] \cap [y] = \emptyset$.

Because of this we say that an equivalence relation R *partitions its field* X into mutually exclusive, non-empty subsets.

Let R^x be the relation R as restricted to the R-equivalence class $[x]$; i.e. for every $y, z \in X$, yR^xz if and only if $y, z \in [x]$.

(*c*) Prove that within $[x]$ the relation R – and hence R^x – is universal; i.e. that for any $y, z \in [x]$, yRz (and so yR^xz).

The point of this is that an equivalence relation is universal within any of the equivalence classes in the partition of its field.

Let X be the union of a collection of pairwise disjoint non-empty sets, and define the relation R on X by the stipulation that xRy just in case x and y are both members of some one set in the collection.

(*d*) Prove that R is an equivalence relation on X.

The point here is that the union of universal relations on disjoint fields is an equivalence relation.

3.3. The schema $G^{k,l,m,n}$

When $\alpha R \beta$ in a standard model $\mathcal{M} = \langle W, R, P \rangle$, we say that β is once removed from α. Let us write

$$\alpha R^n \beta$$

to mean that β is n times removed from α – i.e. that β can be reached from α, so to speak, by n steps in the relation R. So $\alpha R^n \beta$ just in case there are worlds $\gamma_1, \ldots, \gamma_{n-1}$ in $\mathcal{M}$ such that $\alpha R \gamma_1, \ldots, \gamma_{n-1} R \beta$. Pictorially:

$$\underset{\alpha}{\bullet} \xrightarrow{R} \underset{\gamma_1}{\bullet} \xrightarrow{R} \cdots \xrightarrow{R} \underset{\gamma_{n-1}}{\bullet} \xrightarrow{R} \underset{\beta}{\bullet}$$

Of course, that β is n worlds away from α does not preclude the possibility that β is also removed from α by some other number of worlds.

The relation R^n is called the nth *relative product* of R with itself, and it is defined inductively as follows.

DEFINITION 3.6. Let α and β be worlds in a standard model $\mathcal{M} = \langle W, R, P \rangle$.

(1) $\alpha R^0 \beta$ iff $\alpha = \beta$.

(2) For $n > 0$, $\alpha R^n \beta$ iff for some γ in $\mathcal{M}$, $\alpha R \gamma$ and $\gamma R^{n-1} \beta$.

Thus R^0 is the relation of identity, R^1 is the relation R itself, and, for example, $\alpha R^3 \beta$ if and only if there exist γ_1 and γ_2 such that $\alpha R \gamma_1$, $\gamma_1 R \gamma_2$, and $\gamma_2 R \beta$.

Using R^n we can state truth conditions for sentences of the forms $\Box^n$A and $\Diamond^n$A. For just as $\Box$A is true at a world exactly when A is true at all worlds once removed from it, so $\Box^n$A holds at a world just in case A holds at all worlds n times removed; and similarly for $\Diamond$A and $\Diamond^n$A.

THEOREM 3.7. *Let α be a world in a standard model $\mathcal{M} = \langle W, R, P \rangle$. Then, for $n \geqslant 0$:*

(1) $\models^{\mathcal{M}}_{\alpha} \Box^n$A *iff for every β in $\mathcal{M}$ such that $\alpha R^n \beta$, $\models^{\mathcal{M}}_{\beta}$* A.

(2) $\models^{\mathcal{M}}_{\alpha} \Diamond^n$A *iff for some β in $\mathcal{M}$ such that $\alpha R^n \beta$, $\models^{\mathcal{M}}_{\beta}$* A.

Proof. The proof in each case is by induction on n. We carry it out for (1) and leave (2) to the reader as an exercise.

Base case:

$\models^{\mathcal{M}}_{\alpha} \Box^0$A iff $\models^{\mathcal{M}}_{\alpha}$ A

– definition 2.3 (1);

iff for every β in $\mathcal{M}$ such that $\alpha = \beta$, $\models^{\mathcal{M}}_{\beta}$ A;

iff for every β in $\mathcal{M}$ such that $\alpha R^0 \beta$, $\models^{\mathcal{M}}_{\beta}$ A

– definition 3.6 (1).

Inductive case. For $n > 0$, assume as an inductive hypothesis that the theorem holds for all $k < n$. Then in particular, for every α in $\mathcal{M}$,

$$\models^{\mathcal{M}}_{\alpha} \Box^{n-1}\text{A iff for every } \beta \text{ in } \mathcal{M} \text{ such that } \alpha R^{n-1}\beta, \models^{\mathcal{M}}_{\beta} \text{A}.$$

Therefore:

$\models^{\mathcal{M}}_{\alpha} \Box^{n}$A iff $\models^{\mathcal{M}}_{\alpha} \Box\Box^{n-1}$A
– definition 2.3 (2);

iff for every γ in $\mathcal{M}$ such that $\alpha R\gamma$, $\models^{\mathcal{M}}_{\gamma} \Box^{n-1}$A
– definition 3.2 (1);

iff for every γ in $\mathcal{M}$ such that $\alpha R\gamma$, and for every β in $\mathcal{M}$ such that $\gamma R^{n-1}\beta$, $\models^{\mathcal{M}}_{\beta}$ A
– inductive hypothesis;

iff for every β in $\mathcal{M}$, if there is a γ in $\mathcal{M}$ such that $\alpha R\gamma$ and $\gamma R^{n-1}\beta$, then $\models^{\mathcal{M}}_{\beta}$ A;

iff for every β in $\mathcal{M}$, if $\alpha R^{n}\beta$, then $\models^{\mathcal{M}}_{\beta}$ A
– definition 3.6 (2).

We are now in a position to generalize the results in the preceding section concerning the validity of the schemas D, T, B, 4, and 5. For example, for $n \geqslant 0$ the schema

$$4^{n}. \quad \Box\text{A} \rightarrow \Box^{n}\text{A}$$

is valid in the class of *n-transitive* standard models, i.e. models $\mathcal{M} = \langle W, R, P\rangle$ such that for every α and β in $\mathcal{M}$,

$$\text{if } \alpha R^{n}\beta, \text{ then } \alpha R\beta.$$

(Proof: Let α be a world in such a model $\mathcal{M}$ and suppose that $\models^{\mathcal{M}}_{\alpha} \Box$A, i.e. that for every β in $\mathcal{M}$ such that $\alpha R\beta$, $\models^{\mathcal{M}}_{\beta}$ A. Then $\models^{\mathcal{M}}_{\beta}$ A whenever β in $\mathcal{M}$ is such that $\alpha R^{n}\beta$; i.e. $\models^{\mathcal{M}}_{\alpha} \Box^{n}$A.) As an instance of this result we have theorem 3.5 (4) – that the schema 4,

$$\Box\text{A} \rightarrow \Box^{2}\text{A},$$

is valid where R satisfies the condition that for every α and β,

$$\text{if } \alpha R^{2}\beta, \text{ then } \alpha R\beta.$$

For this is simply another way of expressing the transitivity of R – that if β is twice removed from α it is also once removed (compare (*c*) in exercise 3.30). More generally, for $m, n \geqslant 0$ the schema

$$4^{m,n}. \quad \Box^{m}\text{A} \rightarrow \Box^{n}\text{A}$$

is valid in the class of standard models in which for all worlds α and β,

if $\alpha R^n \beta$, then $\alpha R^m \beta$.

The reader may verify this as an exercise. Note that the schema 4 is also an instance of $4^{m,n}$, and that $T = 4^{1,0}$.

But we are after even more generality. Let us consider the schema

$$G^{k,l,m,n}. \quad \Diamond^k \Box^l A \to \Box^m \Diamond^n A,$$

where $k, l, m, n \geqslant 0$. Note that all of the schemas D, T, B, 4, and 5 are special cases of $G^{k,l,m,n}$:

$$D = G^{0,1,0,1}.$$

$$T = G^{0,1,0,0}.$$

$$B = G^{0,0,1,1}.$$

$$4 = G^{0,1,2,0}.$$

$$5 = G^{1,0,1,1}.$$

The schema $G^{k,l,m,n}$ is a generalization of the schema

$$G. \quad \Diamond \Box A \to \Box \Diamond A,$$

discussed at the end of the last section. There we saw that G is valid in the class of incestual standard models, where $\mathcal{M} = \langle W, R, P \rangle$ is such that for every α, β, and γ in $\mathcal{M}$,

if $\alpha R \beta$ and $\alpha R \gamma$, then for some δ in $\mathcal{M}$, $\beta R \delta$ and $\gamma R \delta$.

Let us generalize this condition and say that R (and also $\mathcal{M}$ itself) is *k, l, m, n-incestual* if and only if for every α, β, and γ in $\mathcal{M}$,

if $\alpha R^k \beta$ and $\alpha R^m \gamma$, then for some δ in $\mathcal{M}$, $\beta R^l \delta$ and $\gamma R^n \delta$.

THEOREM 3.8. *The schema* $G^{k,l,m,n}$ *is valid in the class of* k, l, m, n-*incestual standard models.*

Proof. The argument is an elaborate version of that for the validity of the schema G in the class of incestual standard models. We give it here in detail.

Let α be a world in a k, l, m, n-incestual standard model $\mathcal{M} = \langle W, R, P \rangle$, and suppose that $\models_\alpha^\mathcal{M} \Diamond^k \Box^l A$. This means that

(*a*) for some β in $\mathcal{M}$ such that $\alpha R^k \beta$, every δ in $\mathcal{M}$ such that $\beta R^l \delta$ is such that $\models_\delta^\mathcal{M} A$.

We wish to prove that $\models^{\mathcal{M}}_{\alpha} \Box^m \Diamond^n A$, i.e. that

(*b*) for every γ in $\mathcal{M}$ such that $\alpha R^m \gamma$, there is a δ in $\mathcal{M}$ such that $\gamma R^n \delta$ and $\models^{\mathcal{M}}_{\delta} A$.

To show this, we suppose γ to be a world in $\mathcal{M}$ such that $\alpha R^m \gamma$, and argue that there is a world δ in $\mathcal{M}$ such that $\gamma R^n \delta$ and $\models^{\mathcal{M}}_{\delta} A$.

By our assumptions, then, β and γ are such that

$$\alpha R^k \beta \text{ and } \alpha R^m \gamma.$$

So by the *k, l, m, n*-incestuality of R, there is a world δ in $\mathcal{M}$ such that

$$\beta R^l \delta \text{ and } \gamma R^n \delta.$$

From the first half of this and (*a*) it follows that $\models^{\mathcal{M}}_{\delta} A$, so that indeed there is a world δ in $\mathcal{M}$ such that $\gamma R^n \delta$ and $\models^{\mathcal{M}}_{\delta} A$.

As a corollary to theorem 3.8 we have theorem 3.5 – that the schemas D, T, B, 4, and 5 are valid in the classes of serial, reflexive, symmetric, transitive, and euclidean standard models, respectively. To see this it is enough to notice that the properties of seriality, reflexivity, symmetry, transitivity, and euclideanness are the same, respectively, as 0, 1, 0, 1-, 0, 1, 0, 0-, 0, 0, 1, 1-, 0, 1, 2, 0-, and 1, 0, 1, 1-incestuality. Let us show this for the cases of reflexivity and symmetry, and leave the others as exercises.

Let R be the relation in a standard model $\mathcal{M}$. For reflexivity:

R is 0, 1, 0, 0-incestual iff for every α, β, and γ in $\mathcal{M}$, if $\alpha R^0 \beta$ and $\alpha R^0 \gamma$, then for some δ in $\mathcal{M}$, $\beta R^1 \delta$ and $\gamma R^0 \delta$;

iff for every α, β, and γ in $\mathcal{M}$, if $\alpha = \beta$ and $\alpha = \gamma$, then for some δ in $\mathcal{M}$, $\beta R \delta$ and $\gamma = \delta$ – definition 3.6;

iff for every α and β in $\mathcal{M}$, if $\alpha = \beta$, then $\alpha R \beta$;

iff for every α in $\mathcal{M}$, $\alpha R \alpha$;

iff R is reflexive.

For symmetry:

R is 0, 0, 1, 1-incestual iff for every α, β, and γ in $\mathcal{M}$, if $\alpha R^0 \beta$ and $\alpha R^1 \gamma$, then for some δ in $\mathcal{M}$, $\beta R^0 \delta$ and $\gamma R^1 \delta$;

iff for every α, β, and γ in $\mathcal{M}$, if $\alpha = \beta$ and $\alpha R \gamma$, then for some δ in $\mathcal{M}$, $\beta = \delta$, and $\gamma R \delta$
– definition 3.6;

iff for every α and β in $\mathcal{M}$, if $\alpha R \beta$, then $\beta R \alpha$;

iff R is symmetric.

EXERCISES

3.33. Give an example of a standard model $\mathcal{M} = \langle W, R, P\rangle$ in which $\alpha R^m \beta$ and $\alpha R^n \beta$ for some worlds α and β in $\mathcal{M}$ and some natural numbers m and n such that $m \neq n$.

3.34. Prove part (2) of theorem 3.7.

3.35. Prove that the schema $4^{m,n}$ is valid in the class of standard models $\mathcal{M} = \langle W, R, P\rangle$ such that for all α and β in $\mathcal{M}$, if $\alpha R^n \beta$, then $\alpha R^m \beta$.

3.36. Check that the schemas D, T, B, 4, and 5 are all (the advertised) special cases of the schema $G^{k,l,m,n}$.

3.37. Prove that the properties of seriality, transitivity, and euclideanness are the same, respectively, as 0,1,0,1-, 0,1,2,0-, and 1,0,1,1-incestuality.

3.38. Prove that 1,0,1,0-, 0,0,1,0-, and 0,2,1,0-incestuality are the same as the following properties.

partial functionality: if $\alpha R \beta$ and $\alpha R \gamma$, then $\beta = \gamma$
vacuity: if $\alpha R \beta$, then $\alpha = \beta$
(*weak*) *density:* if $\alpha R \beta$, then for some γ, $\alpha R \gamma$ and $\gamma R \beta$

We may infer from this that the schemas D_c ($= G^{1,0,1,0}$), T_c ($= G^{0,0,1,0}$), and 4_c ($= G^{0,2,1,0}$) are valid respectively in classes of partially functional, vacuous, and dense standard models. Prove this directly.

3.39. Let $\mathcal{M} = \langle W, R, P\rangle$ be a standard model. Prove:

(*a*) The schema B_c, $\Box\Diamond A \to A$, is true in $\mathcal{M}$ when for every α in $\mathcal{M}$ there is a β in $\mathcal{M}$ such that both $\alpha R \beta$ and for every γ in $\mathcal{M}$ such that $\beta R \gamma$, $\alpha = \gamma$.

(*b*) The schema 5_c, $\Box\Diamond A \to \Diamond A$, is true in $\mathcal{M}$ when for every α in $\mathcal{M}$ there is a β in $\mathcal{M}$ such that both $\alpha R \beta$ and for every γ in $\mathcal{M}$ such that $\beta R \gamma$, $\beta = \gamma$.

3.40. Prove that P, $\Diamond\top$, is valid in the class of serial standard models.

3.41. Prove that $\overline{\text{P}}$, $\neg\Diamond\top$, is valid in the class of standard models in which the relation is empty (the class of standard models $\mathcal{M} = \langle W, R, P\rangle$ such that for every α and β in $\mathcal{M}$, not $\alpha R \beta$).

3.42. The following express the properties of seriality, euclideanness, incestuality, partial functionality, vacuity, density, and emptiness of a relation R (see exercises 3.30, 3.38, and 3.41). Which are which?

(*a*) $R \subseteq R|R$.

(*b*) $\breve{R}|R \subseteq R$.

(*c*) $R \subseteq \emptyset$.

(*d*) $I \subseteq R|\breve{R}$.

(*e*) $\breve{R}|R \subseteq I$.

(*f*) $\breve{R}|R \subseteq R|\breve{R}$.

(*g*) $R \subseteq I$.

3.43. Consider the condition

$$\breve{R}^k|R^m \subseteq R^l|\breve{R}^n.$$

Prove that this is the same as k, l, m, n-incestuality.

3.44. Consider the condition

$$\breve{R}^m|R^k \subseteq R^n|\breve{R}^l$$

on a standard model $\mathcal{M} = \langle W, R, P\rangle$. Prove:

(*a*) The condition validates the dual of $\text{G}^{k,l,m,n}$, viz. $\Diamond^m\Box^n\text{A}\to\Box^k\Diamond^l\text{A}$.

(*b*) The condition is the same as k, l, m, n-incestuality.

3.45. Identify a single condition on standard models to validate the following schemas.

(*a*) $\Box(\Box\text{A}\vee\text{B})\to(\Box\text{A}\vee\Box\text{B})$

(*b*) $(\Diamond\text{A}\wedge\Diamond\text{B})\to\Diamond(\Diamond\text{A}\wedge\text{B})$

3.46. Identify a single condition on standard models to validate the following schemas.

(*a*) $\Box(\text{A}\vee\text{B})\to(\Box\text{A}\vee\Box\text{B})$

(*b*) $(\Diamond\text{A}\wedge\Diamond\text{B})\to\Diamond(\text{A}\wedge\text{B})$

3.47. Identify conditions on standard models to validate the following schemas.

$$\text{H}^{++}.\quad \Box(\Diamond A \vee \Diamond B) \to (\Box\Diamond A \vee \Box\Diamond B)$$

$$\text{H}^{+}.\quad (\Box(\Box A \vee B) \wedge \Box(A \vee \Box B)) \to (\Box A \vee \Box B)$$

$$\text{H}.\quad (\Box(A \vee B) \wedge \Box(\Box A \vee B) \wedge \Box(A \vee \Box B)) \to (\Box A \vee \Box B)$$

3.48. Show that the schemas in the preceding exercise are true in every euclidean standard model. Then falsify instances of each of the schemas in standard models in which the relation is a quasi-ordering (see exercise 3.23).

3.49. Identify a condition on standard models to validate

$$\text{P}^{k}.\quad \Diamond^{k}\top$$

for every $k > 0$.

3.50. Try to identify a condition on standard models to validate the schema G_{c}, $\Box\Diamond A \to \Diamond\Box A$.

3.51. Consider the following conditions on a standard model $\mathcal{M} = \langle W, R, P\rangle$.

secondary reflexivity: if $\alpha R\beta$, then $\beta R\beta$

reverse secondary reflexivity: if $\beta R\alpha$, then $\beta R\beta$

Prove that the following schemas are valid respectively in classes of secondarily reflexive and reverse secondarily reflexive standard models.

$$\Box(\Box A \to A) \qquad \Diamond\top \to (\Box A \to A)$$

3.52. Consider the following conditions on a standard model $\mathcal{M} = \langle W, R, P\rangle$.

(*a*) if $\alpha R^{j}\beta$, then if $\beta R^{k}\gamma$ and $\beta R^{m}\delta$, then for some ϵ in $\mathcal{M}$, $\gamma R^{l}\epsilon$ and $\delta R^{n}\epsilon$

(*b*) if $\beta R^{j}\alpha$, then if $\beta R^{k}\gamma$ and $\beta R^{m}\delta$, then for some ϵ in $\mathcal{M}$, $\gamma R^{l}\epsilon$ and $\delta R^{n}\epsilon$

Prove that the following schemas are valid respectively in classes of standard models satisfying (*a*) and (*b*).

$$\Box^{j}(\Diamond^{k}\Box^{l}A \to \Box^{m}\Diamond^{n}A) \qquad \Diamond^{j}\top \to (\Diamond^{k}\Box^{l}A \to \Box^{m}\Diamond^{n}A)$$

3.53. Consider again the models $\mathcal{M} = \langle W, f, P\rangle$ in exercises 3.13 and 3.28. The following conditions are equivalent to the properties of partial functionality, vacuity, emptiness, incestuality, secondary reflexivity, and reverse secondary reflexivity (see exercises 3.38, 3.41, and 3.51). Which are which?

(*a*) if $\alpha \in f(\beta)$, then $\beta \in f(\beta)$

(*b*) $f(\alpha) \subseteq \{\alpha\}$

(*c*) if $\beta, \gamma \in f(\alpha)$, then $f(\beta) \cap f(\gamma) \neq \emptyset$

(*d*) $f(\alpha)$ is at most a singleton

(*e*) if $\beta \in f(\alpha)$, then $\beta \in f(\beta)$

(*f*) $f(\alpha) \subseteq \emptyset$

3.54. Identify conditions on standard models to validate the following schemas.

D!. $\Box A \leftrightarrow \Diamond A$

T!. $\Box A \leftrightarrow A$

3.55. Consider models $\mathcal{M} = \langle W, f, P \rangle$ in which W and P are as usual and f is a function from W to W (not a point to set function as in exercise 3.13); i.e. for each world α in $\mathcal{M}$, $f(\alpha)$ is some world in $\mathcal{M}$. In models of this sort truth conditions for necessitations are given by:

$$\models_\alpha^\mathcal{M} \Box A \text{ iff } \models_{f(\alpha)}^\mathcal{M} A.$$

(*a*) State truth conditions for possibilitations so that Df$\Diamond$ is valid in the class of such models.

(*b*) Show that the rule RK (see chapter 1) preserves validity in the class of such models.

(*c*) Show that the schema D!, $\Box A \leftrightarrow \Diamond A$, is valid in the class of such models.

(*d*) Define a class of standard models equivalent to this class of models.

3.56. Let $\mathcal{M} = \langle W, R, P \rangle$ be a standard model. By an R-sequence we mean a finite sequence $\langle \alpha_1, \ldots, \alpha_n \rangle$ of worlds in $\mathcal{M}$ such that $\alpha_i R \alpha_{i+1}$ for $i = 1, \ldots, n-1$. We define the standard model $\mathcal{M}' = \langle W', R', P' \rangle$ as follows.

(1) W' = the set of all R-sequences.

(2) $\langle \alpha_1, \ldots, \alpha_n \rangle R' \langle \beta_1, \ldots, \beta_m \rangle$ iff $\langle \beta_1, \ldots, \beta_m \rangle = \langle \alpha_1, \ldots, \alpha_n, \beta_m \rangle$.

(3) $\langle \alpha_1, \ldots, \alpha_n \rangle \in P'_k$ iff $\alpha_n \in P_k$, for $k = 0, 1, 2, \ldots$.

Prove that for every world $\langle \alpha_1, \ldots, \alpha_n \rangle$ in $\mathcal{M}'$,

$$\models_{\alpha_n}^\mathcal{M} A \text{ iff } \models_{\langle \alpha_1, \ldots, \alpha_n \rangle}^{\mathcal{M}'} A.$$

The proof is by induction on the complexity of A. Give it at least for the cases in which (*a*) A is atomic, $\mathbb{P}_k$, (*b*) A is the falsum, $\bot$, (*c*) A is a conditional, $B \to C$, and (*d*) A is a necessitation, $\Box B$.

3.57. Consider the following conditions on a standard model $\mathscr{M} = \langle W, R, P\rangle$.

irreflexivity: not $\alpha R \alpha$

asymmetry: if $\alpha R \beta$, then not $\beta R \alpha$

antisymmetry: if $\alpha R \beta$ and $\beta R \alpha$, then $\alpha = \beta$

intransitivity: if $\alpha R \beta$ and $\beta R \gamma$, then not $\alpha R \gamma$

(*a*) Prove that the models $\mathscr{M}'$ in the preceding exercise have all these properties.

(*b*) Let C_1, C_2, C_3, and C_4 be the classes of irreflexive, asymmetric, antisymmetric, and intransitive standard models, and let C be the class of all standard models. Prove that each of these classes determines the same modal logic – i.e. that for any sentence A,

$\vDash_{\mathsf{C}} A$ iff $\vDash_{\mathsf{C}n} A$,

for $n = 1, 2, 3, 4$. (This comes down to showing – via part (*a*) and the preceding exercise – that if a sentence is false in any standard model it is also rejected by models that have the properties in question.)

3.58. Prove:

(*a*) Every reflexive relation is serial, secondarily reflexive, and reverse secondarily reflexive.

(*b*) Every euclidean relation is secondarily reflexive and incestual.

(*c*) Every secondarily reflexive relation is dense.

(*d*) Every symmetric relation is incestual.

(*e*) Every symmetric relation is transitive if and only if it is euclidean.

(*f*) Every serial relation is reflexive that is vacuous or reverse secondarily reflexive.

(*g*) Every vacuous relation is symmetric, transitive, euclidean, and partially functional.

(*h*) Every reflexive partially functional relation is vacuous.

(*i*) The empty relation is vacuous.

(*j*) Every asymmetric relation is irreflexive and antisymmetric.

(*k*) Every intransitive relation is irreflexive.

3.4. Generated models

Let α be a possible world in a standard model $\mathcal{M} = \langle W, R, P \rangle$. By an *R-descendant* of α we mean any world in $\mathcal{M}$ that can be reached from α via the relation R in some finite number of steps (including zero, so that α is an R-descendant of itself). In other words, a world β in $\mathcal{M}$ is an R-descendant of α if and only if $\alpha R^n \beta$ for some $n \geqslant 0$.

As the reader may well have noticed already, the truth value of a sentence A at a possible world α in a standard model depends only on the subsentences of A (including A) and the R-descendants of α (including α). That is to say, sentences not involved in the structure of A and worlds not descended from α are irrelevant to the question of what truth value A has at α. We can make this precise in terms of a theorem about *generated models*.

DEFINITION 3.9. Let α be a world in a standard model $\mathcal{M} = \langle W, R, P \rangle$. Then $\mathcal{M}^\alpha = \langle W^\alpha, R^\alpha, P^\alpha \rangle$ is *the standard model generated by* α *from* $\mathcal{M}$ iff:

(1) $W^\alpha = \{\beta \text{ in } \mathcal{M} : \alpha R^n \beta, \text{ for some } n \geqslant 0\}$.

(2) $R^\alpha = R \cap (W^\alpha \times W^\alpha)$.

(3) $P^\alpha_n = P_n \cap W^\alpha$, for each $n \geqslant 0$.

Thus the set of worlds W^α is the set of R-descendants of α, and the relation R^α and sets P^α_n are just the restrictions of R and P_n to the set of R-descendants of α.

THEOREM 3.10. *Let* $\mathcal{M}^\alpha = \langle W^\alpha, R^\alpha, P^\alpha \rangle$ *be the standard model generated by* α *from* $\mathcal{M} = \langle W, R, P \rangle$. *Then for every* β *in* $\mathcal{M}^\alpha$:

$\models^{\mathcal{M}}_\beta$ A *iff* $\models^{\mathcal{M}^\alpha}_\beta$ A.

Proof. The proof is by induction on the complexity of A. We give it for the cases in which A is (*a*) atomic, $\mathbb{P}_n$, (*b*) the falsum, $\perp$, (*c*) a conditional, B$\rightarrow$C, and (*d*) a necessitation, $\Box$B. We suppose throughout that β is a world in $\mathcal{M}^\alpha$.

For (*a*):

$\models^{\mathcal{M}}_\beta \mathbb{P}_n$ iff $\beta \in P_n$
– definition 2.5 (1);

iff $\beta \in P_n \cap W^\alpha$
– since $\beta \in W^\alpha$;

iff $\beta \in P_n^\alpha$
  – definition 3.9 (3);
iff $\models_\beta^{\mathscr{M}^\alpha} \mathbb{P}_n$
  – definition 2.5 (1).

So the theorem holds when A is atomic.

For (*b*). By definition 2.5 (3) $\perp$ is not true at any world in any model. So $\models_\beta^{\mathscr{M}} \perp$ if and only if $\models_\beta^{\mathscr{M}^\alpha} \perp$, and we see that the theorem holds when A is the falsum.

For the inductive cases (*c*) and (*d*) we make the hypothesis that the theorem holds for all sentences shorter than A.

For (*c*):

$\models_\beta^{\mathscr{M}} \mathrm{B} \rightarrow \mathrm{C}$ iff if $\models_\beta^{\mathscr{M}} \mathrm{B}$ then $\models_\beta^{\mathscr{M}} \mathrm{C}$
  – definition 2.5 (7);
iff if $\models_\beta^{\mathscr{M}^\alpha} \mathrm{B}$ then $\models_\beta^{\mathscr{M}^\alpha} \mathrm{C}$
  – inductive hypothesis;
iff $\models_\beta^{\mathscr{M}^\alpha} \mathrm{B} \rightarrow \mathrm{C}$
  – definition 2.5 (7).

Thus the theorem holds when A is a conditional.

For (*d*). For left-to-right, suppose that $\models_\beta^{\mathscr{M}} \Box\mathrm{B}$. Then by definition 3.2, for every γ in $\mathscr{M}$ such that $\beta R \gamma$, $\models_\gamma^{\mathscr{M}} \mathrm{B}$. To show from this that $\models_\beta^{\mathscr{M}^\alpha} \Box\mathrm{B}$, it is enough (by definition 3.2) to suppose that γ is a world in $\mathscr{M}^\alpha$ such that $\beta R^\alpha \gamma$ and then argue that $\models_\gamma^{\mathscr{M}^\alpha} \mathrm{B}$. But if $\beta R^\alpha \gamma$, then also $\beta R \gamma$. So $\models_\gamma^{\mathscr{M}} \mathrm{B}$, and – by the inductive hypothesis – $\models_\gamma^{\mathscr{M}^\alpha} \mathrm{B}$.

For right-to-left, assume that $\models_\beta^{\mathscr{M}^\alpha} \Box\mathrm{B}$, so that for every world γ in $\mathscr{M}^\alpha$ such that $\beta R^\alpha \gamma$, $\models_\gamma^{\mathscr{M}^\alpha} \mathrm{B}$. Now let γ be a world in $\mathscr{M}$ for which it holds that $\beta R \gamma$. Since $\beta \in W^\alpha$, γ is also an R-descendant of α; and so $\gamma \in W^\alpha$. Hence the pair $\langle \beta, \gamma \rangle$ is in the relation $R \cap (W^\alpha \times W^\alpha)$, which means by definition 3.9 that $\beta R^\alpha \gamma$. So $\models_\gamma^{\mathscr{M}^\alpha} \mathrm{B}$, and – by the inductive hypothesis – $\models_\gamma^{\mathscr{M}} \mathrm{B}$, which is what we wished to prove.

Thus the theorem holds when A is a necessitation. This concludes the proof.

As an immediate corollary to theorem 3.10, a sentence is true in a standard model just in case it is true in every model generated from (any world in) the model.

THEOREM 3.11. *Let* $\mathscr{M}$ *be an stadard model. Then:*

$$\models^{\mathscr{M}} \mathrm{A} \text{ iff for every } \alpha \text{ in } \mathscr{M}, \models^{\mathscr{M}^\alpha} \mathrm{A}.$$

Proof. Suppose that $\models^{\mathcal{M}} A$, i.e. by definition 2.6, $\models^{\mathcal{M}}_{\beta} A$ for every β in $\mathcal{M}$. Let $\mathcal{M}^{\alpha}$ be generated from $\mathcal{M}$. Then by theorem 3.10 it follows that $\models^{\mathcal{M}^{\alpha}}_{\beta} A$ for every β in $\mathcal{M}^{\alpha}$, i.e. that $\models^{\mathcal{M}^{\alpha}} A$. For the reverse, suppose that not $\models^{\mathcal{M}} A$, so that for some α in $\mathcal{M}$, not $\models^{\mathcal{M}}_{\alpha} A$. Then by theorem 3.10, not $\models^{\mathcal{M}^{\alpha}}_{\alpha} A$. So not $\models^{\mathcal{M}^{\alpha}} A$, which is what we wished to show.

Let us call a standard model *generated* just in case it is generated by some world in some standard model. The following theorem relates classes of generated models to the classes of models from which they are generated.

THEOREM 3.12. *Let* C *be a class of standard models, and let* $\mathcal{G}(\mathsf{C})$ *be the class of models generated from the models in* C. *Then:*

$$\models_{\mathsf{C}} A \textit{ iff } \models_{\mathcal{G}(\mathsf{C})} A.$$

Proof. This is an immediate consequence of theorem 3.11. For if $\models_{\mathsf{C}} A$ and $\mathcal{M}$ is a model in $\mathcal{G}(\mathsf{C})$ – i.e. $\mathcal{M}$ is generated from some model in C – then clearly A is true in $\mathcal{M}$. Conversely, suppose $\models_{\mathcal{G}(\mathsf{C})} A$, and let $\mathcal{M}$ be a model in C. Then since A is true in every model generated from $\mathcal{M}$, it is true in $\mathcal{M}$.

As we remarked in section 2.4, the set of sentences valid in a class of models is a system of modal logic (see exercise 2.32(*d*)). Thus theorem 3.12 tells us that a system of modal logic that is determined by a class of standard models is also determined by the associated class of generated models.

This is often very interesting. For example, let us use theorem 3.12 to show that the system of modal logic determined by the class E of standard models in which the relation is an equivalence is also determined by the class U of standard models in which the relation is universal (compare exercises 1.10, 3.12, and 3.32).

THEOREM 3.13. $\models_{\mathsf{E}} A$ *iff* $\models_{\mathsf{U}} A$.

Proof. The result follows via theorem 3.12 and the fact that U is precisely the class $\mathcal{G}(\mathsf{E})$ of models generated from members of E. The point is, roughly, that an equivalence relation can be taken apart into a collection of universal relations, and any collection of universal relations (with disjoint fields) can be patched together to form an equivalence relation.

For the proof proper, suppose, first, that $\mathcal{M}$ is a model in U. Then the relation in $\mathcal{M}$ is an equivalence, since it is universal (see exercise 3.12). So $\mathcal{M}$ is in E, too. But $\mathcal{M}$ is generated (by any of its worlds) from itself, so it is in $\mathcal{G}$(E), as we wished to show. Conversely, let $\mathcal{M}^\alpha = \langle W^\alpha, R^\alpha, P^\alpha \rangle$ be a model in $\mathcal{G}$(E) generated by α from $\mathcal{M} = \langle W, R, P \rangle$ in E. Then W^α, the set of R-descendants of α in $\mathcal{M}$, is an equivalence class (see exercise 3.32) within which R – and hence R^α – is universal. So $\mathcal{M}^\alpha$ is in U, and the proof is ended.

Universal models are, essentially, the models of chapter 1, i.e. models without relations (see exercise 3.12). So we see that the same modal logic is determined both by that class of models and by the class of equivalence – i.e. reflexive euclidean – standard models. Of course, this modal logic is the system *S5*; but the proof that this is so awaits us in chapter 5.

EXERCISES

3.59. Give the proof of theorem 3.9 for the cases in which A = $\top$, $\neg$B, B $\wedge$ C, B $\vee$ C, B $\leftrightarrow$ C, $\Diamond$B.

3.60. *The p-morphism theorem.* Let $\mathcal{M} = \langle W, R, P \rangle$ and $\mathcal{M}^\circ = \langle W^\circ, R^\circ, P^\circ \rangle$ be standard models, and let f be a function from W to W° satisfying the following conditions.

(1) f is onto.

(2) For every α and β in $\mathcal{M}$:

(*a*) if $\alpha R \beta$, then $f(\alpha) R^\circ f(\beta)$;

(*b*) if $f(\alpha) R^\circ f(\beta)$, then there is a γ in $\mathcal{M}$ such that $f(\beta) = f(\gamma)$ and $\alpha R \gamma$.

(3) For every α in $\mathcal{M}$ and every $n = 0, 1, 2, \ldots$, $\alpha \in P_n$ if and only if $f(\alpha) \in P^\circ_n$.

The function f is said to be a p-morphism ('p' for *pseudo-epi*) from $\mathcal{M}$ to $\mathcal{M}^\circ$, reliable on the atomic sentences. Prove that for every α in $\mathcal{M}$,

$$\models^{\mathcal{M}}_{\alpha} \text{A iff } \models^{\mathcal{M}^\circ}_{f(\alpha)} \text{A}.$$

The proof is by induction on the complexity of A. Give it at least for the cases in which A is atomic, the falsum, a conditional, and a necessitation.

3.61. Use the p-morphism theorem in the preceding exercise to give another proof of the theorem in exercise 3.56. That is, define a function from the worlds of one model to those of the other, in exercise 3.56, and prove that the function is a p-morphism reliable on the atomic sentences.

3.62. *The safe extension theorem.* Let $\mathcal{M} = \langle W, R, P \rangle$ and $\mathcal{M}^{\#} = \langle W^{\#}, R^{\#}, P^{\#} \rangle$ be standard models satisfying the following conditions.

(1) $W^{\#} = W \cup X$, where X is a set disjoint from W.

(2) $R^{\#} = R \cup S$, where S is a binary relation on $W^{\#}$ for which it holds that $\alpha \in X$ whenever $\alpha S \beta$.

(3) $P^{\#} = P$.

$\mathcal{M}^{\#}$ is said to be a safe extension of $\mathcal{M}$ – safe in the sense that, since $R^{\#}$ does not lead from W into X, truth values of sentences at worlds common to $\mathcal{M}$ and $\mathcal{M}^{\#}$ are the same in both models. That is, for every α in $\mathcal{M}$,

$$\models_{\alpha}^{\mathcal{M}} \mathrm{A} \text{ iff } \models_{\alpha}^{\mathcal{M}^{\#}} \mathrm{A}.$$

The proof of this is by induction on the complexity of A. Give it at least for the cases in which A is atomic, the falsum, a conditional, and a necessitation. (For the modal cases it may be helpful first to prove that $\alpha R \beta$ if and only if $\alpha R^{\#} \beta$, for every α in $\mathcal{M}$.)

3.63. Let us consider the following principle.

If $\models_{\mathsf{C}} \Box \mathrm{A}$, then $\models_{\mathsf{C}} \mathrm{A}$.

This principle holds for some classes of standard models, but not for others. For example, it holds when C is any of the following classes.

(*a*) all
(*b*) serial
(*c*) reflexive
(*d*) transitive
(*e*) serial symmetric
(*f*) serial transitive
(*g*) reflexive symmetric
(*h*) reflexive transitive
(*i*) reflexive euclidean

Cases (*c*), (*g*), (*h*), and (*i*) are trivial: if all the models in C are reflexive, then $\Box \mathrm{A} \to \mathrm{A}$ is valid in C (theorem 3.5); and so if $\models_{\mathsf{C}} \Box \mathrm{A}$, then $\models_{\mathsf{C}} \mathrm{A}$.

For cases (*a*), (*b*), (*d*), and (*f*) we can use the safe extension theorem of the preceding exercise. For example, let us show that the principle holds in case (*a*), i.e. when C is the class of all standard models. We argue contrapositively. Suppose that not $\models_{\mathsf{C}} \mathrm{A}$, so that for some α in some $\mathcal{M} = \langle W, R, P \rangle$ in C, not $\models_{\alpha}^{\mathcal{M}} \mathrm{A}$. Now, for a world α' not in $\mathcal{M}$, we define the standard model $\mathcal{M}^{\#} = \langle W^{\#}, R^{\#}, P^{\#} \rangle$ by the following conditions.

(1) $W^{\#} = W \cup \{\alpha'\}$.

(2) $R^{\#} = R \cup \{\langle \alpha', \alpha \rangle\}$.

(3) $P^{\#} = P$.

It is clear that $\mathcal{M}^{\#}$ is a safe extension of $\mathcal{M}$, and so by the theorem, not $\models_{\alpha}^{\mathcal{M}^{\#}}$ A. But $\alpha' R^{\#} \alpha$, and so by definition 3.2 not $\models_{\alpha'}^{\mathcal{M}^{\#}} \Box$A. So not $\models_{\mathsf{C}} \Box$A, which is what we wished to prove.

Cases (*b*), (*d*), and (*f*) are for the reader. The problem in these cases is to define safe extensions that have the right properties. We also leave case (*e*) for the reader; the argument is simple (remember theorem 3.5) but does not use the safe extension theorem.

The principle we are considering does not hold when **C** is any of the following classes of standard models.

(*j*) symmetric
(*k*) euclidean
(*l*) serial euclidean
(*m*) transitive euclidean
(*n*) serial transitive euclidean
(*o*) symmetric transitive

For the cases in which the models in **C** are euclidean – i.e. (*k*)–(*o*) (for (*o*) see exercise 3.58(*e*)) – it is enough to note that the schema $\Box(\Box A \rightarrow A)$ is valid in each class (see exercises 3.51 and 3.58(*b*)), but that instances of $\Box A \rightarrow A$ have countermodels in each class (as the reader should verify).

Case (*j*) is left for the reader (consider the schema $\Box(A \rightarrow \Diamond\Diamond A)$).

The principle we have been considering is a special case ($n = 1$) of the following principle.

$$\text{If} \models_{\mathsf{C}} \Box A_1 \vee \ldots \vee \Box A_n, \text{ then} \models_{\mathsf{C}} A_i \text{ for some } i = 1, \ldots, n \ (n > 0).$$

This 'rule of disjunction' holds for cases (*a*), (*b*), (*c*), (*d*), (*f*), and (*h*) above, but fails for the rest. The reader may wish to demonstrate the new failures, cases (*e*), (*g*), and (*i*). We return to this matter in chapter 5.

3.5. Filtrations

We explained in section 2.3 the basis of what it means to say that a model $\mathcal{M}^*$ is a filtration through Γ of a model $\mathcal{M}$. To wit, Γ is a set of sentences closed under subsentences, the worlds in $\mathcal{M}^*$ are the equivalence classes of worlds in $\mathcal{M}$ that agree on the truth values of the sentences in Γ, and a world in $\mathcal{M}^*$ verifies exactly those atomic sentences in Γ that are verified in $\mathcal{M}$ by any (hence all) of its members. (Recall the definitions of $\equiv$, $[\alpha]$, and $[X]$ in 2.3.) To define this idea, now, for standard models it is necessary only to insure that the relation in the filtration $\mathcal{M}^*$ is in a certain sense consistent with the relation in $\mathcal{M}$. The following definition suffices for our purposes.

DEFINITION 3.14. Let $\mathcal{M} = \langle W, R, P \rangle$ be a standard model, and let Γ be a set of sentences closed under subsentences. Then a *filtration of* $\mathcal{M}$ *through* Γ is any standard model $\mathcal{M}^* = \langle W^*, R^*, P^* \rangle$ such that:

(1) $W^* = [W]$.

(2) For every α and β in $\mathcal{M}$:

- (*a*) if $\alpha R \beta$, then $[\alpha] R^* [\beta]$;
- (*b*) if $[\alpha] R^* [\beta]$, then for every sentence $\Box \mathrm{A} \in \Gamma$, if $\models^{\mathcal{M}}_{\alpha} \Box \mathrm{A}$, then $\models^{\mathcal{M}}_{\beta} \mathrm{A}$;
- (*c*) if $[\alpha] R^* [\beta]$, then for every sentence $\Diamond \mathrm{A} \in \Gamma$, if $\models^{\mathcal{M}}_{\beta} \mathrm{A}$, then $\models^{\mathcal{M}}_{\alpha} \Diamond \mathrm{A}$.

(3) $P^*_n = [P_n]$, for each n such that $\mathbb{P}_n \in \Gamma$.

Thus R^* is consistent with R in the sense that (*a*) R^* imitates in $\mathcal{M}^*$ the behavior of R in $\mathcal{M}$, (*b*) R^* does not relate worlds $[\alpha]$ and $[\beta]$ in $\mathcal{M}^*$ for which it happens that some $\Box \mathrm{A}$ in Γ is true at α in $\mathcal{M}$ while A is false at β in $\mathcal{M}$, and (*c*) R^* does not relate $[\alpha]$ and $[\beta]$ in $\mathcal{M}^*$ for which it happens that some $\Diamond \mathrm{A}$ in Γ is false at α in $\mathcal{M}$ while A is true at β in $\mathcal{M}$. We give some examples of filtrations after the following three theorems.

THEOREM 3.15. *Let* $\mathcal{M}^* = \langle W^*, R^*, P^* \rangle$ *be a* Γ*-filtration of a standard model* $\mathcal{M} = \langle W, R, P \rangle$. *Then for every* $\mathrm{A} \in \Gamma$*:*

$\models^{\mathcal{M}}_{\alpha} \mathrm{A}$ *iff* $\models^{\mathcal{M}^*}_{[\alpha]} \mathrm{A}$.

In other words, $[\|\mathrm{A}\|^{\mathcal{M}}] = \|\mathrm{A}\|^{\mathcal{M}^*}$, *for every* $\mathrm{A} \in \Gamma$.

Proof. The proof is by induction on the complexity of $\mathrm{A} \in \Gamma$. The non-modal cases were discussed in section 2.3 (exercise 2.27). Of the modal cases, let us treat only that in which A is a necessitation, $\Box \mathrm{B}$. As an inductive hypothesis we assume that the theorem holds for sentences in Γ that are shorter than A. Since Γ is closed under subsentences, $\mathrm{B} \in \Gamma$.

So it follows from the inductive hypothesis that for every α in $\mathcal{M}$,

$$\vDash_{\alpha}^{\mathcal{M}} \mathrm{B} \text{ iff } \vDash_{[\alpha]}^{\mathcal{M}^*} \mathrm{B}.$$

Now let us show that this result holds for $\Box$B.

For left-to-right, suppose that $\vDash_{\alpha}^{\mathcal{M}} \Box \mathrm{B}$. To show that $\vDash_{[\alpha]}^{\mathcal{M}^*} \Box \mathrm{B}$ we suppose that $[\beta]$ is a world in $\mathcal{M}^*$ such that $[\alpha] R^*[\beta]$, and argue from this to the conclusion that $\vDash_{[\beta]}^{\mathcal{M}^*} \mathrm{B}$. But by clause (2)(*b*) of definition 3.14 it follows that if $\vDash_{\alpha}^{\mathcal{M}} \Box \mathrm{B}$, then $\vDash_{\beta}^{\mathcal{M}} \mathrm{B}$. So $\vDash_{\beta}^{\mathcal{M}} \mathrm{B}$ – whence, by the inductive hypothesis, $\vDash_{[\beta]}^{\mathcal{M}^*} \mathrm{B}$. For right-to-left, suppose that $\vDash_{[\alpha]}^{\mathcal{M}^*} \Box \mathrm{B}$, so that for every $[\beta]$ in $\mathcal{M}^*$ such that $[\alpha] R^*[\beta]$, $\vDash_{[\beta]}^{\mathcal{M}^*} \mathrm{B}$. Let β be a world in $\mathcal{M}$ for which it holds that $\alpha R \beta$; it is enough now to show that $\vDash_{\beta}^{\mathcal{M}} \mathrm{B}$. But by clause (2)(*a*) of definition 3.14 it follows that $[\alpha] R^*[\beta]$. So $\vDash_{[\beta]}^{\mathcal{M}^*} \mathrm{B}$ – whence, by the inductive hypothesis, $\vDash_{\beta}^{\mathcal{M}} \mathrm{B}$. This completes the proof.

The next two theorems are corollaries of theorem 3.15. For their proofs see the remarks in section 2.3.

THEOREM 3.16. *Let $\mathcal{M}^*$ be a Γ-filtration of a standard model $\mathcal{M}$. Then $\mathcal{M}$ and $\mathcal{M}^*$ are equivalent modulo Γ – i.e. for every $\mathrm{A} \in \Gamma$:*

$$\vDash^{\mathcal{M}} \mathrm{A} \textit{ iff } \vDash^{\mathcal{M}^*} \mathrm{A}.$$

THEOREM 3.17. *Let* C *be a class of standard models and let $\Gamma(\mathsf{C})$ be the class of Γ-filtrations of models in* C. *Then for every $\mathrm{A} \in \Gamma$:*

$$\vDash_{\mathsf{C}} \mathrm{A} \textit{ iff } \vDash_{\Gamma(\mathsf{C})} \mathrm{A}.$$

Definition 3.14 provides for the possibility of a number of filtrations of a standard model $\mathcal{M} = \langle W, R, P \rangle$ through a given set of sentences Γ. For example, consider $\mathcal{M}^* = \langle W^*, R^*, P^* \rangle$ in which (with W^* and P^* as usual) for every α and β,

$$[\alpha] R^*[\beta] \text{ iff for some } \alpha' \in [\alpha] \text{ and some } \beta' \in [\beta], \alpha' R \beta'.$$

To see that $\mathcal{M}^*$ is a filtration we must check that it satisfies conditions (*a*), (*b*), and (*c*) in clause (2) of the definition. We leave (*a*) and (*c*) as exercises. To prove that (*b*) is satisfied, suppose that $[\alpha] R^*[\beta]$, $\Box \mathrm{A} \in \Gamma$, and $\vDash_{\alpha}^{\mathcal{M}} \Box \mathrm{A}$. Then $\alpha' R \beta'$ for some $\alpha' \in [\alpha]$ and some $\beta' \in [\beta]$. So $\vDash_{\alpha'}^{\mathcal{M}} \Box \mathrm{A}$, since α and α' agree on Γ, from which it follows that $\vDash_{\beta'}^{\mathcal{M}} \mathrm{A}$. But $\mathrm{A} \in \Gamma$, since Γ is closed under subsentences. Therefore, since β and β' agree on Γ, $\vDash_{\beta}^{\mathcal{M}} \mathrm{A}$, which is what we wished to show.

We call $\mathcal{M}^*$ a *finest* filtration of $\mathcal{M}$ through Γ. By a *coarsest* Γ-filtration of $\mathcal{M}$ we mean a case in which for every α and β in $\mathcal{M}$,

$$[\alpha] R^*[\beta] \text{ iff both for every } \Box \mathrm{A} \in \Gamma, \text{ if } \vDash_{\alpha}^{\mathcal{M}} \Box \mathrm{A}, \text{ then } \vDash_{\beta}^{\mathcal{M}} \mathrm{A}, \\ \text{and for every } \Diamond \mathrm{A} \in \Gamma, \text{ if } \vDash_{\beta}^{\mathcal{M}} \mathrm{A}, \text{ then } \vDash_{\alpha}^{\mathcal{M}} \Diamond \mathrm{A}.$$

We leave it as an exercise for the reader to check that R^*, so defined, meets the conditions for a filtration.

We explained in section 2.8 how filtrations may be used to show that a system of modal logic is determined by a class of finite models. The point is that a filtration through a finite set of sentences is always a finite model: if Γ contains n sentences, then a Γ-filtration of a model $\mathcal{M}$ has at most 2^n worlds (that being the maximum number of ways the worlds in $\mathcal{M}$ can agree on the sentences in Γ); and 2^n is finite if n is. Thus, in particular, a model will be finite if it is a filtration through the set of subsentences of a sentence, since such a set is always finite. For the most part we reserve theorems about determination by classes of finite standard models until chapter 5, in connection with decidability results. But as an example let us show here that the modal logic determined by the class U of universal standard models is also determined by the class $\mathsf{U}_{\mathrm{FIN}}$ of finite universal standard models.

THEOREM 3.18. $\models_{\mathsf{U}} \mathrm{A}$ *iff* $\models_{\mathsf{U}_{\mathrm{FIN}}} \mathrm{A}$.

Proof. Left-to-right is trivial, since $\mathsf{U}_{\mathrm{FIN}} \subseteq \mathsf{U}$. For right-to-left, suppose that $\models_{\mathsf{U}_{\mathrm{FIN}}} \mathrm{A}$. To prove that $\models_{\mathsf{U}} \mathrm{A}$ it is enough, in virtue of theorem 3.17, to prove that $\models_{\Gamma(\mathsf{U})} \mathrm{A}$, where Γ is the set of subsentences of A (which of course contains A). So let $\mathcal{M}^* = \langle W^*, R^*, P^* \rangle$ be a model in $\Gamma(\mathsf{U})$, i.e. a Γ-filtration of some model $\mathcal{M} = \langle W, R, P \rangle$ in U. We wish to argue that $\models^{\mathcal{M}^*} \mathrm{A}$, and for this it is sufficient to show that $\mathcal{M}^*$ is in $\mathsf{U}_{\mathrm{FIN}}$. Since $\mathcal{M}^*$ is finite, this amounts to showing that $\mathcal{M}^*$ is universal. But it is: By clause (2)(*a*) of definition 3.14, $[\alpha] R^* [\beta]$ whenever α and β are worlds in $\mathcal{M}$ for which it holds that $\alpha R \beta$; and, since $\mathcal{M}$ is universal – i.e. $\alpha R \beta$ for every α and β in $\mathcal{M}$ – it follows that $[\alpha] R^* [\beta]$ for every $[\alpha]$ and $[\beta]$ in $\mathcal{M}^*$ – i.e. $\mathcal{M}^*$ is universal. (Indeed, this argument can be generalized; *any* filtration of a universal standard model is itself universal.) To sum up, if A is valid in the class of finite universal standard models, then A is true in every (finite) filtration of any universal model through its set of subsentences and so, by theorem 3.17, is true in every universal model. Put contrapositively, if A fails in some universal model it fails in some finite universal model – to wit, any filtration, through its set of subsentences, of the rejecting universal model.

Of course, as we have remarked, the set of sentences valid in the class of universal standard models is the modal logic *S5*. So theorem 3.18 tells us that *S5* has the finite model property: every non-theorem of *S5* is false in some finite model for the system. This leads directly to the result that *S5* is decidable, by the reasoning explained in section 2.8, since *S5* can be axiomatized by finitely many schemas.

The proof of theorem 3.18 turned on the fact that any filtration of a universal model is itself universal. This result does not generalize, however. For example, not every filtration of a transitive standard model is itself transitive (exercise 3.67); so the proof that the modal logic determined by the class of transitive standard models is also determined by the class of finite transitive standard models cannot be so easily made. The moral is that it is often necessary to select a more limited class of filtrations of the models in a given class.

With an eye to the proofs of finite determination and decidability in chapter 5, we devote the next section to the problem of finding appropriate filtrations of serial, reflexive, symmetric, transitive, and euclidean models.

EXERCISES

3.64. Give the proof of theorem 3.15 for the case in which $A = \Diamond B$.

3.65. Check that finest and coarsest filtrations, described in section 3.5, are indeed filtrations; i.e. check that the relations R^* in these structures satisfy the conditions in clause (2) of definition 3.14. Prove that in each of these cases R^* satisfies the following.

$$\text{If } \alpha_1 \equiv \alpha_2 \text{ and } \beta_1 \equiv \beta_2, \text{ then } [\alpha_1]\, R^*[\beta_1] \text{ iff } [\alpha_2]\, R^*[\beta_2].$$

3.66. Prove that any filtration of a reflexive standard model is itself reflexive.

3.67. Give examples of each of the following.

(*a*) a non-symmetric filtration of a symmetric standard model
(*b*) a non-transitive filtration of a transitive standard model
(*c*) a non-euclidean filtration of a euclidean standard model

3.68. Let Γ be a set of sentences closed under subsentences. By $\mathscr{B}(\Gamma)$ – the *boolean closure* of Γ – we mean the result of closing Γ with respect to the (boolean) operations $\top$, $\bot$, $\neg$, $\wedge$, $\vee$, $\rightarrow$, and $\leftrightarrow$. In other words, $\mathscr{B}(\Gamma)$ is defined as follows.

(1) $\Gamma \subseteq \mathscr{B}(\Gamma)$.

(2) $\top \in \mathscr{B}(\Gamma)$.

(3) $\bot \in \mathscr{B}(\Gamma)$.

(4) $\neg A \in \mathscr{B}(\Gamma)$ iff $A \in \mathscr{B}(\Gamma)$.

(5) $A \wedge B \in \mathscr{B}(\Gamma)$ iff $A, B \in \mathscr{B}(\Gamma)$.

(6) $A \vee B \in \mathscr{B}(\Gamma)$ iff $A, B \in \mathscr{B}(\Gamma)$.

(7) $A \rightarrow B \in \mathscr{B}(\Gamma)$ iff $A, B \in \mathscr{B}(\Gamma)$.

(8) $A \leftrightarrow B \in \mathscr{B}(\Gamma)$ iff $A, B \in \mathscr{B}(\Gamma)$.

Note that $\mathscr{B}(\Gamma)$ is closed under subsentences.

(*a*) Let $\mathscr{M}^* = \langle W^*, R^*, P^* \rangle$ be a Γ-filtration of a standard model $\mathscr{M} = \langle W, R, P \rangle$. Then for every $A \in \mathscr{B}(\Gamma)$,

$\models_{\alpha}^{\mathscr{M}} A$ iff $\models_{[\alpha]}^{\mathscr{M}^*} A$.

That is, $[\|A\|^{\mathscr{M}}] = \|A\|^{\mathscr{M}^*}$, for every $A \in \mathscr{B}(\Gamma)$.

The proof of this improvement of theorem 3.15 is by induction on the complexity of $A \in \mathscr{B}(\Gamma)$. Give it at least for the cases in which A is atomic, the falsum, a conditional, and a necessitation.

(*b*) Prove or disprove: $\mathscr{B}(\Gamma)$ is logically finite relative to a model $\mathscr{M}$ if Γ is.

3.6. Filtrations, continued

THEOREM 3.19. *Let $\mathscr{M}^* = \langle W^*, R^*, P^* \rangle$ be a filtration of a standard model $\mathscr{M} = \langle W, R, P \rangle$. Then:*

(1) *$\mathscr{M}^*$ is serial if $\mathscr{M}$ is.*

(2) *$\mathscr{M}^*$ is reflexive if $\mathscr{M}$ is.*

Proof. Part (1) we leave as an exercise. For (2), suppose that $\mathscr{M}$ is reflexive, i.e. that $\alpha R \alpha$ for every α in $\mathscr{M}$. By (2)(*a*) of definition 3.14 it follows that $[\alpha] R^* [\alpha]$ for every $[\alpha]$ in $\mathscr{M}^*$, i.e. that $\mathscr{M}^*$ is reflexive.

There are no analogous results for arbitrary filtrations of symmetric, transitive, and euclidean models. But we can define filtrations that do have certain combinations of these properties.

Let Γ be a set of sentences closed under subsentences, and let $\mathscr{M} = \langle W, R, P \rangle$ be a standard model. We consider the following conditions on worlds α and β in $\mathscr{M}$.

c_1 for every $\Box A \in \Gamma$, if $\models_{\alpha}^{\mathscr{M}} \Box A$, then $\models_{\beta}^{\mathscr{M}} A$
for every $\Diamond A \in \Gamma$, if $\models_{\beta}^{\mathscr{M}} A$, then $\models_{\alpha}^{\mathscr{M}} \Diamond A$

c_2 for every $\Box A \in \Gamma$, if $\models_{\beta}^{\mathscr{M}} \Box A$, then $\models_{\alpha}^{\mathscr{M}} A$
for every $\Diamond A \in \Gamma$, if $\models_{\alpha}^{\mathscr{M}} A$, then $\models_{\beta}^{\mathscr{M}} \Diamond A$

c_3 for every $\Box A \in \Gamma$, if $\models^{\mathcal{M}}_{\alpha} \Box A$, then $\models^{\mathcal{M}}_{\beta} \Box A$
for every $\Diamond A \in \Gamma$, if $\models^{\mathcal{M}}_{\beta} \Diamond A$, then $\models^{\mathcal{M}}_{\alpha} \Diamond A$

c_4 for every $\Box A \in \Gamma$, if $\models^{\mathcal{M}}_{\beta} \Box A$, then $\models^{\mathcal{M}}_{\alpha} \Box A$
for every $\Diamond A \in \Gamma$, if $\models^{\mathcal{M}}_{\alpha} \Diamond A$, then $\models^{\mathcal{M}}_{\beta} \Diamond A$

As their display and designations suggest, these conditions come in pairs – one for $\Box$, and one for $\Diamond$. Where W^* and P^* are defined as usual, we can obtain a Γ-filtration $\mathcal{M}^* = \langle W^*, R^*, P^*\rangle$ of $\mathcal{M}$ by defining R^* in terms of various combinations of the pairs. The details emerge in the following theorem.

THEOREM 3.20. *Let $\mathcal{M}^* = \langle W^*, R^*, P^*\rangle$ be a standard model in which W^* and P^* are defined as in a Γ-filtration of a standard model $\mathcal{M} = \langle W, R, P\rangle$. Then:*

(1) *If R^* is defined by c_1 and c_2, then (a) $\mathcal{M}^*$ is symmetric, and (b) $\mathcal{M}^*$ is a Γ-filtration of $\mathcal{M}$ if $\mathcal{M}$ is symmetric.*

(2) *If R^* is defined by c_1 and c_3, then (a) $\mathcal{M}^*$ is transitive, and (b) $\mathcal{M}^*$ is a Γ-filtration of $\mathcal{M}$ if $\mathcal{M}$ is transitive.*

(3) *If R^* is defined by c_1, c_2, and c_3, then (a) $\mathcal{M}^*$ is symmetric and transitive, and (b) $\mathcal{M}^*$ is a Γ-filtration of $\mathcal{M}$ if $\mathcal{M}$ is symmetric and transitive.*

(4) *If R^* is defined by c_1, c_3, and c_4, then (a) $\mathcal{M}^*$ is transitive and euclidean, and (b) $\mathcal{M}^*$ is a Γ-filtration of $\mathcal{M}$ if $\mathcal{M}$ is transitive and euclidean.*

Proof. The complete proof is very long and involved. We choose to give it in detail for part (2), hoping thereby to illuminate the argument for the other parts, which we leave as exercises for the reader.

We assume that R^* is defined by c_1 and c_3. First we show that $\mathcal{M}^*$ is transitive. Suppose $[\alpha] R^*[\beta]$ and $[\beta] R^*[\gamma]$, for worlds α, β, and γ in $\mathcal{M}$. This means that the following conditions obtain.

$c_1(\alpha, \beta)$ for every $\Box A \in \Gamma$, if $\models^{\mathcal{M}}_{\alpha} \Box A$, then $\models^{\mathcal{M}}_{\beta} A$
for every $\Diamond A \in \Gamma$, if $\models^{\mathcal{M}}_{\beta} A$, then $\models^{\mathcal{M}}_{\alpha} \Diamond A$

$c_3(\alpha, \beta)$ for every $\Box A \in \Gamma$, if $\models^{\mathcal{M}}_{\alpha} \Box A$, then $\models^{\mathcal{M}}_{\beta} \Box A$
for every $\Diamond A \in \Gamma$, if $\models^{\mathcal{M}}_{\beta} \Diamond A$, then $\models^{\mathcal{M}}_{\alpha} \Diamond A$

$c_1(\beta, \gamma)$ for every $\Box A \in \Gamma$, if $\models^{\mathcal{M}}_{\beta} \Box A$, then $\models^{\mathcal{M}}_{\gamma} A$
for every $\Diamond A \in \Gamma$, if $\models^{\mathcal{M}}_{\gamma} A$, then $\models^{\mathcal{M}}_{\beta} \Diamond A$

$c_3(\beta, \gamma)$ for every $\Box A \in \Gamma$, if $\models^{\mathcal{M}}_{\beta} \Box A$, then $\models^{\mathcal{M}}_{\gamma} \Box A$
for every $\Diamond A \in \Gamma$, if $\models^{\mathcal{M}}_{\gamma} \Diamond A$, then $\models^{\mathcal{M}}_{\beta} \Diamond A$

We wish to conclude from this that $[\alpha] R^*[\gamma]$, which means that we must argue for the following conditions.

$$c_1(\alpha, \gamma) \quad \begin{array}{l} \text{for every } \Box A \in \Gamma, \text{ if } \models_\alpha^{\mathcal{M}} \Box A, \text{ then } \models_\gamma^{\mathcal{M}} A \\ \text{for every } \Diamond A \in \Gamma, \text{ if } \models_\gamma^{\mathcal{M}} A, \text{ then } \models_\alpha^{\mathcal{M}} \Diamond A \end{array}$$

$$c_3(\alpha, \gamma) \quad \begin{array}{l} \text{for every } \Box A \in \Gamma, \text{ if } \models_\alpha^{\mathcal{M}} \Box A, \text{ then } \models_\gamma^{\mathcal{M}} \Box A \\ \text{for every } \Diamond A \in \Gamma, \text{ if } \models_\gamma^{\mathcal{M}} \Diamond A, \text{ then } \models_\alpha^{\mathcal{M}} \Diamond A \end{array}$$

For $c_1(\alpha, \gamma)$. Suppose first that $\Box A \in \Gamma$ and $\models_\alpha^{\mathcal{M}} \Box A$, to show that $\models_\gamma^{\mathcal{M}} A$. Then $\models_\beta^{\mathcal{M}} \Box A$, by $c_3(\alpha, \beta)$, and so $\models_\gamma^{\mathcal{M}} A$, by $c_1(\beta, \gamma)$. Next, suppose that $\Diamond A \in \Gamma$ and $\models_\gamma^{\mathcal{M}} A$. Then $\models_\beta^{\mathcal{M}} \Diamond A$, by $c_1(\beta, \gamma)$, and so $\models_\alpha^{\mathcal{M}} \Diamond A$, by $c_3(\alpha, \beta)$, which is what we wished to show.

For $c_3(\alpha, \gamma)$. Suppose again that $\Box A \in \Gamma$ and $\models_\alpha^{\mathcal{M}} \Box A$. Then $\models_\beta^{\mathcal{M}} \Box A$, by $c_3(\alpha, \beta)$, and so $\models_\gamma^{\mathcal{M}} \Box A$, by $c_3(\beta, \gamma)$. Similarly, suppose that $\Diamond A \in \Gamma$ and $\models_\gamma^{\mathcal{M}} \Diamond A$. Then $\models_\beta^{\mathcal{M}} \Diamond A$, by $c_3(\beta, \gamma)$, and so $\models_\alpha^{\mathcal{M}} \Diamond A$, by $c_3(\alpha, \beta)$.

This takes care of the transitivity of $\mathcal{M}^*$. It remains to be shown that $\mathcal{M}^*$ is a Γ-filtration of $\mathcal{M}$ if $\mathcal{M}$ is transitive.

Assume that $\mathcal{M}$ is a transitive standard model. To see that $\mathcal{M}^*$ is a Γ-filtration of $\mathcal{M}$ we must check that R^* satisfies the conditions in clause (2) of definition 3.14. Parts (2)(*b*) and (2)(*c*) are just the condition c_1, however, so the question reduces to (2)(*a*). Thus suppose that $\alpha R \beta$, for worlds α and β in $\mathcal{M}$. We wish to argue from this that $[\alpha] R^*[\beta]$, i.e. that conditions $c_1(\alpha, \beta)$ and $c_3(\alpha, \beta)$ are met.

For $c_1(\alpha, \beta)$. Suppose first that $\Box A \in \Gamma$ and $\models_\alpha^{\mathcal{M}} \Box A$. Then for every β in $\mathcal{M}$ such that $\alpha R \beta$, $\models_\beta^{\mathcal{M}} A$. So $\models_\beta^{\mathcal{M}} A$, since $\alpha R \beta$. Next, suppose that $\Diamond A \in \Gamma$ and $\models_\beta^{\mathcal{M}} A$. Then there is a β in $\mathcal{M}$ such that $\alpha R \beta$ and $\models_\beta^{\mathcal{M}} A$, which means that $\models_\alpha^{\mathcal{M}} \Diamond A$.

For $c_3(\alpha, \beta)$. Suppose first that $\Box A \in \Gamma$ and $\models_\alpha^{\mathcal{M}} \Box A$. Because $\mathcal{M}$ is transitive, it follows by theorem 3.5 that $\models_\alpha^{\mathcal{M}} \Box A \rightarrow \Box\Box A$, and hence that $\models_\alpha^{\mathcal{M}} \Box\Box A$. This means that for every β in $\mathcal{M}$ such that $\alpha R \beta$, $\models_\beta^{\mathcal{M}} \Box A$. So $\models_\beta^{\mathcal{M}} \Box A$, since $\alpha R \beta$, which is what we wished to prove. Now suppose that $\Diamond A \in \Gamma$ and that $\models_\beta^{\mathcal{M}} \Diamond A$. Then there is a β in $\mathcal{M}$ such that $\alpha R \beta$ and $\models_\beta^{\mathcal{M}} \Diamond A$, which means that $\models_\alpha^{\mathcal{M}} \Diamond\Diamond A$. By the transitivity of $\mathcal{M}$ and exercise 3.21 it follows that $\models_\alpha^{\mathcal{M}} \Diamond\Diamond A \rightarrow \Diamond A$, and hence that $\models_\alpha^{\mathcal{M}} \Diamond A$, which is what we wished to prove.

This concludes the proof of the theorem.

By putting together the contents of theorems 3.19 and 3.20 we manage to cover most of the properties, and combinations of properties, of standard filtrations in which we are interested. For example, if $\mathcal{M}$ is a reflexive symmetric transitive model, we can find a filtration $\mathcal{M}^*$ of $\mathcal{M}$

that is symmetric and transitive by defining the relation in $\mathscr{M}^*$ as in part (3) of theorem 3.20, and $\mathscr{M}^*$ will also be reflexive, by theorem 3.19. Thus, reasoning as we did for theorem 3.18, we can prove a number of finite determination theorems. We suggest some of these results in the exercises.

But we have not yet shown how to deal with filtrations of models that are simply euclidean, or serial and euclidean. The case of reflexive euclidean models is the same as, for example, that of reflexive symmetric transitive models (see exercise 3.31); part (3) of theorem 3.20 takes care of models that are symmetric and transitive and hence symmetric and euclidean (see exercise 3.58(*e*)); and part (4) of the theorem only deals with models that are transitive and euclidean. Since by theorem 3.19 any way of constructing euclidean filtrations from euclidean models yields a solution for the serial euclidean case, we may concentrate solely on the simpler question.

In fact, it is impossible in general to produce euclidean filtrations along the lines of theorem 3.20, i.e. by stating conditions on pairs of worlds in the filtrated model with respect to an arbitrary set of sentences closed under subsentences. The proof of this is worth setting out here.

Let Γ be the set $\{\mathbb{P}_0, \Box\mathbb{P}_0\}$, which is closed under subsentences, and let $\mathscr{M} = \langle W, R, P\rangle$ be a standard model containing five distinct worlds, α, β, γ, δ, and ϵ, related by R as indicated by the arrows in figure 3.7. We leave it as an exercise for the reader to verify that $\mathscr{M}$ is euclidean and that the distribution of the sentences in Γ and their negations is

Figure 3.7

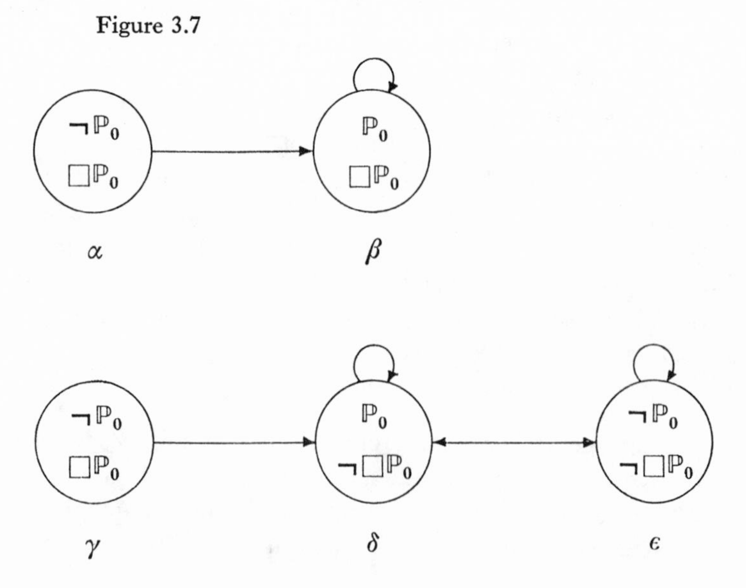

coherent. Note that α and γ are the only worlds in $\mathcal{M}$ that agree on all the sentences in Γ. Thus in a Γ-filtration $\mathcal{M}^* = \langle W^*, R^*, P^* \rangle$ of $\mathcal{M}$ there are four worlds: $[\alpha](= [\gamma])$, $[\beta]$, $[\delta]$, and $[\epsilon]$. $\mathcal{M}^*$ is pictured in figure 3.8, where the arrows reflect the minimal pairings under R^* demanded by clause (2)(*a*) of definition 3.14. Note that $\mathcal{M}^*$ is not euclidean, since $[\alpha]\, R^*[\beta]$ and $[\alpha]\, R^*[\delta]$, but not $[\beta]\, R^*[\delta]$. Furthermore, $\mathcal{M}^*$ cannot be made to be euclidean; for this would mean adding to R^* all the pairs of worlds represented by drawing double-headed arrows between $[\beta]$ and $[\delta]$ and between $[\beta]$ and $[\epsilon]$, which in turn would mean that the sentence $\Box \mathbb{P}_0$ is false at $[\beta]$, contrary to the fact of its truth at that world. In short, this Γ-filtration cannot be made to be euclidean on pain of contradiction.

So something more special is required for the construction of euclidean filtrations; we cannot in general deal with arbitrary sets of sentences closed under subsentences. One approach to the problem uses the idea

Figure 3.8

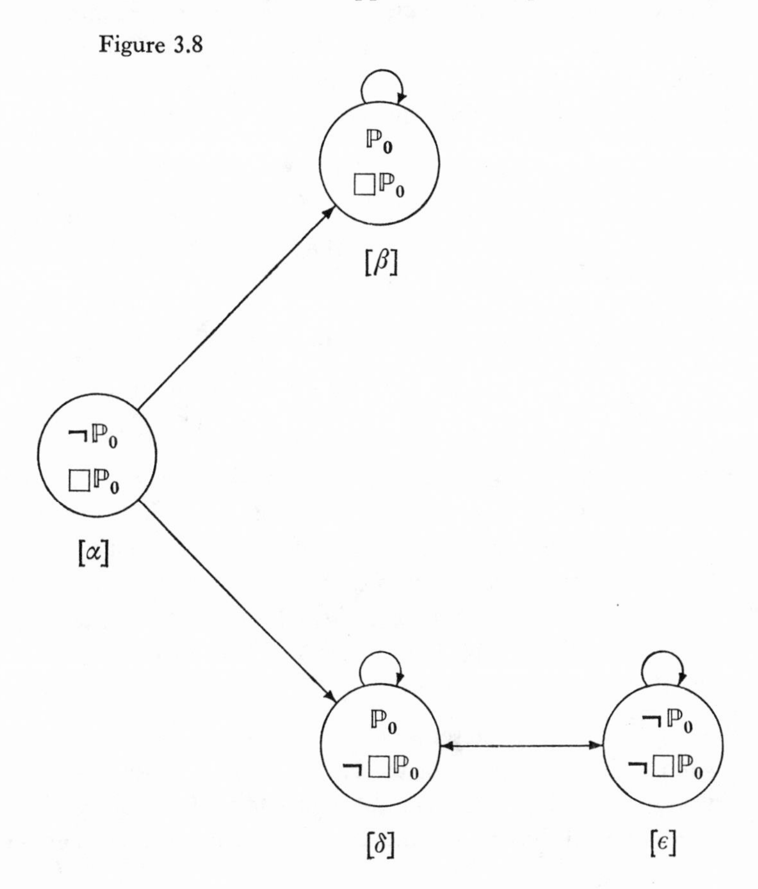

of the modal closure of a set of sentences. Because it is applicable also to models that are symmetric or transitive we state the following theorem for models of all three kinds.

THEOREM 3.21. *Let Γ be a modally closed set of sentences closed under subsentences, and let $\mathcal{M}^* = \langle W^*, R^*, P^* \rangle$ be a coarsest Γ-filtration of a standard model $\mathcal{M} = \langle W, R, P \rangle$. Then:*

(1) $\mathcal{M}^*$ is symmetric if $\mathcal{M}$ is.

(2) $\mathcal{M}^*$ is transitive if $\mathcal{M}$ is.

(3) $\mathcal{M}^*$ is euclidean if $\mathcal{M}$ is.

Proof. We give the proof for part (3) only. Recall that the modal closure of a set of sentences is the result of adding to the set every modalization ϕA of any sentence A in the set. Suppose that $\mathcal{M}^*$ is a coarsest Γ-filtration of a euclidean model $\mathcal{M}$. That is to say, R^* is defined by conditions (*b*) and (*c*) of clause (2) in definition 3.14. To show that $\mathcal{M}^*$ is euclidean, let α, β, and γ be worlds in $\mathcal{M}$ such that $[\alpha] R^* [\beta]$ and $[\alpha] R^* [\gamma]$. In other words, we assume that the following conditions are satisfied.

$$c_1(\alpha, \beta) \quad \begin{array}{l} \text{for every } \Box A \in \Gamma, \text{ if } \models^{\mathcal{M}}_{\alpha} \Box A, \text{ then } \models^{\mathcal{M}}_{\beta} A \\ \text{for every } \Diamond A \in \Gamma, \text{ if } \models^{\mathcal{M}}_{\beta} A, \text{ then } \models^{\mathcal{M}}_{\alpha} \Diamond A \end{array}$$

$$c_1(\alpha, \gamma) \quad \begin{array}{l} \text{for every } \Box A \in \Gamma, \text{ if } \models^{\mathcal{M}}_{\alpha} \Box A, \text{ then } \models^{\mathcal{M}}_{\gamma} A \\ \text{for every } \Diamond A \in \Gamma, \text{ if } \models^{\mathcal{M}}_{\gamma} A, \text{ then } \models^{\mathcal{M}}_{\alpha} \Diamond A \end{array}$$

We wish to argue that $[\beta] R^* [\gamma]$, i.e. that the following conditions are met.

$$c_1(\beta, \gamma) \quad \begin{array}{l} \text{for every } \Box A \in \Gamma, \text{ if } \models^{\mathcal{M}}_{\beta} \Box A, \text{ then } \models^{\mathcal{M}}_{\gamma} A \\ \text{for every } \Diamond A \in \Gamma, \text{ if } \models^{\mathcal{M}}_{\gamma} A, \text{ then } \models^{\mathcal{M}}_{\beta} \Diamond A \end{array}$$

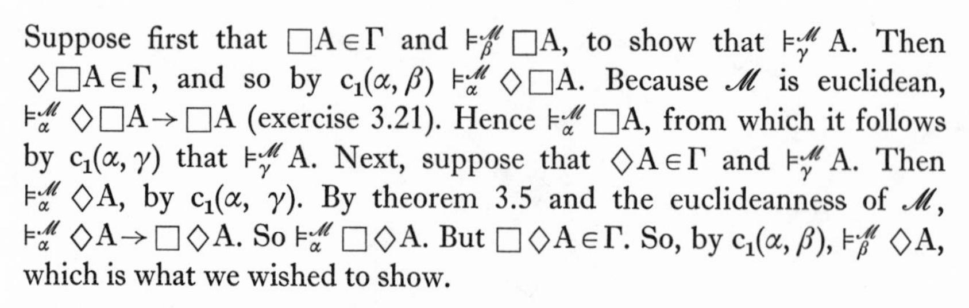

Suppose first that $\Box A \in \Gamma$ and $\models^{\mathcal{M}}_{\beta} \Box A$, to show that $\models^{\mathcal{M}}_{\gamma} A$. Then $\Diamond\Box A \in \Gamma$, and so by $c_1(\alpha, \beta)$ $\models^{\mathcal{M}}_{\alpha} \Diamond\Box A$. Because $\mathcal{M}$ is euclidean, $\models^{\mathcal{M}}_{\alpha} \Diamond\Box A \to \Box A$ (exercise 3.21). Hence $\models^{\mathcal{M}}_{\alpha} \Box A$, from which it follows by $c_1(\alpha, \gamma)$ that $\models^{\mathcal{M}}_{\gamma} A$. Next, suppose that $\Diamond A \in \Gamma$ and $\models^{\mathcal{M}}_{\gamma} A$. Then $\models^{\mathcal{M}}_{\alpha} \Diamond A$, by $c_1(\alpha, \gamma)$. By theorem 3.5 and the euclideanness of $\mathcal{M}$, $\models^{\mathcal{M}}_{\alpha} \Diamond A \to \Box\Diamond A$. So $\models^{\mathcal{M}}_{\alpha} \Box\Diamond A$. But $\Box\Diamond A \in \Gamma$. So, by $c_1(\alpha, \beta)$, $\models^{\mathcal{M}}_{\beta} \Diamond A$, which is what we wished to show.

Noticing that modally closed sets of sentences are infinite the reader may wonder about the point of the last theorem: it does not seem to provide the finite filtrations wanted for proofs of decidability. But, as we shall see, modal closures of finite sets of sentences are logically finite relative to models of certain kinds. In particular, such sets are logically finite relative to any euclidean model. And, as we observed in section 2.3, logical finiteness is sufficient to yield finite filtrations. We return to these matters in chapter 5.

EXERCISES

3.69. Prove part (1) of theorem 3.19.

3.70. Prove parts (1), (3), and (4) of theorem 3.20.

3.71. Check that the model pictured in figure 3.7 is euclidean, and that the distribution of the sentences to the worlds in the model is coherent.

3.72. Show that defining a filtration by conditions c_1 and c_4 (in section 3.6) does not always result in a euclidean model.

3.73. Prove parts (1) and (2) of theorem 3.21.

3.74. Let C be a class of standard models, and let C_{FIN} be the class of all finite models in C. Using filtration theorems we can prove for a number of cases that these classes determine the same modal logic, i.e. that for every A,

$$\models_{\mathsf{C}} A \text{ iff } \models_{\mathsf{C}_{FIN}} A.$$

The reasoning in each case is analogous to that for theorem 3.18, where C is the class of universal standard models. Give the proofs for the cases in which C is any one of the following classes.

(*a*) all
(*b*) serial
(*c*) reflexive
(*d*) symmetric
(*e*) transitive
(*f*) serial symmetric
(*g*) serial transitive
(*h*) transitive euclidean
(*i*) serial transitive euclidean
(*j*) symmetric transitive
(*k*) reflexive symmetric

(*l*) reflexive transitive
(*m*) reflexive euclidean

Except for (*a*), the proofs use theorems 3.19 and 3.20, in addition to the results in section 3.5. For (*m*), note exercise 3.31.

3.75. Give an example of a non-incestual filtration of an incestual standard model.

4

NORMAL SYSTEMS OF MODAL LOGIC

This chapter is devoted to studying, from a purely deductive standpoint, a class of systems of modal logic we call *normal*.

In section 4.1 we first define the class of normal systems. Then we derive a number of theorems and rules of inference common to all normal modal logics and use some of them to formulate alternative deductive characterizations of such systems. Theorems on replacement, negation, and duality are proved in section 4.2 for normal modal logics (they hold more generally for all classical systems, as we discover in chapter 8). These results provide rules and theorems that serve to facilitate derivations.

The smallest normal system of modal logic we call K. Thus every normal system of modal logic is a K-system. (The converse is false; not all K-systems are normal.) To simplify naming normal systems we write

$$KS_1 \ldots S_n$$

to denote the normal modal logic obtained by taking the schemas $S_1, \ldots, S_n$ as theorems. In other words:

$$KS_1 \ldots S_n = \text{the smallest normal system of modal logic containing (every instance of) the schemas } S_1, \ldots, S_n.$$

So, for example, *KT4* is the smallest normal system produced by treating the schemas T and 4 as theorems in a normal modal logic. (It is also denoted by *K4T*; the order of the schema names is irrelevant.) As the limiting case, where there are no schemas, the definition yields K as the smallest normal system.

In section 4.3 we begin a survey of the normal extensions of K containing various combinations of the schemas D, T, B, 4, and 5. This continues in section 4.4 with an account of the numbers of distinct modalities present in certain of these systems.

The chapter concludes with section 4.5, which contains some theorems about maximal sets of sentences in normal modal logics. These results figure importantly with regard to some theorems in chapter 5.

4.1. Normal systems

As we learned in chapter 2, a system of modal logic is a set of sentences containing all tautologies and closed under the rule of inference RPL. We characterize normal systems of modal logic in terms of the schema

$$\text{Df}\Diamond.\quad \Diamond A \leftrightarrow \neg\Box\neg A$$

and the rule of inference

$$\text{RK.}\quad \frac{(A_1 \wedge \ldots \wedge A_n) \rightarrow A}{(\Box A_1 \wedge \ldots \wedge \Box A_n) \rightarrow \Box A} \quad (n \geqslant 0).$$

DEFINITION 4.1. A system of modal logic is *normal* iff it contains Df$\Diamond$ and is closed under RK.

Beginning with theorem 4.2 and continuing with theorem 4.4 we register some of the more important rules and theorems present in all normal systems of modal logic. Many of these are familiar from chapter 1. In theorems 4.3 and 4.5 we record some alternative ways of characterizing normal modal logics.

THEOREM 4.2. *Every normal system of modal logic has the following rules of inference and theorems.*

$$\text{RN.}\quad \frac{A}{\Box A}$$

$$\text{RM.}\quad \frac{A \rightarrow B}{\Box A \rightarrow \Box B}$$

$$\text{RR.}\quad \frac{(A \wedge B) \rightarrow C}{(\Box A \wedge \Box B) \rightarrow \Box C}$$

$$\text{RE.}\quad \frac{A \leftrightarrow B}{\Box A \leftrightarrow \Box B}$$

N. $\Box\top$

M. $\Box(A \wedge B) \rightarrow (\Box A \wedge \Box B)$

C. $(\Box A \wedge \Box B) \rightarrow \Box(A \wedge B)$

R. $\Box(A \wedge B) \leftrightarrow (\Box A \wedge \Box B)$

K. $\Box(A \rightarrow B) \rightarrow (\Box A \rightarrow \Box B)$

Proof. Let Σ be a normal system of modal logic. By theorem 2.13 propositional logic is a part of Σ, a fact we take advantage of frequently and casually.

For RN, RM, and RR. These rules of inference are simply RK for $n = 0$, 1, and 2, respectively.

For RE. Suppose that $\vdash_\Sigma A \leftrightarrow B$. Then by *PL* both $\vdash_\Sigma A \rightarrow B$ and $\vdash_\Sigma B \rightarrow A$. By RM in each case, $\vdash_\Sigma \Box A \rightarrow \Box B$ and $\vdash_\Sigma \Box B \rightarrow \Box A$. Hence by *PL* again, $\vdash_\Sigma \Box A \leftrightarrow \Box B$.

For N. By *PL*, $\vdash_\Sigma \top$. Hence by RN, $\vdash_\Sigma \Box \top$.

For M. By *PL*, $\vdash_\Sigma (A \wedge B) \rightarrow A$ and $\vdash_\Sigma (A \wedge B) \rightarrow B$. So by RM, $\vdash_\Sigma \Box (A \wedge B) \rightarrow \Box A$ and $\vdash_\Sigma \Box (A \wedge B) \rightarrow \Box B$. By *PL* again, $\vdash_\Sigma \Box (A \wedge B) \rightarrow (\Box A \wedge \Box B)$.

For C. By *PL*, $\vdash_\Sigma (A \wedge B) \rightarrow (A \wedge B)$. Hence by RR, $\vdash_\Sigma (\Box A \wedge \Box B) \rightarrow \Box (A \wedge B)$.

For R. This is just the biconditional of M and C.

For K. By *PL*, $\vdash_\Sigma ((A \rightarrow B) \wedge A) \rightarrow B$. So by RR, $\vdash_\Sigma (\Box (A \rightarrow B) \wedge \Box A) \rightarrow \Box B$. Therefore by *PL*, $\vdash_\Sigma \Box (A \rightarrow B) \rightarrow (\Box A \rightarrow \Box B)$.

As in chapter 1, proofs like these for theorem 4.2 can often be stated more perspicuously as annotated sequences of theorems. For example, the proof above for K can be presented thus:

1.	$((A \rightarrow B) \wedge A) \rightarrow B$	*PL*
2.	$(\Box (A \rightarrow B) \wedge \Box A) \rightarrow \Box B$	1, RR
3.	$\Box (A \rightarrow B) \rightarrow (\Box A \rightarrow \Box B)$	2, *PL*

On top of propositional logic, the schema Df$\Diamond$ and the rule RK provide an axiomatic basis for normal systems of modal logic. Together with Df$\Diamond$ the rules and theorems listed in theorem 4.2 provide a number of alternative bases for – i.e. alternative ways of characterizing – normal systems. We select just four for attention in the next theorem; some others appear in the exercises.

THEOREM 4.3. *Let* Σ *be a system of modal logic containing* Df$\Diamond$. *Then:*

(1) Σ *is normal iff it contains* K *and is closed under* RN.

(2) Σ *is normal iff it contains* N *and is closed under* RR.

(3) Σ *is normal iff it contains* N *and* C *and is closed under* RM.

(4) Σ *is normal iff it contains* N, C, *and* M *and is closed under* RE.

Proof. Let Σ be a system containing Df$\diamondsuit$. Theorem 4.2 takes care of left-to-right in each case, so we show only right-to-left.

For (1). We need to show that if Σ contains K and is closed under RN, then Σ is closed under RK; i.e. that for $n \geqslant 0$,

$$\text{if } \vdash_\Sigma (A_1 \wedge \ldots \wedge A_n) \to A,$$
$$\text{then } \vdash_\Sigma (\Box A_1 \wedge \ldots \wedge \Box A_n) \to \Box A.$$

The proof is by induction on n and is like that for theorem 3.3 (2) (recall lemmas 1 and 2 there). With this hint we leave the details to the reader.

For (2). Suppose Σ contains N and is closed under RR. In view of (1) it is enough to show that Σ contains K and is closed under RN. As to K, see the proof of theorem 4.2. For RN:

1. A	hypothesis
2. $(\top \wedge \top) \to A$	1, *PL*
3. $(\Box\top \wedge \Box\top) \to \Box A$	2, RR
4. $\Box\top$	N
5. $\Box A$	3, 4, *PL*

Note that line 1 means that $\vdash_\Sigma A$, so that RR is applicable at line 2.

For (3). Suppose Σ contains N and C and is closed under RM. Given (2), we need only show that Σ is closed under RR. Thus:

1. $(A \wedge B) \to C$	hypothesis
2. $\Box(A \wedge B) \to \Box C$	1, RM
3. $(\Box A \wedge \Box B) \to \Box(A \wedge B)$	C
4. $(\Box A \wedge \Box B) \to \Box C$	2, 3, *PL*

For (4). If Σ contains N, C, and M and is closed under RE, it is sufficient, given (3), to show that Σ is closed under RM. We leave this as an exercise.

With the exception of Df$\diamondsuit$, the rules and theorems so far have featured the necessity operator. The next theorem catalogues some rules and theorems of normal systems in which the possibility operator predominates.

THEOREM 4.4. *Every normal system of modal logic has the following rules of inference and theorems.*

$$\text{RK}\diamondsuit. \quad \frac{A \to (A_1 \vee \ldots \vee A_n)}{\diamondsuit A \to (\diamondsuit A_1 \vee \ldots \vee \diamondsuit A_n)} \quad (n \geqslant 0)$$

RN$\Diamond$. $\dfrac{\neg A}{\neg\Diamond A}$

RM$\Diamond$. $\dfrac{A\to B}{\Diamond A\to\Diamond B}$

RR$\Diamond$. $\dfrac{A\to(B\vee C)}{\Diamond A\to(\Diamond B\vee\Diamond C)}$

RE$\Diamond$. $\dfrac{A\leftrightarrow B}{\Diamond A\leftrightarrow\Diamond B}$

Df$\Box$. $\Box A\leftrightarrow\neg\Diamond\neg A$

N$\Diamond$. $\neg\Diamond\bot$

M$\Diamond$. $(\Diamond A\vee\Diamond B)\to\Diamond(A\vee B)$

C$\Diamond$. $\Diamond(A\vee B)\to(\Diamond A\vee\Diamond B)$

R$\Diamond$. $\Diamond(A\vee B)\leftrightarrow(\Diamond A\vee\Diamond B)$

K$\Diamond$. $(\neg\Diamond A\wedge\Diamond B)\to\Diamond(\neg A\wedge B)$

Proof. Let Σ be a normal system.

For RK$\Diamond$. Suppose that $\vdash_\Sigma A\to(A_1\vee\ldots\vee A_n)$. Then by *PL*, $\vdash_\Sigma(\neg A_1\wedge\ldots\wedge\neg A_n)\to\neg A$. By applying RK, $\vdash_\Sigma(\Box\neg A_1\wedge\ldots\wedge\Box\neg A_n)\to\Box\neg A$. Hence by *PL* again, $\vdash_\Sigma\neg\Box\neg A\to(\neg\Box\neg A_1\vee\ldots\vee\neg\Box\neg A_n)$. Therefore by Df$\Diamond$ and *PL*, $\vdash_\Sigma\Diamond A\to(\Diamond A_1\vee\ldots\vee\Diamond A_n)$.

For RN$\Diamond$, RM$\Diamond$, and RR$\Diamond$. These are the rule RK$\Diamond$ for $n = 0$, 1, and 2, respectively. (For RN$\Diamond$, recall that when $n = 0$ the conditionals in RK$\Diamond$ are identified with the negations of their antecedents.)

For RE$\Diamond$. The proof uses RM$\Diamond$ and is like that for RE in theorem 4.3. Exercise.

For Df$\Box$. Compare the proof of this in section 1.2, and note that it uses only *PL*, Df$\Diamond$, and RE.

For N$\Diamond$. By *PL*, $\vdash_\Sigma\neg\bot$. So by RN$\Diamond$, $\vdash_\Sigma\neg\Diamond\bot$.

For M$\Diamond$. By *PL*, $\vdash_\Sigma A\to(A\vee B)$ and $\vdash_\Sigma B\to(A\vee B)$. Hence by RM$\Diamond$, $\vdash_\Sigma\Diamond A\to\Diamond(A\vee B)$ and $\vdash_\Sigma\Diamond B\to\Diamond(A\vee B)$. By *PL*, $\vdash_\Sigma(\Diamond A\vee\Diamond B)\to\Diamond(A\vee B)$.

For C$\Diamond$. The proof uses RR$\Diamond$ and the tautology $(A\vee B)\to(A\vee B)$. Exercise.

For R$\Diamond$. This is the biconditional of M$\Diamond$ and C$\Diamond$.

For K$\Diamond$:

1. $B\to(A\vee(\neg A\wedge B))$ — *PL*
2. $\Diamond B\to(\Diamond A\vee\Diamond(\neg A\wedge B))$ — 1, RR$\Diamond$
3. $(\neg\Diamond A\wedge\Diamond B)\to\Diamond(\neg A\wedge B)$ — 2, *PL*

The reader should appreciate the parallels between the proofs above and those for the corresponding rules and theorems in theorem 4.2. We have developed this analogy intentionally, for the sake of simplicity and also to enhance the reader's ability to create such proofs on his own. There are of course other ways of doing this. As an example, let us prove again that R$\Diamond$ is a theorem of all normal modal logics, as follows.

1. $\Box(\neg A \wedge \neg B) \leftrightarrow (\Box\neg A \wedge \Box\neg B)$	R
2. $(\neg A \wedge \neg B) \leftrightarrow \neg(A \vee B)$	*PL*
3. $\Box(\neg A \wedge \neg B) \leftrightarrow \Box\neg(A \vee B)$	2, RE
4. $\Box\neg(A \vee B) \leftrightarrow (\Box\neg A \wedge \Box\neg B)$	1, 3, *PL*
5. $\neg\Box\neg(A \vee B) \leftrightarrow (\neg\Box\neg A \vee \neg\Box\neg B)$	4, *PL*
6. $\Diamond(A \vee B) \leftrightarrow (\Diamond A \vee \Diamond B)$	5, Df$\Diamond$ and *PL*

Other alternative proofs of rules and theorems are suggested in the exercises.

The characterization of normal systems of modal logic in terms of Df$\Diamond$ and RK and in theorem 4.3 may be said to be *necessity-based*, inasmuch as $\Box$ is treated as though it were primitive and $\Diamond$ is introduced only definitionally, through Df$\Diamond$. In the next theorem we turn this around by using rules and theorems from theorem 4.4 to give five characterizations of normal systems that are *possibility-based* and introduce necessity definitionally via Df$\Box$.

THEOREM 4.5. *Let* Σ *be a system of modal logic containing* Df$\Box$. *Then:*

(1) Σ *is normal iff it is closed under* RK$\Diamond$.

(2) Σ *is normal iff it contains* K$\Diamond$ *and is closed under* RN$\Diamond$.

(3) Σ *is normal iff it contains* N$\Diamond$ *and is closed under* RR$\Diamond$.

(4) Σ *is normal iff it contains* N$\Diamond$ *and* C$\Diamond$ *and is closed under* RM$\Diamond$.

(5) Σ *is normal iff it contains* N$\Diamond$, C$\Diamond$, *and* M$\Diamond$ *and is closed under* RE$\Diamond$.

Proof. Let Σ be a system containing Df$\Box$. The left-to-right cases are covered by theorem 4.4, so we need show only the converses.

For (1). Suppose that Σ is closed under RK$\Diamond$. We wish to prove first that Σ is closed under RK. The argument is analogous to that given for RK$\Diamond$, using RK, in theorem 4.4. Thus:

1. $(A_1 \wedge \ldots \wedge A_n) \rightarrow A$ — hypothesis
2. $\neg A \rightarrow (\neg A_1 \vee \ldots \vee \neg A_n)$ — 1, *PL*
3. $\Diamond\neg A \rightarrow (\Diamond\neg A_1 \vee \ldots \vee \Diamond\neg A_n)$ — 2, RK$\Diamond$
4. $(\neg\Diamond\neg A_1 \wedge \ldots \wedge \neg\Diamond\neg A_n) \rightarrow \neg\Diamond\neg A$ — 3, *PL*
5. $(\Box A_1 \wedge \ldots \wedge \Box A_n) \rightarrow \Box A$ — 4, Df$\Box$ and *PL*

Next we must show that Σ contains Df$\Diamond$. The argument for this is like that suggested for Df$\Box$ in theorem 4.4, if as a lemma it is shown first that Σ is closed under RE$\Diamond$. This is left to the reader as an exercise. If thus Σ contains Df$\Diamond$ and is closed under RK, then by definition 4.1 it is normal.

The proofs for parts (2)–(5) parallel those for (1)–(4) in theorem 4.3.

For (2). Suppose that Σ contains K$\Diamond$ and is closed under RN$\Diamond$. In view of (1), just proved, it is enough to show that Σ is closed under RK$\Diamond$, i.e. that for $n \geqslant 0$,

$$\text{if } \vdash_\Sigma A \rightarrow (A_1 \vee \ldots \vee A_n), \text{ then } \vdash_\Sigma \Diamond A \rightarrow (\Diamond A_1 \vee \ldots \vee \Diamond A_n).$$

The proof is by induction on n. Where $n = 0$, we need to show that if $\vdash_\Sigma \neg A$, then $\vdash_\Sigma \neg\Diamond A$. This is just RN$\Diamond$. So suppose as an inductive hypothesis that the rule holds for $k < n$. Then we reason as follows.

1. $A \rightarrow (A_1 \vee \ldots \vee A_n)$ — hypothesis
2. $(\neg A_1 \wedge A) \rightarrow (A_2 \vee \ldots \vee A_n)$ — 1, *PL*
3. $\Diamond(\neg A_1 \wedge A) \rightarrow (\Diamond A_2 \vee \ldots \vee \Diamond A_n)$ — 2, inductive hypothesis
4. $(\neg\Diamond A_1 \wedge \Diamond A) \rightarrow \Diamond(\neg A_1 \wedge A)$ — K$\Diamond$
5. $(\neg\Diamond A_1 \wedge \Diamond A) \rightarrow (\Diamond A_2 \vee \ldots \vee \Diamond A_n)$ — 3, 4, *PL*
6. $\Diamond A \rightarrow (\Diamond A_1 \vee \ldots \vee \Diamond A_n)$ — 5, *PL*

For (3). Suppose Σ contains N$\Diamond$ and is closed under RR$\Diamond$. Given (2), we need only show that Σ contains K$\Diamond$ and is closed under RN$\Diamond$. The proof of K$\Diamond$ appears in the proof of theorem 4.4. For RN$\Diamond$:

1. $\neg A$ — hypothesis
2. $A \rightarrow (\bot \vee \bot)$ — 1, *PL*
3. $\Diamond A \rightarrow (\Diamond\bot \vee \Diamond\bot)$ — 2, RR$\Diamond$
4. $\neg\Diamond\bot$ — N$\Diamond$
5. $\neg\Diamond A$ — 3, 4, *PL*

For (4). Suppose Σ contains N$\Diamond$ and C$\Diamond$ and is closed under RM$\Diamond$. Given (3), it is sufficient to prove that Σ is closed under RR$\Diamond$. Exercise (compare the proof of theorem 4.3 (3)).

For (5). Suppose that Σ contains N$\Diamond$, C$\Diamond$, and M$\Diamond$ and is closed under RE$\Diamond$. In view of (4), it will do just to show that Σ is closed under RM$\Diamond$. Thus:

1. $A \to B$	hypothesis
2. $(A \vee B) \leftrightarrow B$	1, *PL*
3. $\Diamond(A \vee B) \leftrightarrow \Diamond B$	2, RE$\Diamond$
4. $(\Diamond A \vee \Diamond B) \to \Diamond(A \vee B)$	M$\Diamond$
5. $\Diamond A \to \Diamond B$	3, 4, *PL*

Many principles of normal systems can be generalized modally. For example, for every $k \geqslant 0$ every normal modal logic is closed under the rule of inference

$$\text{RK}^k. \quad \frac{(A_1 \wedge \ldots \wedge A_n) \to A}{(\Box^k A_1 \wedge \ldots \wedge \Box^k A_n) \to \Box^k A} \quad (n \geqslant 0).$$

This should be evident, for the conclusion of the rule will follow from the hypothesis by k applications of the rule RK. More formally, it may be proved quite simply by induction on k. When $k = 0$ the hypothesis and the conclusion of the rule are the same, so of course the inference is good in this case. And from an inductive hypothesis that the rule holds whenever it has fewer than k $\Box$s, it follows by RK that it holds also when the number is k. That is to say, we may argue the inductive part of the proof as follows.

1. $(A_1 \wedge \ldots \wedge A_n) \to A$	hypothesis
2. $(\Box^{k-1} A_1 \wedge \ldots \wedge \Box^{k-1} A_n) \to \Box^{k-1} A$	1, inductive hypothesis
3. $(\Box\Box^{k-1} A_1 \wedge \ldots \wedge \Box\Box^{k-1} A_n) \to \Box\Box^{k-1} A$	2, RK
4. $(\Box^k A_1 \wedge \ldots \wedge \Box^k A_n) \to \Box^k A$	3, definition 2.3

Therefore, the rule RK^k holds in any normal system, for every $k \geqslant 0$.

The schema Df$\Diamond$ likewise generalizes along the modal dimension. For every $k \geqslant 0$ the schema

$$\text{Df}\Diamond^k. \quad \Diamond^k A \leftrightarrow \neg\Box^k \neg A$$

is a theorem of any normal modal logic. Here, too, a simple inductive argument suffices. For the basis, note that Df$\Diamond^k$ is a tautology when

$k = 0$. For the inductive part, assume that the schema is a theorem whenever the number of $\Box$s and $\Diamond$s is less than k. Then we argue as follows.

1. $\Diamond^{k-1}A \leftrightarrow \neg\Box^{k-1}\neg A$	inductive hypothesis
2. $\Diamond\Diamond^{k-1}A \leftrightarrow \Diamond\neg\Box^{k-1}\neg A$	1, RE$\Diamond$
3. $\Box\Box^{k-1}\neg A \leftrightarrow \neg\Diamond\neg\Box^{k-1}\neg A$	Df$\Box$
4. $\Diamond\Diamond^{k-1}A \leftrightarrow \neg\Box\Box^{k-1}\neg A$	2, 3, *PL*
5. $\Diamond^{k}A \leftrightarrow \neg\Box^{k}\neg A$	4, definition 2.3

It should be apparent, given RK^k and Df$\Diamond^k$, that similar generalizations of all the principles in theorems 4.2 and 4.4 are part of any normal system of modal logic. More precisely, the results of putting $\Box^k$ and $\Diamond^k$ for $\Box$ and $\Diamond$ throughout these principles yield theorems and rules of inference that belong to every normal system, for every $k \geqslant 0$. Because we will need some of these principles later on (especially in chapter 5), we record this formally.

THEOREM 4.6. *Every normal system of modal logic has the principles* RK^k, Df$\Diamond^k$, RN^k, RM^k, RR^k, RE^k, N^k, M^k, C^k, R^k, K^k, RK$\Diamond^k$, RN$\Diamond^k$, RM$\Diamond^k$, RR$\Diamond^k$, RE$\Diamond^k$, Df$\Box^k$, N$\Diamond^k$, M$\Diamond^k$, C$\Diamond^k$, R$\Diamond^k$, *and* K$\Diamond^k$, *for every* $k \geqslant 0$.

Given the proofs above for RK^k and Df$\Diamond^k$, the reader can easily construct proofs for the remaining principles by attending to the proofs of theorems 4.2 and 4.4. Separate inductive proofs are also possible in each case.

EXERCISES

Where appropriate, freely make use of theorems and rules of inference established in section 4.1 and, farther along, the results of previous exercises.

4.1. Complete the proof of theorem 4.3 (parts (1) and (4)). (For (4), note that $A \to B$ is *PL*-equivalent to $A \leftrightarrow (A \wedge B)$.)

4.2. Complete the proof of theorem 4.4 by showing that every normal system has the rule RE$\Diamond$ and the theorem C$\Diamond$.

4.3. Complete the proof of theorem 4.5 (parts (1) and (4)).

4.4. Prove some of the parts of theorem 4.6.

4.5. Let Σ be a system of modal logic containing Df$\Diamond$. Prove:

(*a*) Σ is normal iff it is closed under RR and RN.

(*b*) Σ is normal iff it contains C and is closed under RM and RN.

(*c*) Σ is normal iff it contains N and K and is closed under RM.

(*d*) Σ is normal iff it contains C and M and is closed under RE and RN.

(*e*) Σ is normal iff it contains N and R and is closed under RE.

(*f*) Σ is normal iff it contains R and is closed under RE and RN.

(*g*) Σ is normal iff it contains N and K and is closed under RE.

4.6. Let Σ be a system of modal logic containing Df$\Box$. Prove:

(*a*) Σ is normal iff it is closed under RR$\Diamond$ and RN$\Diamond$.

(*b*) Σ is normal iff it contains C$\Diamond$ and is closed under RM$\Diamond$ and RN$\Diamond$.

(*c*) Σ is normal iff it contains N$\Diamond$ and K$\Diamond$ and is closed under RM$\Diamond$.

(*d*) Σ is normal iff it contains C$\Diamond$ and M$\Diamond$ and is closed under RE$\Diamond$ and RN$\Diamond$.

(*e*) Σ is normal iff it contains N$\Diamond$ and R$\Diamond$ and is closed under RE$\Diamond$.

(*f*) Σ is normal iff it contains R$\Diamond$ and is closed under RE$\Diamond$ and RN$\Diamond$.

(*g*) Σ is normal iff it contains N$\Diamond$ and K$\Diamond$ and is closed under RE$\Diamond$.

4.7. Prove that the following schemas are theorems of any normal system.

(*a*) $\Box A \to \Box(B \to A)$

(*b*) $\Box\neg A \to \Box(A \to B)$

(*c*) $\Diamond\top \leftrightarrow \neg\Box\bot$

(*d*) $\Box(A \to B) \to (\Diamond A \to \Diamond B)$

(*e*) $\Box(A \leftrightarrow B) \to (\Box A \leftrightarrow \Box B)$

(*f*) $\Box(A \leftrightarrow B) \to (\Diamond A \leftrightarrow \Diamond B)$

(*g*) $(\Box A \vee \Box B) \to \Box(A \vee B)$

(*h*) $\Diamond(A \wedge B) \to (\Diamond A \wedge \Diamond B)$

(*i*) $(\Box A \wedge \Diamond B) \to \Diamond(A \wedge B)$

(*j*) $\Box(A \vee B) \to (\Diamond A \vee \Box B)$
(*k*) $\Diamond(A \to B) \vee \Box(B \to A)$
(*l*) $\Diamond(A \to B) \leftrightarrow (\Box A \to \Diamond B)$
(*m*) $\Diamond\top \leftrightarrow (\Box A \to \Diamond A)$
(*n*) $(\Diamond A \to \Box B) \to \Box(A \to B)$
(*o*) $(\Diamond A \to \Box B) \to (\Box A \to \Box B)$
(*p*) $(\Diamond A \to \Box B) \to (\Diamond A \to \Diamond B)$

4.8. Prove that the following schemas are theorems of any normal system (for any $n \geqslant 2$).

(*a*) $\Box(A_1 \wedge \ldots \wedge A_n) \leftrightarrow (\Box A_1 \wedge \ldots \wedge \Box A_n)$
(*b*) $\Diamond(A_1 \vee \ldots \vee A_n) \leftrightarrow (\Diamond A_1 \vee \ldots \vee \Diamond A_n)$
(*c*) $(\Box A_1 \vee \ldots \vee \Box A_n) \to \Box(A_1 \vee \ldots \vee A_n)$
(*d*) $\Diamond(A_1 \wedge \ldots \wedge A_n) \to (\Diamond A_1 \wedge \ldots \wedge \Diamond A_n)$
(*e*) $(\Box A_1 \wedge \ldots \wedge \Box A_{n-1} \wedge \Diamond A_n) \to \Diamond(A_1 \wedge \ldots \wedge A_n)$
(*f*) $\Box(A_1 \vee \ldots \vee A_n) \to (\Diamond A_1 \vee \ldots \vee \Diamond A_{n-1} \vee \Box A_n)$

4.9. Prove that the following sentences are theorems of any normal system whenever $m \leqslant n$.

(*a*) $\Diamond^n\top \to \Diamond^m\top$ (*b*) $\Box^m\bot \to \Box^n\bot$

4.10. Let Σ be any system of modal logic containing Df$\Diamond$ and satisfying the conditions that, for every $n \geqslant 0$,

(*a*) $\Box^n A \in \Sigma$ if $\vdash_{PL} A$,
(*b*) $\Box^n(\Box(A \to B) \to (\Box A \to \Box B)) \in \Sigma$,
(*c*) Σ is closed under the rule MP.

Prove that Σ is normal. (This boils down to a proof, by induction on n, that Σ is closed under the rule RN.)

4.11. Prove that every normal system has the following rule of inference, for any $k, m, n \geqslant 0$.

$$\frac{(A_1 \wedge \ldots \wedge A_m \wedge \Diamond^k B_1 \wedge \ldots \wedge \Diamond^k B_n) \to \bot}{(\Box A_1 \wedge \ldots \wedge \Box A_m) \to \Box^{k+1}\neg(B_1 \wedge \ldots \wedge B_n)}$$

4.12. Use the erasure transformation ϵ from exercise 1.27 to prove the consistency of the system *K*. (Alternatively, consider the mappings τ in exercises 1.11 and 3.16.)

4.13. Consider the following rules of inference.

$$(a)\ \frac{\Box A}{A} \qquad (d)\ \frac{\Diamond A}{A}$$

$$(b)\ \frac{\Box A \to \Box B}{A \to B} \qquad (e)\ \frac{\Diamond A \to \Diamond B}{A \to B}$$

$$(c)\ \frac{\Box A \leftrightarrow \Box B}{A \leftrightarrow B} \qquad (f)\ \frac{\Diamond A \leftrightarrow \Diamond B}{A \leftrightarrow B}$$

These rules hold for some normal systems, but not for all. To prove that they hold for the system *K* we first define the mapping σ, as follows.

(1) $\sigma(\mathbb{P}_n) = \mathbb{P}_n$, for $n = 0, 1, 2, \ldots$.
(2) $\sigma(\top) = \top$.
(3) $\sigma(\bot) = \bot$.
(4) $\sigma(\neg A) = \neg\sigma(A)$.
(5) $\sigma(A \wedge B) = \sigma(A) \wedge \sigma(B)$.
(6) $\sigma(A \vee B) = \sigma(A) \vee \sigma(B)$.
(7) $\sigma(A \to B) = \sigma(A) \to \sigma(B)$.
(8) $\sigma(A \leftrightarrow B) = \sigma(A) \leftrightarrow \sigma(B)$.
(9) $\sigma(\Box A) = A$.
(10) $\sigma(\Diamond A) = A$.

So to speak, σ searches through a sentence – or schema – for its first, or outermost, occurrences of $\Box$ and $\Diamond$, and 'erases' them. Thus $\sigma(\Box A \to \Diamond A)$ is $A \to A$. Note that σ is not the same as ϵ in exercises 1.27 and 4.12: σ does not delete all occurrences of $\Box$ and $\Diamond$. For example, $\sigma(\Box A \to \Box\Box A)$ is $A \to \Box A$, not $A \to A$.

Now consider *K* as axiomatized by Df$\Diamond$, RK, and RPL. Prove by induction on the length (number of lines) of a proof, relative to this axiomatization, the following lemma.

If $\vdash_K A$, then $\vdash_K \sigma(A)$.

That is, prove that if A appears on the first line of a proof, then $\sigma(A)$ is also a *K*-theorem (this is the basis of the induction), and that, assuming that the result holds for all lines $k < n$, it holds as well for line n (this is the inductive step). (Take it for granted that the result holds with respect to RPL, i.e. that if A is a tautological consequence of $A_1, \ldots, A_n$, then $\sigma(A)$ is a tautological consequence of $\sigma(A_1), \ldots, \sigma(A_n)$.)

It follows from this lemma that *K* has rules (*a*)–(*f*). For example, for (*a*) we argue as follows. If $\vdash_K \Box A$, then by the lemma $\vdash_K \sigma(\Box A)$, and so – by the definition of σ – $\vdash_K A$.

Give the arguments for cases (*b*)–(*f*).

4.2. Replacement and duality

In this section we pause to state and prove some simple theorems about replacement and duality in normal modal logics. These principles function as theorems and rules of inference in every normal system, and where possible we present them as such. Their usefulness is illustrated by means of several examples.

THEOREM 4.7. *Every normal system of modal logic has the rule of replacement:*

$$\text{REP.} \quad \frac{B \leftrightarrow B'}{A \leftrightarrow A[B/B']}$$

(Recall from section 2.1 that $A[B/B']$ is any sentence that results from A by replacing zero or more occurrences of B, in A, by B'.)

Proof. Let Σ be a normal system, and suppose (throughout the proof) that $\vdash_\Sigma B \leftrightarrow B'$. Then what we wish to prove is that $\vdash_\Sigma A \leftrightarrow A[B/B']$.

We consider first the possibility that A and B are the same sentence. Then $A[B/B']$ is either A (when there is no replacement) or B' (when A, i.e. B, is replaced by B'). In either case, $\vdash_\Sigma A \leftrightarrow A[B/B']$. For in the first case this is just $\vdash_\Sigma A \leftrightarrow A$, which is trivial; and in the second it is $\vdash_\Sigma B \leftrightarrow B'$, which is the assumption.

Thus we may assume henceforth that A and B are distinct.

The proof proceeds now by induction on the complexity of A. We give it for the cases in which A is (*a*) atomic, $\mathbb{P}_n$, (*b*) the falsum, $\perp$, (*c*) a conditional, $C \to D$, and (*d*) a necessitation, $\Box C$; the rest are left for the reader.

For (*a*). Given that $\mathbb{P}_n$ and B are distinct, $\mathbb{P}_n[B/B'] = \mathbb{P}_n$. So, $\vdash_\Sigma \mathbb{P}_n \leftrightarrow \mathbb{P}_n[B/B']$, trivially. So the theorem holds when A is atomic.

For (*b*). The argument is the same as for (*a*).

For the inductive cases (*c*) and (*d*) we make the hypothesis that the result holds for all sentences shorter than A.

For (*c*). By the inductive hypothesis, $\vdash_\Sigma C \leftrightarrow C[B/B']$ and $\vdash_\Sigma D \leftrightarrow D[B/B']$. It follows (by *PL*; the proof is left to the reader) that $\vdash_\Sigma (C \to D)$

$\leftrightarrow(C[B/B'] \to D[B/B'])$. But note that $(C \to D)[B/B'] = C[B/B'] \to D[B/B']$. Therefore, $\vdash_\Sigma (C \to D) \leftrightarrow (C \to D)[B/B']$. So the theorem holds when A is a conditional.

For (*d*). By the inductive hypothesis, $\vdash_\Sigma C \leftrightarrow C[B/B']$. By the rule RE it follows that $\vdash_\Sigma \Box C \leftrightarrow \Box(C[B/B'])$. However, $(\Box C)[B/B'] = \Box(C[B/B'])$. Therefore, $\vdash_\Sigma \Box C \leftrightarrow (\Box C)[B/B']$. So the theorem holds when A is a necessitation.

This ends the proof of theorem 4.7.

The use of the rule REP is illustrated in the following proof that the schema

$$\Diamond(A \to B) \leftrightarrow (\Box A \to \Diamond B)$$

is a theorem in any normal system of modal logic.

1.	$\Diamond(A \to B) \leftrightarrow \Diamond(\neg A \lor B)$	*PL* and REP
2.	$\leftrightarrow(\Diamond\neg A \lor \Diamond B)$	1, R$\Diamond$ and REP
3.	$\leftrightarrow(\neg\Diamond\neg A \to \Diamond B)$	2, *PL* and REP
4.	$\leftrightarrow(\Box A \to \Diamond B)$	3, Df$\Box$ and REP

This highly abbreviated proof needs some explanation. The justification of line 1 indicates a tacit use of REP in which, since $(A \to B) \leftrightarrow (\neg A \lor B)$ is a tautology, $\neg A \lor B$ replaces $A \to B$ in the tautology $\Diamond(A \to B) \leftrightarrow \Diamond(A \to B)$. In line 2, $\Diamond\neg A \lor \Diamond B$ replaces $\Diamond(\neg A \lor B)$ in line 1, in virtue of the theorem R$\Diamond$. Then in line 3, $\neg\Diamond\neg A \to \Diamond B$ replaces the tautologically equivalent $\Diamond\neg A \lor \Diamond B$ in line 2. Finally, in line 4 the theorem Df$\Box$ is used in replacing $\neg\Diamond\neg A$ by $\Box A$ in line 3.

Use of the rule REP is further illustrated in the proofs of theorems 4.8 and 4.10 below.

Let us turn now to the subject of duality (recall definition 2.4).

THEOREM 4.8. *Every normal system of modal logic has the following theorems and rules of inference, all referred to as* DUAL.

(1) $A \leftrightarrow \neg A^*$

(2) $\dfrac{A}{\neg A^*} \quad \dfrac{\neg A}{A^*}$

(3) $\dfrac{A \to B}{B^* \to A^*}$

(4) $\dfrac{A \leftrightarrow B}{A^* \leftrightarrow B^*}$

Proof. We assume throughout that Σ is a normal modal logic.

For (1). The proof here is by induction on the complexity of A. Let us treat the cases in which A is (*a*) atomic, $\mathbb{P}_n$, (*b*) the falsum, $\bot$, (*c*) a conjunction, $B \land C$, and (*d*) a necessitation, $\Box B$.

For (*a*):

1. $\mathbb{P}_n \leftrightarrow \neg\neg\mathbb{P}_n$ *PL*
2. $\leftrightarrow \neg\mathbb{P}_n^*$ 1, definition 2.4(1)

So the theorem holds when A is atomic.

For (*b*):

1. $\bot \leftrightarrow \neg\top$ *PL*
2. $\leftrightarrow \neg\bot^*$ 1, definition 2.4(3)

So the theorem holds when A is the falsum.

For the inductive cases (*c*) and (*d*), we make the hypothesis that the theorem holds for sentences shorter than A. Thus, $\vdash_\Sigma B \leftrightarrow \neg B^*$ and $\vdash_\Sigma C \leftrightarrow \neg C^*$.

For (*c*):

1. $(B \land C) \leftrightarrow (\neg B^* \land \neg C^*)$ inductive hypothesis and REP
2. $\leftrightarrow \neg(B^* \lor C^*)$ 1, *PL* and REP
3. $\leftrightarrow \neg(B \land C)^*$ 2, definition 2.4(5)

So the theorem holds when A is a conjunction.

For (*d*):

1. $\Box B \leftrightarrow \Box(\neg B^*)$ inductive hypothesis and REP
2. $\leftrightarrow \neg\Diamond(B^*)$ 1, Df$\Diamond$ and REP
3. $\leftrightarrow \neg(\Box B)^*$ 2, definition 2.4(9)

So the theorem holds when A is a necessitation.

This concludes the proof of (1). Parts (2)–(4) are corollaries.

For (2). It follows at once from (1) that if $\vdash_\Sigma A$, then $\vdash_\Sigma \neg A^*$. So Σ is closed under the first rule DUAL in (2). For the second, it is enough to note that (1) means that $\neg A \leftrightarrow A^*$ is always a theorem of Σ.

For (3). If $\vdash_\Sigma A \to B$, then $\vdash_\Sigma \neg A^* \to \neg B^*$, by (1) and REP. Hence by *PL*, $\vdash_\Sigma B^* \to A^*$. So Σ is closed under the rule DUAL in (3).

For (4). We leave this as an exercise.

This concludes the proof of theorem 4.8.

As an example of the use of DUAL, let us see that $\Diamond\top \leftrightarrow \neg\Box\bot$ is a theorem of every normal system. For by DUAL(1), $\Diamond\top \leftrightarrow \neg(\Diamond\top)^*$ is, and by definition 2.4, $(\Diamond\top)^* = \Box\bot$.

Similarly, using DUAL(2) one can show that every normal system Σ has $\neg(\Box A \wedge \Box\neg A)$ as a theorem if it has $\Diamond A \vee \Diamond\neg A$. (Note that neither schema is a theorem of every normal system, however.) For suppose that $\vdash_\Sigma \Diamond A \vee \Diamond\neg A$. Then also, $\vdash_\Sigma \Diamond\neg A \vee \Diamond\neg\neg A$. Hence by DUAL(2), $\vdash_\Sigma \neg(\Diamond\neg A \vee \Diamond\neg\neg A)^*$. But by definition 2.4 this means that $\vdash_\Sigma \neg(\Box(\neg A^*) \wedge \Box\neg(\neg A^*))$. So by DUAL(1) and REP, $\vdash_\Sigma \neg(\Box A \wedge \Box\neg A)$.

Finally, let us show that since $(\Box A \vee \Box B) \rightarrow \Box(A \vee B)$ is always a theorem of a normal system, so is $\Diamond(A \wedge B) \rightarrow (\Diamond A \wedge \Diamond B)$. The proof uses DUAL(3):

1.	$(\Box\neg A \vee \Box\neg B) \rightarrow \Box(\neg A \vee \neg B)$	theorem
2.	$(\Box(\neg A \vee \neg B))^* \rightarrow (\Box\neg A \vee \Box\neg B)^*$	1, DUAL(3)
3.	$\Diamond(\neg A^* \wedge \neg B^*) \rightarrow (\Diamond(\neg A^*) \wedge \Diamond(\neg B^*))$	2, definition 2.4
4.	$\Diamond(A \wedge B) \rightarrow (\Diamond A \wedge \Diamond B)$	3, DUAL(1) and REP

In our last theorems of this section we state some simple principles concerning *duals of modalities*. Recall that a modality ϕ is a finite (possibly null) sequence of the operators $\neg$, $\Box$, and $\Diamond$, and that the dual of a modality ϕ – written ϕ^* – is the result of interchanging $\Box$ and $\Diamond$ throughout ϕ (see section 2.1).

THEOREM 4.9. *Let Σ be a normal system of modal logic. Then:*

(1) $\vdash_\Sigma \phi A \leftrightarrow \neg\phi^*\neg A$.

(2) $\vdash_\Sigma \phi A$ *iff* $\vdash_\Sigma \neg\phi^*\neg A$.

(3) $\vdash_\Sigma \phi A \rightarrow \psi A$, *for every* A, *iff* $\vdash_\Sigma \psi^* A \rightarrow \phi^* A$, *for every* A.

(4) $\vdash_\Sigma \phi A \leftrightarrow \psi A$, *for every* A, *iff* $\vdash_\Sigma \phi^* A \leftrightarrow \psi^* A$, *for every* A.

Proof. Let Σ be a normal system. The result is a corollary to theorem 4.8.

For (1):

1.	$\phi A \leftrightarrow \neg(\phi A)^*$	DUAL(1)
2.	$\leftrightarrow \neg\phi^*(A^*)$	1, definition of *
3.	$\leftrightarrow \neg\phi^*\neg A$	2, DUAL(1) and REP

For (2). This follows easily from (1). Exercise.

For (3):

$\vdash_\Sigma \phi A \rightarrow \psi A$, for every A, iff $\vdash_\Sigma \neg\phi^*\neg A \rightarrow \neg\psi^*\neg A$, for every A
– (1) and REP;

iff $\vdash_\Sigma \psi^* \neg A \to \phi^* \neg A$, for every A
– *PL*;

iff $\vdash_\Sigma \psi^* A \to \phi^* A$, for every A
– *PL* and REP.

For (4). This follows from (3). Exercise.

As an application of theorem 4.9 we see that a normal modal logic contains the schema 4, $\Box A \to \Box\Box A$, as a theorem just in case it contains the dual schema 4$\Diamond$, $\Diamond\Diamond A \to \Diamond A$. Similarly, the schema B, $A \to \Box\Diamond A$, is a theorem of a normal system if and only if its dual B$\Diamond$, $\Diamond\Box A \to A$, is. (Of course not every normal system contains these schemas as theorems.)

Our final theorem is a rather obvious consequence of the preceding one. We set it out primarily in order to simplify the discussion in the next few sections. Recall that an affirmative modality contains an even number of occurrences of $\neg$.

THEOREM 4.10. *Let Σ be a normal system of modal logic, and let ϕ and ψ be affirmative modalities. Then Σ has the schema*

S. $\phi A \to \psi A$

as a theorem iff Σ has any one of the following theorem and rules of inference.

S$\Diamond$. $\psi^* A \to \phi^* A$

RS. $\dfrac{A \to B}{\phi A \to \psi B}$

RS$\Diamond$. $\dfrac{A \to B}{\psi^* A \to \phi^* B}$

Proof. Let Σ be a normal modal logic, and let ϕ and ψ be affirmative modalities. For the sake of simplicity we assume that ϕ and ψ are in fact composed solely of the operators $\Box$ and $\Diamond$, so that $\neg$ does not appear.

For S$\Diamond$. This follows from theorem 4.9 (3).

For RS. Suppose that $\vdash_\Sigma A \to B$. Then by repeated applications of the rules RM and RM$\Diamond$, $\vdash_\Sigma \psi A \to \psi B$. So if $\vdash_\Sigma \phi A \to \psi A$, then by *PL*, $\vdash_\Sigma \phi A \to \psi B$. Thus Σ has RS if it has S. Conversely, suppose Σ is closed under RS. Then $\vdash_\Sigma \phi A \to \psi A$, by RS on the tautology $A \to A$. So Σ has S if it has RS.

For RS$\Diamond$. Exercise.

Theorem 4.10 is illustrated by the fact that a normal system of modal logic has the schema 5, $\Diamond A \to \Box\Diamond A$, as a theorem if and only if it has its dual 5$\Diamond$, $\Diamond\Box A \to \Box A$, or either of these rules of inference:

R5. $$\frac{A \to B}{\Diamond A \to \Box\Diamond B}$$

R5$\Diamond$. $$\frac{A \to B}{\Diamond\Box A \to \Box B}$$

The theorems in this section afford the reader a handy means of recognizing theorems of normal systems. It is not so important, at this point, that the details of the several proofs be mastered and absorbed. It is worth remarking, however, that the proofs of theorems 4.7, 4.8, and 4.9 (but not 4.10) all depend ultimately only on *PL* and the presence of RE and Df$\Diamond$ (or RE$\Diamond$ and Df$\Box$) in normal systems of modal logic. This becomes important in chapter 8, where we return to these results.

EXERCISES

4.14. Complete the proof of theorem 4.7 (for the cases in which $A = \top$, $\neg C$, $C \wedge D$, $C \vee D$, $C \leftrightarrow D$, $\Diamond C$).

4.15. Complete the proof of theorem 4.8 (part (1) – for the cases in which $A = \top$, $\neg B$, $B \vee C$, $B \to C$, $B \leftrightarrow C$, $\Diamond B$ – and part (4)).

4.16. Complete the proof of theorem 4.9 (parts (2) and (4)).

4.17. Give the proof of theorem 4.10 for RS$\Diamond$.

4.18. Prove that if Σ is a system of modal logic closed under the rule REP, then Σ contains Df$\Diamond$ if and only if Σ contains Df$\Box$.

4.19. Prove that a system of modal logic is normal if it contains Df$\Diamond$, N, K, and is closed under REP.

4.20. Prove that $A \leftrightarrow A^{**}$ is a theorem of any normal modal logic.

4.21. Use REP and DUAL (and perhaps the result in the preceding exercise) to prove that N$\Diamond$, M$\Diamond$, C$\Diamond$, R$\Diamond$, and K$\Diamond$ are theorems of every normal system given that N, M, C, R, and K are. Then prove the reverse of this, i.e. that N etc. are theorems of every normal system given that N$\Diamond$ etc. are.

4.3. The schemas D, T, B, 4, and 5

The smallest normal system of modal logic, *K*, contains as theorems just what comes from Df$\Diamond$, RK, and propositional logic, nothing more. Thus we have canvassed the principal rules and theorems of the system *K* already in the preceding sections.

In this and the next section we are interested in the normal extensions of *K* obtained by adding as theorems the following schemas.

D. $\Box A \to \Diamond A$

T. $\Box A \to A$

B. $A \to \Box\Diamond A$

4. $\Box A \to \Box\Box A$

5. $\Diamond A \to \Box\Diamond A$

Including *K* itself there are just fifteen distinct normal systems produced by taking these schemas as theorems in all possible combinations. These systems appear on the diagram in figure 4.1.

The inclusions among the systems on the diagram are marked by lines: extensions of a system are reached by going in a rightward direction along the lines (for example, *KT* is shown to be an extension of *KD*). Most of the inclusions are obvious; some of those that are not we shall establish, and others are given as exercises. Likewise it is possible to show that each of the seventeen systems apparently not registered on the diagram is identical with one that is. Indeed, many of these identities are obvious from the diagram – for example, that *KDT* is the same as *KT*. The distinctness of the systems listed – and so the properness of the inclusions – is proved in chapter 5.

Historically the most important of these systems are *KD*, *KT*, *KTB*, *KT4*, and *KT5*. The first two are widely regarded as basic deontic and alethic modal logics, respectively, and are sometimes referred to simply as *D* and *T*. The other three systems – *KTB*, *KT4*, and *KT5* – are the well-known *Brouwersche* system (sometimes called *B*) and the Lewis systems *S4* and *S5*. Nevertheless, we approach these logics more analytically, by focusing on the systems *KD*, *KT*, *KB*, *K4*, *K5*, and their normal extensions. We begin with the following theorem about some alternative characterizations of these systems.

THEOREM 4.11. *Let* Σ *be a normal system of modal logic. Then:*

(1) Σ *is a KD-system iff it has* RD.

(2) Σ *is a KT-system iff it has any of* T$\Diamond$, RT, *and* RT$\Diamond$.

(3) Σ *is a KB-system iff it has any of* B◇, RB, *and* RB◇.

(4) Σ *is a K4-system iff it has any of* 4◇, R4, *and* R4◇.

(5) Σ *is a K5-system iff it has any of* 5◇, R5, *and* R5◇.

Proof. The theorem is an immediate consequence of theorem 4.10 and the fact that the modalities in the schemas D, T, B, 4, and 5 are all affirmative.

Figure 4.1

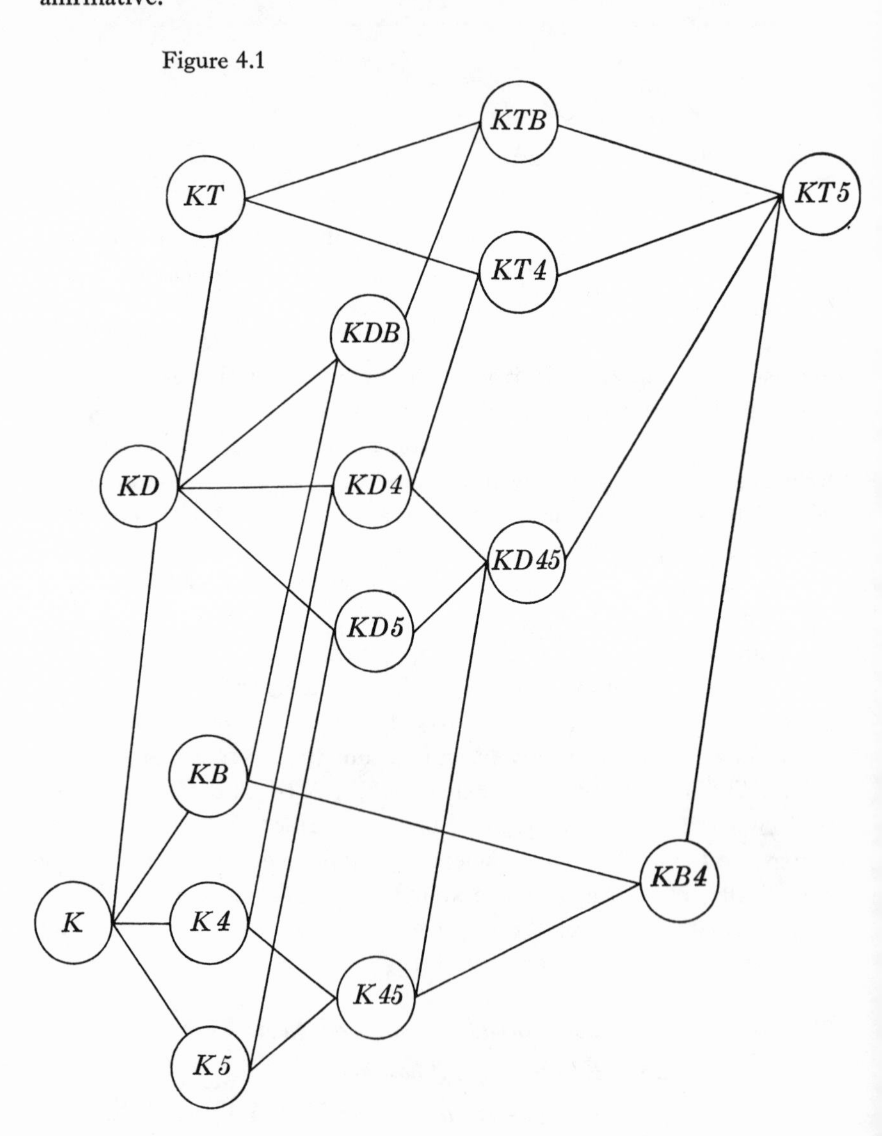

In what follows we freely make use of these theorems and rules of inference wherever appropriate. The reader should consult theorem 4.10 to ascertain their identities.

Now let us examine in turn each of the systems *KD*, *KT*, *KB*, *K4*, and *K5*, and their normal extensions.

Normal *KD*-systems. These come in many guises, as theorems 4.12 and 4.13 reveal.

THEOREM 4.12. *A normal system of modal logic is a KD-system iff it has any of the following theorems and rules of inference.*

RP. $\dfrac{A}{\Diamond A}$

P. $\Diamond\top$

O. $\Diamond A \vee \Diamond\neg A$

RP$\Box$. $\dfrac{\neg A}{\neg\Box A}$

P$\Box$. $\neg\Box\bot$

O$\Box$. $\neg(\Box A \wedge \Box\neg A)$

Proof. Suppose that Σ is a normal system.

For RP:

1. A	hypothesis
2. $\Box A$	1, RN
3. $\Box A \to \Diamond A$	D
4. $\Diamond A$	2, 3, *PL*

Thus Σ is closed under RP if it is a *KD*-system. For the reverse, suppose that Σ has RP. Then $\vdash_\Sigma \Diamond(A \to A)$, by RP on the tautology $A \to A$. To see from this that $\vdash_\Sigma \Box A \to \Diamond A$, it is enough to recall that the schema

$$\Diamond(A \to B) \leftrightarrow (\Box A \to \Diamond B)$$

is a theorem of every normal system (see following the proof of theorem 4.7). So Σ is a *KD*-system if it has the rule RP.

For P. Every normal modal logic has the theorem

$$\Diamond\top \leftrightarrow (\Box A \to \Diamond A)$$

(exercise 4.7(*m*)). So P is a theorem of Σ if and only if D is, which means that Σ is a *KD*-system just in case it contains P.

For O:

$$\vdash_\Sigma \Box A \to \Diamond A \text{ iff } \vdash_\Sigma \neg\Diamond\neg A \to \Diamond A$$
$$- \text{Df}\Box \text{ and REP;}$$
$$\text{iff } \vdash_\Sigma \Diamond A \vee \Diamond\neg A$$
$$- PL.$$

So Σ is a *KD*-system if and only if it contains O. More generally, this follows from the fact that every normal system has the theorem

$$(\Box A \to \Diamond B) \leftrightarrow (\Diamond B \vee \Diamond\neg A).$$

For RP$\Box$, P$\Box$, and O$\Box$. These principles are dual to RP, P, and O, and we leave the proofs as exercises. (See the examples after the proof of theorem 4.8.)

This completes our proof of theorem 4.12.

The theorem D admits of generalization along the modal dimension. To wit, for every $k > 0$ the schema

$$\text{D}^k. \quad \Box^k A \to \Diamond^k A$$

is a theorem of a normal *KD*-system. For D^k is D itself when $k = 1$. And if we suppose (as an inductive hypothesis) that the schema is a theorem when it has fewer than k $\Box$s and $\Diamond$s, then we can argue that D^k is, too. Thus:

1. $\Box^{k-1}A \to \Diamond^{k-1}A$	inductive hypothesis
2. $\Box\Box^{k-1}A \to \Diamond\Diamond^{k-1}A$	1, RD
3. $\Box^k A \to \Diamond^k A$	2, definition 2.3

Therefore, D^k is a theorem of every normal *KD*-system, for every $k > 0$.

From this result one can readily perceive analogous generalizations of the theorems and rules of normal *KD*-systems in theorems 4.11 (1) and 4.12. That is to say, the results of putting $\Box^k$ and $\Diamond^k$ for $\Box$ and $\Diamond$ throughout RD, RP, P, O, RP$\Box$, P$\Box$, and O$\Box$ are all principles of any normal *KD*-system, for every $k > 0$.

Moreover, the reverse is true. If a normal system has D^k or any one of these generalizations – RD^k, RP^k, P^k, O^k, $\text{RP}\Box^k$, $\text{P}\Box^k$, and $\text{O}\Box^k$ – for any $k > 0$, then it is a *KD*-system. We may illustrate this by showing that if

$$\text{P}^k. \quad \Diamond^k \top$$

is a theorem of a normal modal logic, for $k > 0$, then so is P, $\Diamond\top$, and so the modal logic is a *KD*-system. The proof:

1. $\Diamond^{k-1}\top \to \top$	*PL*
2. $\Diamond\Diamond^{k-1}\top \to \Diamond\top$	1, RM$\Diamond$

3. $\Diamond^k\top \to \Diamond\top$ 2, definition 2.3
4. $\Diamond^k\top$ P^k
5. $\Diamond\top$ 3, 4, *PL*

Therefore, a normal modal logic is a *KD*-system whenever it contains P^k, for any $k > 0$.

These proofs should be enough to convince the reader of the correctness of these alternative ways of characterizing normal *KD*-systems of modal logic. We state this formally as a theorem and leave the remaining proofs as exercises.

THEOREM 4.13. *A normal system of modal logic is a KD-system iff it has any of the theorems and rules of inference* D^k, RD^k, RP^k, P^k, O^k, $RP\Box^k$, $P\Box^k$, *and* $O\Box^k$, *for any* $k > 0$.

Normal *KT*-systems. The schema D is a theorem of any modal logic containing T and T$\Diamond$ ($\Box A \to \Diamond A$ follows by *PL* from $\Box A \to A$ and $A \to \Diamond A$). Therefore:

THEOREM 4.14. *Every normal KT-system is a KD-system.*

Thus all the principles mentioned in theorems 4.11 (1), 4.12, and 4.13 are present in any normal *KT*-system of modal logic. (This is not true the other way around, as we shall prove.)

The theorem T can be generalized modally; i.e. the schema

$$T^k. \quad \Box^k A \to A$$

is a theorem of every normal *KT*-system, for every $k > 0$. The inductive proof of this is left to the reader as an exercise. Thus, in virtue of theorem 4.10:

THEOREM 4.15. *Every normal KT-system has the theorems and rules of inference* T^k, $T\Diamond^k$, RT^k, *and* $RT\Diamond^k$, *for every* $k > 0$.

The name *T* for the logic *KT* derives from the designation *logique t* of Feys. The system is also called *M*, following von Wright.

Normal *KB*-systems. We begin by noting some recondite ways of characterizing systems of this kind.

THEOREM 4.16. *A normal system of modal logic is a KB-system iff it has any of the following theorems and rules of inference.*

X. $\Box(\Diamond A \to B) \to (A \to \Box B)$

X$\Diamond$. $\Box(A \to \Box B) \to (\Diamond A \to B)$

RX. $\dfrac{\Diamond A \to B}{A \to \Box B}$

RX$\Diamond$. $\dfrac{A \to \Box B}{\Diamond A \to B}$

Proof
For X:

1. $\Box(\Diamond A \to B) \to (\Box\Diamond A \to \Box B)$ K
2. $A \to \Box\Diamond A$ B
3. $\Box(\Diamond A \to B) \to (A \to \Box B)$ 1, 2, *PL*

So a normal *KB*-system contains X. For the reverse, note that $\Box(\Diamond A \to \Diamond A) \to (A \to \Box\Diamond A)$ is a special case of X and that the antecedent is a theorem by RN on the tautology $\Diamond A \to \Diamond A$. By *PL*, then, B is a theorem. So a normal logic is a *KB*-system if X is a theorem.

For X$\Diamond$. Exercise.

For RX:

1. $\Diamond A \to B$ hypothesis
2. $\Box\Diamond A \to \Box B$ 1, RM
3. $A \to \Box\Diamond A$ B
4. $A \to \Box B$ 2, 3, *PL*

So a normal *KB*-system has the rule RX. (Alternatively, if $\Diamond A \to B$ is a theorem so is $\Box(\Diamond A \to B)$ (by RN) – whence $A \to \Box B$ is a theorem by MP on X.) Conversely, a normal system closed under RX has the theorem $A \to \Box\Diamond A$, by RX on the tautology $\Diamond A \to \Diamond A$. So a normal modal logic is a *KB*-system if it has the rule RX.

For RX$\Diamond$. Exercise.

The theorem B can be generalized modally in two ways. In the first, the operators $\Box$ and $\Diamond$ are each iterated k times, for $k > 0$:

B^k. $A \to \Box^k \Diamond^k A$

To prove that every normal *KB*-system has B^k for every $k > 0$, notice first that B^k = B, for $k = 1$, and then suppose as an inductive hypothesis that the schema is a theorem whenever it has fewer than k $\Box$s and $\Diamond$s. Then:

1. $\Diamond A \to \Box^{k-1}\Diamond^{k-1}\Diamond A$ inductive hypothesis
2. $A \to \Box\Box^{k-1}\Diamond^{k-1}\Diamond A$ 1, RX
3. $A \to \Box^{k}\Diamond^{k}A$ 2, definition 2.3 and exercise 2.6

In the second way of generalizing B, the modality $\Box\Diamond$ itself is iterated k times, for $k > 0$:

$$\text{B}(\ )^k.\quad A \to (\Box\Diamond)^k A$$

We leave it to the reader to prove, inductively, that every normal *KB*-system contains $\text{B}(\ )^k$ for every $k > 0$.

By means of theorem 4.10 the principles in theorem 4.11(3) can similarly be generalized, and so can the rules RX and RX$\Diamond$ in theorem 4.26. We record all these generalizations formally.

THEOREM 4.17. *Every normal KB-system has the theorems and rules of inference* B^k, $\text{B}\Diamond^k$, RB^k, $\text{RB}\Diamond^k$, RX^k, $\text{RX}\Diamond^k$, $\text{B}(\ )^k$, $\text{B}\Diamond(\ )^k$, $\text{RB}(\ )^k$, *and* $\text{RB}\Diamond(\ )^k$, *for every* $k > 0$.

According to the next theorem the schema 4 is a theorem of a normal *KB*-system just in case the schema 5 is.

THEOREM 4.18. *A normal modal logic is a KB4-system iff it is a KB5-system.*

Proof. To show that 5 is a *KB4*-theorem we may argue as follows.

1. $\Diamond\Diamond A \to \Diamond A$ 4$\Diamond$
2. $\Diamond A \to \Box\Diamond A$ 1, RX

And to show that 4 is a *KB5*-theorem we may argue as follows.

1. $\Diamond\Box A \to \Box A$ 5$\Diamond$
2. $\Box A \to \Box\Box A$ 1, RX

In particular, then, the systems *KB4* and *KB5* are identical. (Our choice of the designation *KB4* in the diagram in figure 4.1 is thus somewhat arbitrary.)

The schema B is called the *Brouwersche* axiom for the curious reason that when it is stated equivalently as

$$A \to \neg\Diamond\neg\Diamond A$$

and the modality $\neg\Diamond$ is replaced by the intuitionistic negation sign, $\sim$, the result is

$$A \to \sim\sim A,$$

the intuitionistically valid version of the law of double negation. Brouwer was a leading exponent of intuitionism. So far as is known, however, Brouwer had no concern with the modal schema B; the name *Brouwersche* was given by Becker. The *Brouwersche system*, it should be noted, is *KTB*, not *KB*.

Normal *K4*-systems. The important modal generalization of the schema 4 is

$$4^k. \quad \Box A \to \Box^k \Box A.$$

This is a theorem of every normal *K4*-system, for any $k > 0$. The proof is left as an exercise. Hence by theorem 4.10:

THEOREM 4.19. *Every normal K4-system has the theorems and rules of inference* 4^k, $4\Diamond^k$, $R4^k$, *and* $R4\Diamond^k$, *for every* $k > 0$.

An interesting feature of normal *K4*-systems is that in them it is inconsistent to hold that every proposition is at least possibly possible, i.e. that the schema

$$\Diamond\Diamond A$$

is a theorem. For in conjunction with $4\Diamond$ this would lead to

$$\Diamond A,$$

and so in particular to

$$\Diamond\bot,$$

which conflicts with $N\Diamond$, $\neg\Diamond\bot$.

The schema 4 is often called the characteristic theorem of the system *S4*. But note that *S4* is *KT4*, which is stronger than *K4*.

Normal *K5*-systems. These all contain, for every $k > 0$, the schema

$$5^k. \quad \Diamond A \to \Box^k \Diamond A.$$

(Again we leave the proof to the reader.) Hence:

THEOREM 4.20. *Every normal K5-system has the theorems and rules of inference* 5^k, $5\Diamond^k$, $R5^k$, *and* $R5\Diamond^k$, *for every* $k > 0$.

As the diagram in figure 4.1 shows, the strongest normal system that can be formed using the schemas D, T, B, 4, and 5 is *KT5* – better known as the Lewis system *S5* – which we discussed in chapter 1. (Thus the schema 5 or 5◇ is often referred to as the characteristic theorem of *S5*.) There are many ways of axiomatizing *S5*. The next theorem gives the principal axiomatizations of *S5* using D, T, B, 4, and 5; of course duality (for example, putting T◇ for T) provides many more possibilities.

THEOREM 4.21. *A normal modal logic is a KT5-system iff it has as theorems* (1) T, B, *and* 4, (2) D, B, *and* 4, *or* (3) D, B, *and* 5. *In particular, then,* $KT5 = KTB4 = KDB4 = KDB5$.

Proof. Part (1) was established in chapter 1. In light of this and theorems 4.14 and 4.18 it is then sufficient to show that T is a theorem of every normal *KDB4*-system. We leave the details of the reasoning as an exercise.

This is a good place to affirm the correctness of figure 4.1 with respect to the inclusions advertised there. For the most part this is a matter of definition – for example, *KTB* is obviously an extension of *KT*. For the rest, note that $KD \subseteq KT$, and so $KDB \subseteq KTB$ and $KD4 \subseteq KT4$, by theorem 4.15; that $K45 \subseteq KB4$ by theorem 4.18; and that by theorem 4.21 *KT5* is an extension of *KTB*, *KT4*, *KD45*, and *KB4*. Several of the seventeen systems apparently missing from figure 4.1 have already been mentioned, for example, in the alternative axiomatizations of *S5* in theorem 4.21. We leave it as an exercise for the reader to identify all the missing systems and locate them in figure 4.1.

We might also remark here that although the system *KD* results from the addition to *K* of D^k (or any other of the principles listed in theorem 4.13) for any $k > 0$, there is no analogous result with respect to the modal generalizations of T, B, 4, and 5, for $k > 1$. We shall be in a position to prove this in chapter 5.

The point of our analytical approach in this section may by now be apparent. It enables us to see better the individual contributions of the schemas D, T, B, 4, and 5 to more familiar modal logics such as *KTB*, *KT4*, and *KT5*. Two examples will make this clear. First, it is often pointed out that the rules of inference RX and RX◇ are present in the *Brouwersche* system, *KTB*. But as we have seen, these rules are already in the modal logic *KB* (and hence in any normal *KB*-system); the theorem T has no bearing on the matter. Second, the result that the schema $\Diamond\Diamond$A is inconsistent as an addition to the Lewis system *S4* (*KT4*) is frequently mentioned. But our analytical exposition shows this to be so with respect to *K4* and normal *K4*-systems generally. Here again the presence of the theorem T is of no consequence.

EXERCISES

4.22. Use theorem 4.10 to ascertain the identities of the schemas and rules of inference mentioned in theorem 4.11.

4.23. Complete the proof of theorem 4.12 (for RP$\Box$, P$\Box$, and O$\Box$).

4.24. Complete the proof of theorem 4.13.

4.25. Prove by induction that, for any $k > 0$, the schema T^k is a theorem of every normal *KT*-system (for theorem 4.15).

4.26. Complete the proof of theorem 4.16 (for X$\Diamond$ and RX$\Diamond$).

4.27. Prove by induction that, for any $k > 0$, the schema $B(\ )^k$ is a theorem of every normal *KB*-system (for theorem 4.17).

4.28. Prove by induction that, for any $k > 0$, the schema 4^k is a theorem of every normal *K4*-system (for theorem 4.19).

4.29. Prove by induction that, for any $k > 0$, the schema 5^k is a theorem of every normal *K5*-system (for theorem 4.20).

4.30. Complete the proof of theorem 4.21 by proving that the schema T is a theorem of any normal *KDB4*-system.

4.31. Identify and locate on the diagram in figure 4.1 the seventeen systems not already listed there.

4.32. Prove that a normal modal logic is a *KD*-system if and only if it has theorems of the form $\Diamond$A.

4.33. Consider the following schemas.

U.	$\Box(\Box A \to A)$	U$\Diamond$.	$\Box(A \to \Diamond A)$
4_c.	$\Box\Box A \to \Box A$	$4\Diamond_c$.	$\Diamond A \to \Diamond\Diamond A$
5_c.	$\Box\Diamond A \to \Diamond A$	$5\Diamond_c$.	$\Box A \to \Diamond\Box A$

Prove:

(*a*) U is a theorem of a normal system if and only if U$\Diamond$ is.

(*b*) 4_c (and hence $4\Diamond_c$) is a theorem of any normal *KU*-system.

(*c*) U (and hence 4_c) and 5_c (and hence $5\Diamond_c$) are theorems of any normal *KT*-system.

(*d*) D is a theorem of any normal *K5*$_c$-system.

4.34. Prove that the schema G, $\Diamond\Box A \to \Box\Diamond A$, is a theorem of any normal *KB*-system.

4.35. Prove that any normal *KB*-system is closed under the following rules of inference.

$$\frac{\Box A}{\Diamond\top \to A} \qquad \frac{\Diamond\top \to A}{\Box A}$$

4.36. Prove that a normal system of modal logic is a *K4*-system if and only if it has any of the following theorems.

(*a*) $\Box(A \to B) \to \Box(\Box A \to \Box B)$

(*b*) $(\Box A \vee \Box B) \to \Box(\Box A \vee \Box B)$

(*c*) $\Box(\Box(A \to B) \to C) \to \Box(\Box(A \to B) \to \Box C)$

4.37. Prove that a normal system of modal logic is a *K5*-system if and only if it has either of the following theorems.

(*a*) $\Box(\Box A \vee B) \to (\Box A \vee \Box B)$

(*b*) $(\Diamond A \wedge \Diamond B) \to \Diamond(\Diamond A \wedge B)$

4.38. Prove:

(*a*) U (and hence 4_c) is a theorem of any normal *K5*-system (see exercise 4.33).

(*b*) G is a theorem of any normal *K5*-system.

4.39. Prove that every normal *K5*-system contains the following theorems.

(*a*) $\Box(\Box A \leftrightarrow \Box\Box A)$ (*b*) $\Box(\Diamond A \leftrightarrow \Diamond\Diamond A)$

(*c*) $\Box(\Box A \leftrightarrow \Diamond\Box A)$ (*d*) $\Box(\Diamond A \leftrightarrow \Box\Diamond A)$

4.40. Referring to the preceding exercise, prove that every normal *K5*-system contains the following theorems.

(*a*) $\Box\Box A \leftrightarrow \Box\Box\Box A$ (*e*) $\Diamond\Diamond A \leftrightarrow \Diamond\Diamond\Diamond A$

(*b*) $\Box\Box A \leftrightarrow \Box\Diamond\Box A$ (*f*) $\Diamond\Diamond A \leftrightarrow \Diamond\Box\Diamond A$

(*c*) $\Diamond\Box A \leftrightarrow \Diamond\Box\Box A$ (*g*) $\Box\Diamond A \leftrightarrow \Box\Diamond\Diamond A$

(*d*) $\Diamond\Box A \leftrightarrow \Diamond\Diamond\Box A$ (*h*) $\Box\Diamond A \leftrightarrow \Box\Box\Diamond A$

4.41. Notice that the interiors of the four necessitations listed in exercise

4.39 are all theorems of any normal *KT5*-system. This suggests the following result (which leads at once to a solution for exercise 4.39). Where Σ is any normal *K5*-system:

$$\vdash_{\Sigma} \Box A, \text{ whenever A is a theorem of } KT5.$$

Prove this by induction on (the set of theorems of) *KT5*. For the basis, show that the necessitations of the axioms Df$\Diamond$, T, and 5 are theorems of Σ; for the inductive part, show that the set $\{A: \vdash_{\Sigma} \Box A\}$ is closed under the rules RPL and RK.

4.42. Prove that every normal *K5*-system contains the following theorems.

(*a*) $\Diamond\Box A \rightarrow \Diamond A$ (*b*) $\Box A \rightarrow \Box\Diamond A$
(*c*) $\Diamond\Box A \rightarrow \Box\Box A$ (*d*) $\Diamond\Diamond A \rightarrow \Box\Diamond A$

4.43. Prove:

(*a*) 5_c and $5\Diamond_c$ (see exercise 4.33) are theorems of any normal *KD4*-system.

(*b*) The schemas $\Diamond\Box A \leftrightarrow \Diamond\Box\Diamond\Box A$ and $\Box\Diamond A \leftrightarrow \Box\Diamond\Box\Diamond A$ are theorems of any normal *KD4*-system.

4.44. Prove that every normal *KD5*-system contains the following theorems.

(*a*) $\Box\Box A \leftrightarrow \Diamond\Box A$ (*b*) $\Diamond\Diamond A \leftrightarrow \Box\Diamond A$

4.45. Prove that every normal *K45*-system contains the following theorems.

4!. $\Box A \leftrightarrow \Box\Box A$ 4$\Diamond$!. $\Diamond A \leftrightarrow \Diamond\Diamond A$

4.46. Consider the following schemas.

5!. $\Diamond A \leftrightarrow \Box\Diamond A$ 5$\Diamond$!. $\Box A \leftrightarrow \Diamond\Box A$

Prove that 4! and 4$\Diamond$! (see the preceding exercise) are theorems of any normal *K5!*- or normal *K5$\Diamond$!*-system.

4.47. Prove that 4!, 4$\Diamond$!, 5!, and 5$\Diamond$! are theorems of any normal *KD45*-system (see exercises 4.43(*a*), 4.45, and 4.46).

4.48. Let us say that a sentence A is *fully modalized* just in case every atomic sentence in A is within the scope of an occurrence of $\Box$ or $\Diamond$. Show that where Σ is any normal *KD45*-system and A is fully modalized:

$$\vdash_{\Sigma} A \leftrightarrow \Box A \quad \text{and} \quad \vdash_{\Sigma} A \leftrightarrow \Diamond A.$$

The proof is by induction on the complexity of A.

4.49. Consider the following rules of inference.

RFM. $\dfrac{A \to B}{A \to \Box B}$, where A is fully modalized

RFM$\Diamond$. $\dfrac{A \to B}{\Diamond A \to B}$, where A is fully modalized

Using the results in the preceding exercise, prove:

(*a*) A normal system is closed under RFM if and only if it is closed under RFM$\Diamond$.

(*b*) Every normal *KD45*-system is closed under RFM and RFM$\Diamond$.

(*c*) The schemas 4 and 5 are theorems of any normal *KD*-system closed under RFM or RFM$\Diamond$.

4.50. Prove that the schema $\Box(\Diamond A \to B) \leftrightarrow \Box(A \to \Box B)$ is a theorem of any normal *KB4*-system.

4.51. Prove that if any of the schemas B, $\Diamond A \to \Diamond\Box\Diamond A$, and $\Box\Diamond\Box A \to \Box A$ is a theorem of a normal *KT*-system, then so are the others.

4.52. Consider the following schemas.

T_c. $A \to \Box A$ $\qquad$ $T\Diamond_c$. $\Diamond A \to A$

D_c. $\Diamond A \to \Box A$

F. $\Box(A \vee B) \to (\Box A \vee \Box B)$ $\qquad$ F$\Diamond$. $(\Diamond A \wedge \Diamond B) \to \Diamond(A \wedge B)$

Prove:

(*a*) T_c is a theorem of a normal system if and only if $T\Diamond_c$ is.

(*b*) If any one of D_c, $\Box A \vee \Box\neg A$, $\neg(\Diamond A \wedge \Diamond\neg A)$, F, and F$\Diamond$ is a theorem of a normal system, then so are all the others.

(*c*) D_c (and hence the rest in (*b*)) is a theorem of any normal KT_c-system.

(*d*) B, 4, 5, and G are theorems of any normal KT_c-system.

(*e*) T is a theorem of any normal KDT_c-system.

(*f*) T_c is a theorem of any normal KD_cT-system.

(*g*) 4 and 4_c are theorems of any normal KD_c5_c-system.

4.53. Consider the following schemas.

T!. $\Box A \leftrightarrow A$ $\qquad$ T$\Diamond$!. $\Diamond A \leftrightarrow A$

D!. $\Box A \leftrightarrow \Diamond A$

Prove:

(*a*) T! is a theorem of a normal system if and only if T$\Diamond$! is.

(*b*) D! is a theorem of any normal *KT!*-system.

(*c*) If any of 4, 4_c, 5, and 5_c is a theorem of a normal *KD!*-system, then so are the others.

4.54. Consider the sentences $\overline{P}$ ($\neg\Diamond\top$) and $\overline{P\Box}$ ($\Box\bot$). Prove:

(*a*) $\overline{P}$ is a theorem of a normal system if and only if $\overline{P\Box}$ is.

(*b*) T_c (and hence D_c, etc., B, 4, 5, and G) is a theorem of any normal $K\overline{P}$-system.

4.55. Prove that a normal system of modal logic is a *KG*-system if and only if it has either of the following theorems.

(*a*) $\Box\Diamond A \vee \Box\Diamond\neg A$

(*b*) $\neg(\Diamond\Box A \wedge \Diamond\Box\neg A)$

The system *KT4G* is known as *S4.2*. This system (properly) contains *S4* (*KT4*) and is (properly) contained in *S5* (*KT5*); see exercise 4.38.

4.56. Consider the following schemas.

H^{++}. $\Box(\Diamond A \vee \Diamond B) \to (\Box\Diamond A \vee \Box\Diamond B)$

$H^{++}\Diamond$. $(\Diamond\Box A \wedge \Diamond\Box B) \to \Diamond(\Box A \wedge \Box B)$

H^{+}. $(\Box(\Box A \vee B) \wedge \Box(A \vee \Box B)) \to (\Box A \vee \Box B)$

$H^{+}\Diamond$. $(\Diamond A \wedge \Diamond B) \to (\Diamond(\Diamond A \wedge B) \vee \Diamond(A \wedge \Diamond B))$

H. $(\Box(A \vee B) \wedge \Box(\Box A \vee B) \wedge \Box(A \vee \Box B)) \to (\Box A \vee \Box B)$

H$\Diamond$. $(\Diamond A \wedge \Diamond B) \to (\Diamond(A \wedge B) \vee \Diamond(\Diamond A \wedge B) \vee \Diamond(A \wedge \Diamond B))$

L^{++}. $\Box(\Box A \to \Box B) \vee \Box(\Box B \to \Box A)$

$L^{++}\Diamond$. $\Box(\Diamond A \to \Diamond B) \vee \Box(\Diamond B \to \Diamond A)$

L^{+}. $\Box(\Box A \to B) \vee \Box(\Box B \to A)$

$L^{+}\Diamond$. $\Box(A \to \Diamond B) \vee \Box(B \to \Diamond A)$

L. $\Box(A \to (\Box B \to C)) \vee \Box(A \vee (\Box C \to B))$

L$\Diamond$. $\Box(A \vee (B \to \Diamond C)) \vee \Box(A \to (C \to \Diamond B))$

Taken as theorems these schemas are all equivalent additions to any normal *KT4*-system; that is, any one of them is a theorem of a normal *KT4*-system if and only if all the others are. Thus the systems $KT4H^{++}$,

..., *KT4L*$\Diamond$ are identical. This system is known as *S4.3*: it is (properly) contained in *S5* (*KT5*), and it (properly) contains *S4* (*KT4*). Indeed, *S4.3* is a (proper) extension of the system *S4.2* (*KT4G*) mentioned in exercise 4.55. Except for properness, these facts are all consequences of the following results (which are for the reader to prove) in conjunction with exercises 4.33 and 4.38.

(*a*) In any normal system H^{++} is a theorem if and only if $H^{++}\Diamond$ is.

(*b*) In any normal system H^{+} is a theorem if and only if $H^{+}\Diamond$ is.

(*c*) In any normal system H is a theorem if and only if $H\Diamond$ is.

(*d*) In any normal system L^{++} is a theorem if and only if $L^{++}\Diamond$ is.

(*e*) In any normal system L^{+} is a theorem if and only if $L^{+}\Diamond$ is.

(*f*) In any normal system L is a theorem if and only if $L\Diamond$ is.

(*g*) In any normal system H^{++} is a theorem if and only if L^{++} is.

(*h*) In any normal system H^{+} is a theorem if and only if L^{+} is.

(*i*) In any normal system H is a theorem if and only if L is.

(*j*) U is a theorem of any normal *KH*$^{+}$-system.

(*k*) H is a theorem of any normal *KH*$^{+}$-system.

(*l*) H^{+} is a theorem of any normal *KUH*-system.

(*m*) H^{+} is a theorem of any normal *KUH*$^{++}$-system.

(*n*) H^{++} is a theorem of any normal *K4H*-system.

In proving (*j*)–(*n*) it may help to restate some of them using (*a*)–(*i*). In any case, the foregoing yield the identity of *KT4H*$^{++}$, ..., *KT4L*$\Diamond$ – i.e. *S4.3* – as the reader should confirm.

(*o*) H^{++} is a theorem of any normal *K5*-system.

This is enough, given the preceding results, to show that *S5* (*KT5*) is an extension of *S4.3*. See also exercise 4.37 with respect to H^{+} (and hence the rest).

(*p*) G is a theorem of any normal *KH*$^{+}$-system.

This is enough, given the preceding results, to show that *S4.3* is an extension of *S4.2*. But also:

(*q*) G is a theorem of any normal *KDH*$^{++}$-system.

4.57. Consider the following schemas.

$$A \vee (\Diamond A \to \Box\Diamond A) \qquad A \to (\Diamond\Box A \to \Box A)$$
$$(\Diamond A \to A) \vee (\Diamond A \to \Box\Diamond A) \qquad (A \to \Box A) \vee (\Diamond\Box A \to \Box A)$$

The two on the left we indifferently dub 5^-, the two on the right, $5^-\Diamond$. Prove:

(*a*) In any normal system 5^- is a theorem if and only if $5^-\Diamond$ is.

(*b*) 5^- is a theorem of any normal *K5*-system.

(*c*) The schema $\Box(\Box(A \to \Box A) \to A) \to \Box(\Diamond\Box A \to A)$ is a theorem of any normal *K5*$^-$-system.

Added as a theorem to *S4* (*KT4*) the schema 5^- produces a system called *S4.4*. By (*b*) *S4.4* is contained in *S5* (*KT5*) (in fact it is properly contained). *S4.4* does not, however, contain *S4.3* or *S4.2* (in exercises 4.55 and 4.56). Added as a theorem to *S4* the schema mentioned in (*c*) produces a system that Hughes and Cresswell call *S4.1*, which is therefore contained (in fact properly) in *S4.4*. But *S4.1* is not extended by *S4.2* or *S4.3*. N.B. this *S4.1* is not the same as that described by McKinsey; see the following exercise.

4.58. Consider the following schema.

$$\text{G}_c. \quad \Box\Diamond A \to \Diamond\Box A$$

Prove:

(*a*) In any normal system the schema $\Diamond(\Diamond A \to \Box A)$ is a theorem if and only if the schema G_c is.

(*b*) D is a theorem of any normal KG_c-system.

(*c*) In any normal system the schema $\Diamond(\Box A \leftrightarrow \Diamond A)$ is a theorem if and only if the schema G_c is.

$KT4G_c$ is the system called *S4.1* by McKinsey. It is clearly an extension of *S4* (*KT4*) – in fact a proper one (otherwise $\Diamond\Box A \leftrightarrow \Diamond\Box\Diamond A$ and $\Box\Diamond A \leftrightarrow \Box\Diamond\Box A$ would be theorems of *S4*, which they are not). But this *S4.1* is not included in *S5* (*KT5*) (if it were, D_c would be a theorem of *S5*, which it is not). *S4.1* is equivalently axiomatized by adding to *S4* any of the following schemas.

$$\Diamond\Box(A \to \Box A) \qquad \Diamond\Box(\Diamond A \to A)$$

$$\Box(A \vee B) \to (\Diamond\Box A \vee \Diamond\Box B) \qquad (\Box\Diamond A \wedge \Box\Diamond B) \to \Diamond(A \wedge B)$$

This is for the reader to prove.

4.59. Consider the following schema and rule of inference.

$$\text{Gr.} \quad \Box(\Box A \to A) \to \Box A \qquad \text{RGr.} \quad \frac{\Box A \to A}{A}$$

Prove:

(*a*) 4 is a theorem of any normal *KGr*-system.

(*b*) Every normal *KGr*-system is closed under RGr.

(*c*) Gr is a theorem of any normal *K4*-system that has the rule RGr.

Is the system *KGr* an extension of *S4* (*KT4*)? Is it included in *S5* (*KT5*)?

4.60. Use the erasure transformation ϵ from exercise 1.27 to prove the consistency of the fourteen systems beyond *K* on the diagram in figure 4.1. Identify some consistent normal modal logics for which ϵ cannot be deployed to prove consistency. Prove the consistency of these examples.

4.61. Amplify the proof of the lemma in exercise 4.13 to show that the systems *KD*, KD_c, and *KD!* have the rules (*a*)–(*f*) listed there.

4.62. Consider the rule (*a*) from exercise 4.13:

$$\frac{\Box A}{A}$$

Of the fifteen systems in figure 4.1 only *K*, *KD*, *KT*, *K4*, *KDB*, *KD4*, *KTB*, *KT4*, and *KT5* have this rule. Exercise 4.13 and 4.61 cover the first two cases, and it is obvious that any *KT*-system has the rule (if $\Box$A is a theorem of a system containing $\Box A \to A$, then A is a theorem of the system). The cases of *K4* and *KD4* must await the developments in chapter 5. Prove that every normal *KDB*-system has the rule.

4.4. Modalities

A modality, once again, is any sequence of the operators $\neg$, $\Box$, and $\Diamond$, including the empty sequence $\cdot$. Within a system of modal logic two modalities ϕ and ψ are *equivalent* if and only if for every A the sentence

$$\phi A \leftrightarrow \psi A$$

is a theorem; otherwise ϕ and ψ are said to be *distinct*. For example, in the system *S4* we have the theorem $\Box A \leftrightarrow \Box\Box A$, so in *S4* the modalities $\Box$ and $\Box\Box$ are equivalent.

Theorems like $\Box A \leftrightarrow \Box\Box A$ are often called *reduction laws*, since in virtue of them one modality is reducible to another.

In some systems of modal logic it happens that every modality is equivalent to one or another in a finite class. For example, in the system *S5* every modality is equivalent to one of $\cdot$, $\Box$, $\Diamond$, or their negations, $\neg$,

$\neg\Box$, $\neg\Diamond$. Thus *S5* is said to have at most six distinct modalities (three affirmative, three negative). To see this it is sufficient to note that *S5* contains the following reduction laws.

(1) $\Box A \leftrightarrow \Box\Box A$
(2) $\Diamond A \leftrightarrow \Diamond\Diamond A$
(3) $\Box A \leftrightarrow \Diamond\Box A$
(4) $\Diamond A \leftrightarrow \Box\Diamond A$

Because of these every modality in *S5* reduces to one of the specified six.

An example may help to make this clear. Suppose we have the sentence

$$\neg\Box\Box\Diamond\neg\Box\Diamond A.$$

First we put the modality $\neg\Box\Box\Diamond\neg\Box\Diamond$ in a standard form by using Df$\Diamond$, Df$\Box$, and REP to bring the negation signs all to the outside – successively,

$$\neg\Box\Box\neg\Box\Box\Diamond A,\ \neg\Box\neg\Diamond\Box\Box\Diamond A,\ \neg\neg\Diamond\Diamond\Box\Box\Diamond A$$

– and then reducing the number of occurrences of $\neg$ to zero (as in this case) or one by *PL*:

$$\Diamond\Diamond\Box\Box\Diamond A$$

According to reduction law (2) the modality $\Diamond\Diamond$ can be replaced by $\Diamond$ alone, yielding

$$\Diamond\Box\Box\Diamond A$$

By (3), $\Diamond\Box$ becomes $\Box$:

$$\Box\Box\Diamond A$$

By (1), $\Box\Box$ becomes $\Box$:

$$\Box\Diamond A$$

And $\Box\Diamond$ reduces to $\Diamond$ by (4):

$$\Diamond A$$

Thus the modality $\neg\Box\Box\Diamond\neg\Box\Diamond$ is shown to be equivalent to $\Diamond$. In a similar way one can show the modality $\neg\Box\Diamond\neg\Box\Diamond\neg\Box$ to be equivalent to $\neg\Box$; details of the reduction are left to the reader.

Of course the presence of reduction laws can only put an upper bound on the number of distinct modalities in a system. To show that *S5* has at least – and hence exactly – six distinct modalities it is necessary to

establish that there are no further reduction laws in the system (for example, that $\Box A \leftrightarrow \Diamond A$ is not also a theorem). In general other means are required to fix a lower bound (possibly infinite) for the number of distinct modalities in a modal logic. We return to this point in chapter 5.

Systems of modal logic can have the same distinct modalities but differ with respect to the pattern of implications among them. *S5* and the system *KD45* provide an example of this; each has $\cdot$, $\Box$, $\Diamond$, and their negations, but the *S5*-theorems $\Box A \rightarrow A$ and $A \rightarrow \Diamond A$ are absent from *KD45*. The diagrams in figures 4.7 and 4.8 chart the differences (among the affirmative modalities; for their negations, reverse the arrows). The systems *KD5*, *K45*, and *KB4* provide another example; compare figures 4.4, 4.5, and 4.6. Moreover, systems may be different even though they have the same distinct modalities and the same patterns of implications among them. Some examples of this situation will be found in the exercises at the end of the section.

It turns out that of the normal systems that can be formed using D, T, B, 4, and 5, only seven have a finite number of distinct modalities: *KT4*, *K5*, *KD5*, *K45*, *KB4*, *KD45*, and *KT5*. The following theorems give the details.

THEOREM 4.22. *Every normal KT4-system has at most fourteen distinct modalities, viz.* $\cdot$, $\Box$, $\Diamond$, $\Box\Diamond$, $\Diamond\Box$, $\Box\Diamond\Box$, $\Diamond\Box\Diamond$, *and their negations, with implications among the affirmative seven as diagramed in figure 4.2.*

Proof. To show that a normal *KT4*-system has at most the specified

Figure 4.2. Modalities in normal *KT4*-systems.

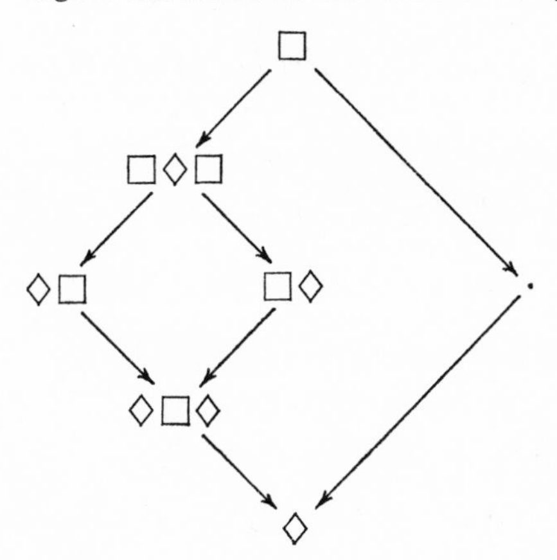

fourteen distinct modalities it is sufficient to show the following reduction laws to be theorems of the system.

$$\Box A \leftrightarrow \Box\Box A \qquad \Diamond A \leftrightarrow \Diamond\Diamond A$$

$$\Diamond\Box A \leftrightarrow \Diamond\Box\Diamond\Box A \qquad \Box\Diamond A \leftrightarrow \Box\Diamond\Box\Diamond A$$

For then every modality will be reducible to one of those specified (as the reader should confirm). The laws on the right are dual to those on the left, and $\Box A \leftrightarrow \Box\Box A$ is obvious in view of T and 4. So it suffices to establish $\Diamond\Box A \leftrightarrow \Diamond\Box\Diamond\Box A$. For left-to-right:

1. $\Box A \rightarrow \Diamond\Box A$ T$\Diamond$
2. $\Box\Box A \rightarrow \Box\Diamond\Box A$ 1, RM
3. $\Box A \rightarrow \Box\Box A$ 4
4. $\Box A \rightarrow \Box\Diamond\Box A$ 2, 3, *PL*
5. $\Diamond\Box A \rightarrow \Diamond\Box\Diamond\Box A$ 4, RM$\Diamond$

And for right-to-left:

1. $\Box\Diamond\Box A \rightarrow \Diamond\Box A$ T
2. $\Diamond\Box\Diamond\Box A \rightarrow \Diamond\Diamond\Box A$ 1, RM$\Diamond$
3. $\Diamond\Diamond\Box A \rightarrow \Diamond\Box A$ 4$\Diamond$
4. $\Diamond\Box\Diamond\Box A \rightarrow \Diamond\Box A$ 2, 3, *PL*

For the eight implications diagramed in figure 4.2 we need consider only the top four; the others are duals. Of these four, two are T and one appears on line 4 of the first proof above. The remaining theorem, $\Box\Diamond\Box A \rightarrow \Box\Diamond A$, follows from T by RM$\Diamond$ and RM.

Thus the system *S4* – i.e. *KT4* itself – has at most fourteen distinct modalities. In chapter 5 we prove it has exactly that many.

THEOREM 4.23. *Every normal K5-system has at most fourteen distinct modalities, viz.* ·, $\Box$, $\Diamond$, $\Box\Box$, $\Diamond\Diamond$, $\Box\Diamond$, $\Diamond\Box$, *and their negations, with implications among the affirmative seven as diagramed in figure 4.3.*

Proof. For this result we require the following reduction laws.

$$\Box\Box A \leftrightarrow \Box\Box\Box A \qquad \Diamond\Diamond A \leftrightarrow \Diamond\Diamond\Diamond A$$

$$\Box\Box A \leftrightarrow \Box\Diamond\Box A \qquad \Diamond\Diamond A \leftrightarrow \Diamond\Box\Diamond A$$

$$\Diamond\Box A \leftrightarrow \Diamond\Box\Box A \qquad \Box\Diamond A \leftrightarrow \Box\Diamond\Diamond A$$

$$\Diamond\Box A \leftrightarrow \Diamond\Diamond\Box A \qquad \Box\Diamond A \leftrightarrow \Box\Box\Diamond A$$

These laws might be summed up by the motto *Drop the middle modality.* The conditional halves of those on the left appear below on lines 2, 3, and 9–14; those on the right follow by duality. The last six lines below make implicit appeal to $(\Diamond A \to \Box B) \to (\Box A \to \Box B)$ and $(\Diamond A \to \Box B) \to (\Diamond A \to \Diamond B)$. Because both are theorems of any normal system (exercise 4.7 (*o*, *p*)) we mark these steps *K*.

1.	$\Diamond\Box A \to \Box A$	$5\Diamond$
2.	$\Box\Diamond\Box A \to \Box\Box A$	1, RM
3.	$\Diamond\Diamond\Box A \to \Diamond\Box A$	1, RM$\Diamond$
4.	$\Diamond\Box A \to \Box\Diamond\Box A$	5
5.	$\Diamond\Box\Box A \to \Box\Box A$	$5\Diamond$
6.	$\Diamond\Box A \to \Box\Box A$	2, 4, *PL*
7.	$\Box\Diamond\Box A \to \Box\Box\Box A$	6, RM
8.	$\Diamond\Box A \to \Box\Box\Box A$	4, 7, *PL*
9.	$\Box\Box A \to \Box\Diamond\Box A$	4, *K*
10.	$\Diamond\Box A \to \Diamond\Diamond\Box A$	4, *K*
11.	$\Box\Box\Box A \to \Box\Box A$	5, *K*
12.	$\Diamond\Box\Box A \to \Diamond\Box A$	5, *K*
13.	$\Box\Box A \to \Box\Box\Box A$	8, *K*
14.	$\Diamond\Box A \to \Diamond\Box\Box A$	8, *K*

As to the six implications pictured in figure 4.3, first note that those on the right are duals of those on the left. Of the latter, $\Diamond\Box A \to \Box\Box A$

Figure 4.3. Modalities in normal *K5*-systems.

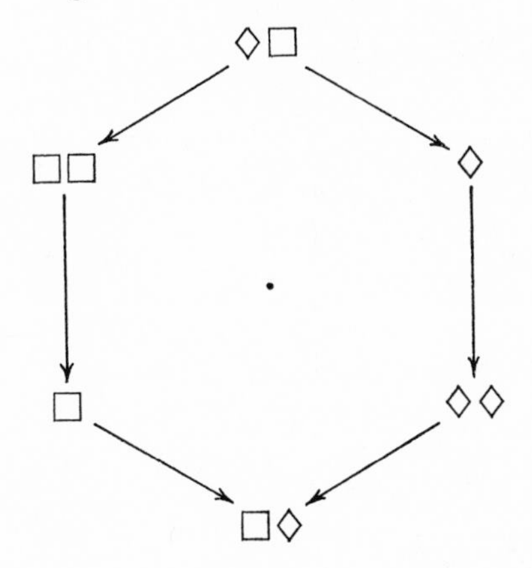

appears on line 6 above, and the remaining two follow from the theorems 5 and 5$\Diamond$ by the *K*-theorems mentioned in the last paragraph.

THEOREM 4.24. *Every normal KD5-system has at most ten distinct modalities, viz.* $\cdot$, $\Box$, $\Diamond$, *one from each of the pairs* $\Box\Box$, $\Diamond\Box$ *and* $\Diamond\Diamond$, $\Box\Diamond$, *and their negations, with implications among the affirmative five as diagramed in figure 4.4.*

Proof. By theorem 4.23 a normal *KD5*-system has at most the distinct modalities, and at least the implications among them, pictured in figure 4.3. But in addition to the reduction laws in the proof above, every normal *KD5*-system contains the laws $\Box\Box A \leftrightarrow \Diamond\Box A$ and $\Diamond\Diamond A \leftrightarrow \Box\Diamond A$; each is the theorem D in one direction, and the converses belong to any normal *K5*-system (as in figure 4.3). Finally, D gives the implication from $\Box$ to $\Diamond$. Hence the modalities and implications in figure 4.4.

THEOREM 4.25. *Every normal K45-system has at most ten distinct modalities, viz.* $\cdot$, $\Box$, $\Diamond$, $\Box\Diamond$, $\Diamond\Box$, *and their negations, with implications among the affirmative five as diagramed in figure 4.5.*

Proof. By theorem 4.23 it is sufficient to point out that a normal *K45*-system contains the reduction laws $\Box A \leftrightarrow \Box\Box A$ and $\Diamond A \leftrightarrow \Diamond\Diamond A$, so that we may delete $\Box\Box$ and $\Diamond\Diamond$ from the diagram in figure 4.3. The result is figure 4.5.

Figure 4.4. Modalities in normal *KD5*-systems.

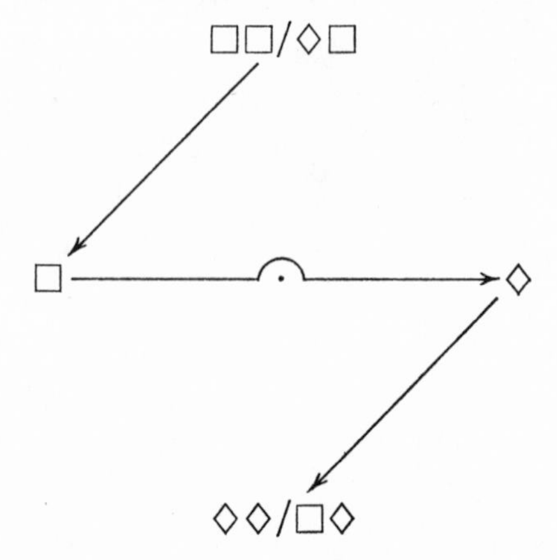

THEOREM 4.26. *Every normal KB4-system has at most ten distinct modalities, viz.* ·, □, ◇, □◇, ◇□, *and their negations, with implications among the affirmative five as diagramed in figure 4.6.*

Proof. Recall (theorem 4.18) that 5 is a theorem of any normal *KB4*-system, so that every such system is an extension of *K45*. By theorem 4.25, then, a normal *KB4*-system has at most the distinct modalities ·, □, ◇, □◇, ◇□, and their negations. The only new elements are the implications involving the modalities · and ¬. Thus figure 4.6.

Figure 4.5. Modalities in normal *K45*-systems.

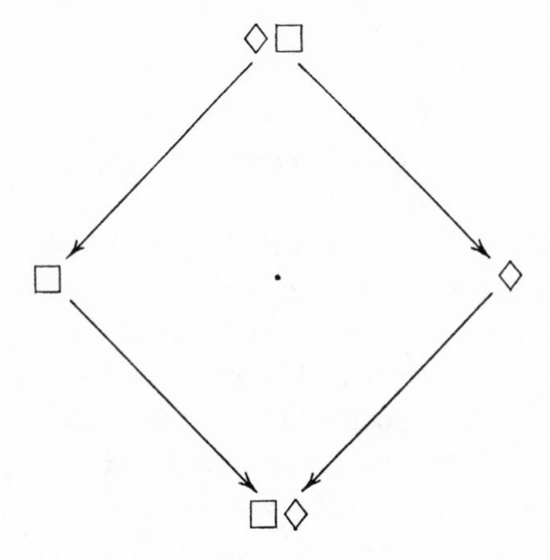

Figure 4.6. Modalities in normal *KB4*-systems.

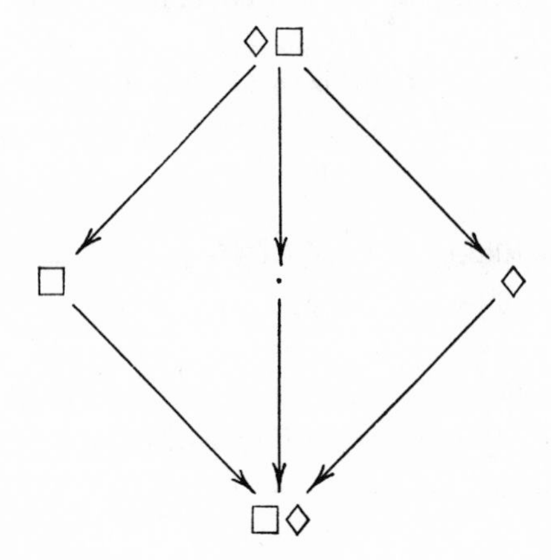

THEOREM 4.27. *Every normal KD45-system has at most six distinct modalities, viz.* $\cdot$, $\Box$, $\Diamond$, *and their negations, with implications among the affirmative three as diagramed in figure 4.7.*

Proof. By the proofs for theorems 4.24 and 4.25 every normal *KD45*-system has the reduction laws

$$\Box A \leftrightarrow \Box\Box A \qquad \Diamond A \leftrightarrow \Diamond\Diamond A$$
$$\Box A \leftrightarrow \Diamond\Box A \qquad \Diamond A \leftrightarrow \Box\Diamond A$$

as well as the implication $\Box A \rightarrow \Diamond A$. Thus a system of this kind has at most the distinct modalities and at least the implication laid out in figure 4.7.

THEOREM 4.28. *Every normal KT5-system has at most six distinct modalities, viz.* $\cdot$, $\Box$, $\Diamond$, *and their negations, with implications among the affirmative three as diagramed in figure 4.8.*

Proof. By theorem 4.21 every normal *KT5*-system contains D and 4. So by theorem 4.27 every such system contains at most the distinct modalities $\cdot$, $\Box$, $\Diamond$, and their negations. In virtue of T and T$\Diamond$ we may add arrows to and from $\cdot$ in figure 4.7. The result is figure 4.8. Alternatively, we may note that a normal *KT5*-system contains B and 4 (theorem 4.21) and so has all the reduction laws and implications had jointly by normal *KT4*- and normal *KB4*-systems. Applying theorems 4.22 and 4.26 – or, combining figures 4.2 and 4.6 – we arrive at the desired result.

Thus, as we said at the beginning of the section, both *KD45* and *KT5* (*S5*) have at most the six modalities $\cdot$, $\Box$, $\Diamond$, $\neg$, $\neg\Box$, and $\neg\Diamond$. The moral about modalities in systems of this sort is that *iteration is vacuous*: any sequence of $\Box$s and $\Diamond$s can always be reduced to its innermost term.

Figure 4.7. Modalities in normal *KD45*-systems.

Figure 4.8. Modalities in normal *KT5*-systems.

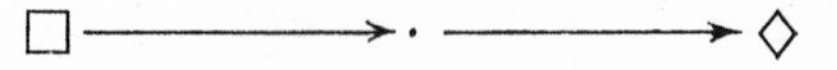

EXERCISES

4.63. Show that the modalities $\neg\Box\Diamond\neg\Box\Diamond\neg\Box$ and $\neg\Box$ are equivalent in *S5* (*KT5*).

4.64. Using the reduction laws mentioned in the proof of theorem 4.22, show that every modality in a normal *KT4*-system is reducible to one of $\cdot$, $\Box$, $\Diamond$, $\Box\Diamond$, $\Diamond\Box$, $\Box\Diamond\Box$, $\Diamond\Box\Diamond$, or the negation of one of these.

4.65. Prove that normal *KD4!*- and normal *KD4U*-systems have at most fourteen distinct modalities, viz. $\cdot$, $\Box$, $\Diamond$, $\Box\Diamond$, $\Diamond\Box$, $\Box\Diamond\Box$, $\Diamond\Box\Diamond$, and their negations, with implications among the affirmative seven as diagramed in figure 4.9. (See exercises 4.33(*b*) and 4.43(*b*)).

4.66. Prove that normal *KT4G*- and normal *KT4H*$^+$-systems have at most ten distinct modalities, viz. $\cdot$, $\Box$, $\Diamond$, $\Box\Diamond$, $\Diamond\Box$, and their negations, with implications among the affirmative five as diagramed in figure 4.10. (See exercises 4.43, 4.55, and 4.56.)

4.67. Prove that normal *KD4H*$^+$-systems have at most ten distinct modalities, viz. $\cdot$, $\Box$, $\Diamond$, $\Box\Diamond$, $\Diamond\Box$, and their negations, with implications among the affirmative five as diagramed in figure 4.11. (See exercises 4.43 and 4.56.)

4.68. Prove that normal *KT4G*$_c$-systems have at most ten distinct modalities, viz. $\cdot$, $\Box$, $\Diamond$, $\Box\Diamond$, $\Diamond\Box$, and their negations, with implications among the affirmative five as diagramed in figure 4.12. (See exercises 4.43 and 4.58.) Note that the modalities $\Box\Diamond\Box$ and $\Diamond\Box\Diamond$ are equi-

Figure 4.9. Modalities in normal *KD4!*- and normal *KD4U*-systems.

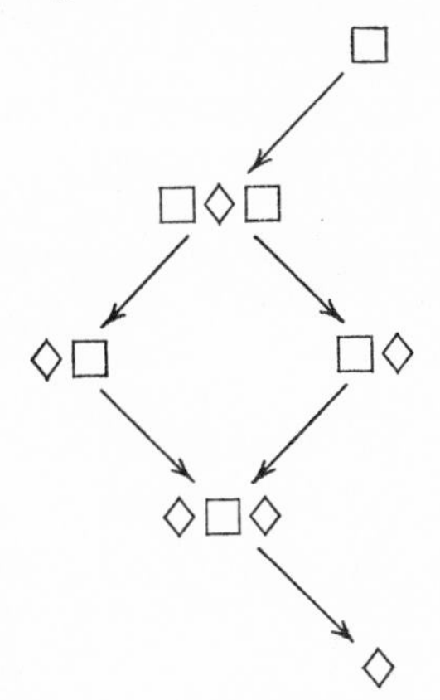

Figure 4.10. Modalities in normal *KT4G*- and normal $KT4H^{+}$-systems.

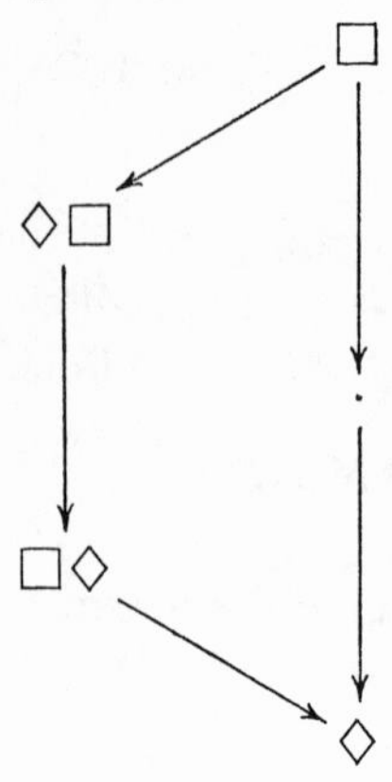

Figure 4.11. Modalities in normal $KD4H^{+}$-systems.

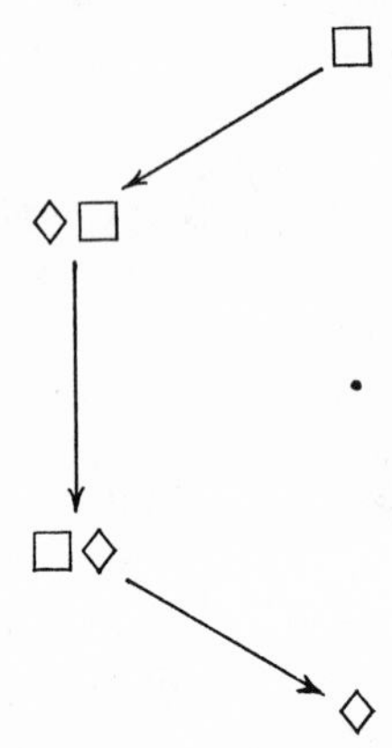

Figure 4.12. Modalities in normal $KT4G_c$-systems.

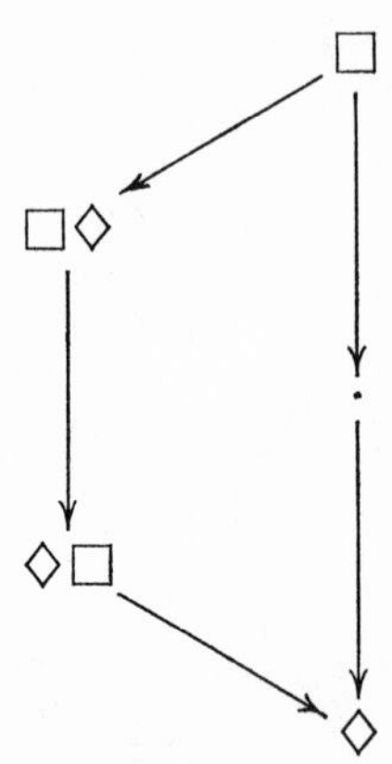

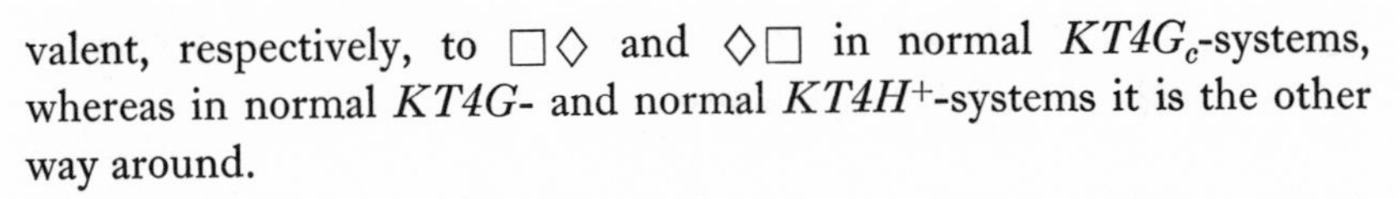

valent, respectively, to $\Box\Diamond$ and $\Diamond\Box$ in normal $KT4G_c$-systems, whereas in normal $KT4G$- and normal $KT4H^+$-systems it is the other way around.

4.69. Prove that normal $K4G_c$-systems have at most ten distinct modalities, viz. $\cdot$, $\Box$, $\Diamond$, $\Box\Diamond$, $\Diamond\Box$, and their negations, with implications among the affirmative five as diagramed in figure 4.13. (See exercises 4.43 and 4.58.) Compare the reduction laws here for $\Box\Diamond\Box$ and $\Diamond\Box\Diamond$ with those in normal $KD4H^+$-systems.

4.70. Prove that normal $K5!$-systems have at most six distinct modalities, viz. $\cdot$, $\Box$, $\Diamond$, and their negations, with implications among the affirmative three as in the diagram in figure 4.7 for normal $KD45$-systems. (See exercise 4.47.)

4.71. Prove that normal extensions of $KD!B$, $KD!B_c$, $KD!4$, $KD!4_c$, $KD!5$, and KD_c5_c have at most four distinct modalities, viz. $\cdot$, one of $\Box$ and $\Diamond$ and their negations. (See exercises 4.33 (*d*), 4.38, 4.52, and 4.53.)

4.72. Prove that normal extensions of $KT!$ (and hence KDT_c and KD_cT) have at most two distinct modalities, viz. $\cdot$ and $\neg$. (See exercises 4.52 and 4.53.)

4.73. Describe a normal modal logic that has just one (distinct!) modality.

4.5. Maximal sets in normal systems

We bring this chapter to a close by stating and proving a few theorems about maximal sets of sentences (section 2.6) in normal systems of modal

Figure 4.13. Modalities in normal $K4G_c$-systems.

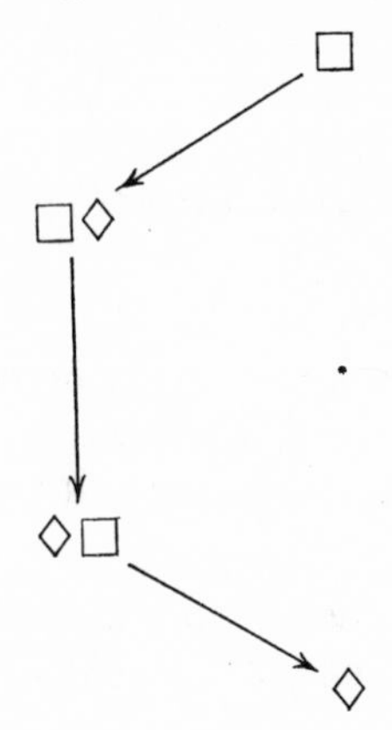

logic. The importance of these theorems will become apparent in chapter 5, where they are useful in several proofs.

THEOREM 4.29. *Let* Γ *and* Δ *be maximal sets of sentences in a normal system* Σ. *Then:*

(1) $\{A: \Box A \in \Gamma\} \subseteq \Delta$ *iff* $\{\Diamond A: A \in \Delta\} \subseteq \Gamma$;

and more generally, for any $k \geqslant 0$,

(2) $\{A: \Box^k A \in \Gamma\} \subseteq \Delta$ *iff* $\{\Diamond^k A: A \in \Delta\} \subseteq \Gamma$.

Proof. Let Γ and Δ be Σ-maximal sets of sentences, and suppose that Σ is normal. We prove (1) only; (2) is a simple generalization. For left-to-right, assume that $\{A: \Box A \in \Gamma\} \subseteq \Delta$ and that A is in Δ. We wish to show that Γ contains $\Diamond A$. By theorem 2.18(5), $\neg A$ is not in Δ. So $\Box\neg A$ is not in Γ, which means that $\neg\Box\neg A$ is a member of Γ. Because Σ is normal, Γ contains Df$\Diamond$, and so – by theorem 2.18(9) – $\Diamond A$ is also in Γ. For right-to-left, assume that $\{\Diamond A: A \in \Delta\} \subseteq \Gamma$ and that $\Box A$ is in Γ. Since Γ contains Df$\Box$, $\neg\Diamond\neg A$ is in Γ, and so $\Diamond\neg A$ is not. Hence $\neg A$ is not a member of Δ, which means that A is, which is what we wished to show.

Note that when $k = 0$ the preceding theorem means that for Σ-maximal sets Γ and Δ, $\Gamma \subseteq \Delta$ if and only if $\Delta \subseteq \Gamma$, and so $\Gamma = \Delta$ just in case $\Gamma \subseteq \Delta$.

The next theorem may be regarded as an extension, for normal modal logics, of theorem 2.18 (particularly parts (3)–(9)).

THEOREM 4.30. *Let* Γ *be a maximal set of sentences in a normal system* Σ. *Then:*

(1) $\Box A \in \Gamma$ *iff for every* $\text{Max}_\Sigma \Delta$ *such that* $\{A: \Box A \in \Gamma\} \subseteq \Delta$, $A \in \Delta$.

(2) $\Diamond A \in \Gamma$ *iff for some* $\text{Max}_\Sigma \Delta$ *such that* $\{\Diamond A: A \in \Delta\} \subseteq \Gamma$, $A \in \Delta$.

Proof. Suppose Σ is a normal system and that Γ is a Σ-maximal set of sentences.

For (1). From left to right the theorem is trivial, for if $\Box A \in \Gamma$ and $\{A: \Box A \in \Gamma\} \subseteq \Delta$, then $A \in \Delta$.

The reverse is thus the interesting direction. Suppose that $A \in \Delta$, for every Σ-maximal set Δ such that $\{A: \Box A \in \Gamma\} \subseteq \Delta$, i.e. that A belongs to every Σ-maximal extension of the set $\{A: \Box A \in \Gamma\}$. By a corollary to Lindenbaum's lemma, theorem 2.20(1), this means that A is Σ-deducible from this set of sentences; i.e.

$$\{A: \Box A \in \Gamma\} \vdash_{\Sigma} A.$$

This in turn means that there are sentences $A_1, \ldots, A_n$ $(n \geqslant 0)$ in the set $\{A: \Box A \in \Gamma\}$ that are such that

$$\vdash_{\Sigma}(A_1 \wedge \ldots \wedge A_n) \rightarrow A.$$

Because Σ is normal, we may infer by RK that

$$\vdash_{\Sigma}(\Box A_1 \wedge \ldots \wedge \Box A_n) \rightarrow \Box A.$$

But Γ contains each of $\Box A_1, \ldots, \Box A_n$, so $\Box A$ is Σ-deducible from Γ; i.e.

$$\Gamma \vdash_{\Sigma} \Box A.$$

By theorem 2.18(1) this means that

$$\Box A \in \Gamma,$$

which was to be proved.

For (2):

$\Diamond A \in \Gamma$ iff $\neg\Box\neg A \in \Gamma$
– Df$\Diamond$ and theorem 2.18(9);
iff $\Box\neg A \notin \Gamma$
– theorem 2.18(5);
iff for some $\text{Max}_{\Sigma}\Delta$ such that $\{A: \Box A \in \Gamma\} \subseteq \Delta$, $\neg A \notin \Delta$
– part (1);
iff for some $\text{Max}_{\Sigma}\Delta$ such that $\{\Diamond A: A \in \Delta\} \subseteq \Gamma$, $\neg A \notin \Delta$
– theorem 4.29;
iff for some $\text{Max}_{\Sigma}\Delta$ such that $\{\Diamond A: A \in \Delta\} \subseteq \Gamma$, $A \in \Delta$
– theorem 2.18(5).

THEOREM 4.31. *Let Γ and Δ be maximal sets of sentences in a normal system Σ. Then for every $k \geqslant 0$:*

$\{A: \Box^{k+1}A \in \Gamma\} \subseteq \Delta$ *iff for some* $\text{Max}_{\Sigma}E$,
$\{A: \Box A \in \Gamma\} \subseteq E$ *and* $\{A: \Box^{k}A \in E\} \subseteq \Delta$.

Proof. We assume that Γ and Δ are maximal sets of sentences in a normal modal logic Σ. From right to left the proof is straightforward, and we leave it as an exercise. For left-to-right we suppose that $\{A: \Box^{k+1}A \in \Gamma\} \subseteq \Delta$, to show that there exists a Σ-maximal set E such that

$$\{A : \Box A \in \Gamma\} \subseteq E \text{ and } \{A : \Box^k A \in E\} \subseteq \Delta,$$

i.e. by theorem 4.29, such that

$$\{A : \Box A \in \Gamma\} \subseteq E \text{ and } \{\Diamond^k A : A \in \Delta\} \subseteq E.$$

In other words, we wish to show that there is a Σ-maximal set of sentences E that includes the set

$$\{A : \Box A \in \Gamma\} \cup \{\Diamond^k A : A \in \Delta\}.$$

By Lindenbaum's lemma (theorem 2.19) this is equivalent to showing that this union is Σ-consistent.

Let us suppose otherwise, and argue to a contradiction. If the set is Σ-inconsistent, then $\bot$ is Σ-deducible from it, and this in turn means that for some $m, n \geqslant 0$ there are sentences $B_1, \ldots, B_m$ in $\{A : \Box A \in \Gamma\}$ and sentences $\Diamond^k C_1, \ldots, \Diamond^k C_n$ in $\{\Diamond^k A : A \in \Delta\}$ such that

$$\vdash_\Sigma (B_1 \wedge \ldots \wedge B_m \wedge \Diamond^k C_1 \wedge \ldots \wedge \Diamond^k C_n) \to \bot.$$

By a rule of inference present in every normal modal logic (see exercise 4.11) we infer that

$$\vdash_\Sigma (\Box B_1 \wedge \ldots \wedge \Box B_m) \to \Box^{k+1} \neg (C_1 \wedge \ldots \wedge C_n).$$

Because each of $\Box B_1, \ldots, \Box B_m$ is in Γ, the consequent $\Box^{k+1}\neg(C_1 \wedge \ldots \wedge C_n)$ is Σ-deducible from Γ, and so belongs to Γ. By our original assumption, then, Δ contains $\neg(C_1 \wedge \ldots \wedge C_n)$. But Δ contains $C_1 \wedge \ldots \wedge C_n$ too, since this is the consequent of the theorem $(C_1 \wedge \ldots \wedge C_n) \to (C_1 \wedge \ldots \wedge C_n)$, for which Δ contains each conjunct of the antecedent. So Δ is Σ-inconsistent, which is a contradiction, and we may consider the proof complete.

EXERCISES

4.74. Prove part (2) of theorem 4.29.

4.75. Give the proof for right-to-left in theorem 4.31.

4.76. Prove the following generalizations of theorem 4.30, for any $k > 0$, where Γ is a maximal set of sentences in a normal system Σ.

(*a*) $\Box^k A \in \Gamma$ iff for every $\text{Max}_\Sigma \Delta$ such that $\{A : \Box^k A \in \Gamma\} \subseteq \Delta$, $A \in \Delta$.

(*b*) $\Diamond^k A \in \Gamma$ iff for some $\text{Max}_\Sigma \Delta$ such that $\{\Diamond^k A : A \in \Delta\} \subseteq \Gamma$, $A \in \Delta$.

4.77. Let Σ be a normal system, and define the relation R on the set of Σ-maximal sets of sentences by:

$$\Gamma R \Delta \text{ iff } \{A : \Box A \in \Gamma\} \subseteq \Delta.$$

(Thus, by theorem 4.29, $\Gamma R \Delta$ if and only if $\{\Diamond A : A \in \Delta\} \subseteq \Gamma$.) Prove:

(*a*) R is serial if Σ contains D.

(*b*) R is reflexive if Σ contains T.

(*c*) R is symmetric if Σ contains B.

(*d*) R is transitive if Σ contains 4.

(*e*) R is euclidean if Σ contains 5.

(See section 3.2 for these properties of R.)

5

DETERMINATION AND DECIDABILITY FOR NORMAL SYSTEMS

In this chapter we present and prove a number of determination theorems for normal modal logics with respect to classes of standard models. Section 5.1 contains the basic theorem for the soundness of such systems and a proof that the fifteen systems on the diagram in figure 4.1 are all distinct. In section 5.2 we return briefly to the topic of modalities in normal systems. In section 5.3 we define the idea of a canonical standard model for a normal system and prove some fundamental theorems about completeness. Section 5.4 contains determination theorems for the logics in figure 4.1, including a theorem to the effect that the system *S5* is determined by the class of models of the sort described in chapter 1. In section 5.5 we generalize the ideas in sections 5.1 and 5.3 to obtain a very large class of determination results in one fell swoop, by proving that the system $KG^{k,l,m,n}$ is determined by the class of k,l,m,n-incestual models. Finally, in section 5.6 we prove the decidability of the fifteen systems in figure 4.1.

5.1. Soundness

The following theorem provides the basis for proofs of soundness for normal modal logics, with respect to classes of standard models.

THEOREM 5.1. *Let* $S_1, \ldots, S_n$ *be schemas valid respectively in classes of standard models* $\mathsf{C}_1, \ldots, \mathsf{C}_n$. *Then the system of modal logic* $KS_1 \ldots S_n$ *is sound with respect to the class* $\mathsf{C}_1 \cap \ldots \cap \mathsf{C}_n$.

Proof. If $S_1, \ldots, S_n$ are valid respectively in classes of standard models $\mathsf{C}_1, \ldots, \mathsf{C}_n$, then each is valid in the intersection of these classes, $\mathsf{C}_1 \cap \ldots \cap \mathsf{C}_n$. It remains to be shown that this class validates Df$\Diamond$, and that validity in this class is preserved by the rules of inference RK and RPL. This follows from theorems 2.8 and 3.3.

Soundness theorems for the smallest normal system, K, are immediate corollaries of theorem 5.1: K is sound with respect to every class of standard models, including the whole class; for this system is axiomatized by Df$\Diamond$, RK, and RPL. The role of the theorem in connection with normal extensions of K should moreover be apparent. For example, by theorem 3.5 the schemas D, T, B, 4, and 5 are valid respectively in classes of serial, reflexive, symmetric, transitive, and euclidean standard models. So it follows by theorem 5.1 that the systems *KD*, *KT*, *KB*, *K4*, and *K5* are sound with respect to these classes of models. For another example, the system *S4* (*KT4*) is sound with respect to classes of quasi-ordered standard models, since T and 4 are valid in classes, respectively, of reflexive and transitive standard models. In general, each of the fourteen normal systems beyond K in figure 4.1 is sound with respect to classes of models of the kinds indicated in figure 5.1.

Some more soundness results appear in the exercises. We shall not state any formally, however.

By means of the theorems in section 4.3 we established there the inclusions among the fifteen systems of modal logic diagramed in figure 4.1. We can employ theorem 5.1 to demonstrate the distinctness of these systems, and hence the properness of the inclusions. This is an appropriate place to do so.

THEOREM 5.2. *The fifteen normal systems in figure 4.1 are all distinct.*

Proof. In general, to show that a system Σ is distinct from a system Σ' it is sufficient to exhibit a model of Σ that falsifies a theorem of Σ', or vice versa. For example, to show that $KD \neq KT$, and so $KD \subset KT$, it is enough to describe a serial standard model (of *KD*, by theorem 5.1) that falsifies an instance of the theorem T of *KT*. This is our method.

Consider the following six standard models $\mathcal{M} = \langle W, R, P\rangle$.

(1) $W = \{\alpha, \beta, \gamma\}$ (distinct); $R = \{\langle\alpha, \alpha\rangle, \langle\beta, \beta\rangle, \langle\gamma, \gamma\rangle, \langle\alpha, \beta\rangle, \langle\beta, \alpha\rangle, \langle\beta, \gamma\rangle, \langle\gamma, \beta\rangle\}$; $P_0 = \{\alpha, \beta\}$ and $P_n = \{\alpha\}$ for $n > 0$.

(2) $W = \{\alpha, \beta\}$ (distinct); $R = \{\langle\alpha, \alpha\rangle, \langle\beta, \beta\rangle, \langle\alpha, \beta\rangle\}$; $P_n = \{\alpha\}$ for $n \geqslant 0$.

(3) $W = \{\alpha, \beta\}$ (distinct); $R = \{\langle\alpha, \beta\rangle, \langle\beta, \beta\rangle\}$; $P_n = \{\beta\}$ for $n \geqslant 0$.

(4) $W = \{\alpha\}$; $R = \varnothing$; $P_n = \varnothing$ for $n \geqslant 0$.

(5) $W = \{\alpha, \beta, \gamma\}$ (distinct); $R = \{\langle\alpha, \beta\rangle, \langle\beta, \beta\rangle, \langle\beta, \gamma\rangle, \langle\gamma, \gamma\rangle\}$; $P_n = \{\beta\}$ for $n \geqslant 0$.

(6) $W = \{\alpha, \beta\}$ (distinct); $R = \{\langle\alpha, \beta\rangle, \langle\beta, \alpha\rangle\}$; $P_n = \{\beta\}$ for $n \geqslant 0$.

Figure 5.1

	serial	reflexive	symmetric	transitive	euclidean
KD	●				
KT		●			
KB			●		
K4				●	
K5					●
KDB	●		●		
KD4	●			●	
KD5	●				●
K45				●	●
KD45	●			●	●
KB4			●	●	
			●		●
KTB		●	●		
KT4		●		●	
KT5		●			●
		●	●	●	
	●		●	●	
	●		●		●

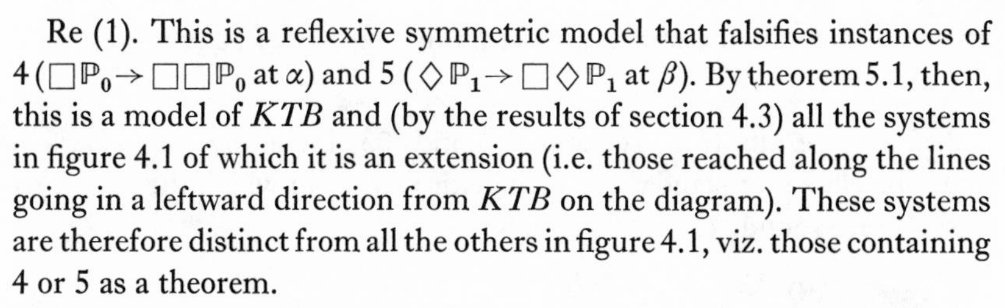

Re (1). This is a reflexive symmetric model that falsifies instances of 4 ($\Box\mathbb{P}_0 \to \Box\Box\mathbb{P}_0$ at α) and 5 ($\Diamond\mathbb{P}_1 \to \Box\Diamond\mathbb{P}_1$ at β). By theorem 5.1, then, this is a model of *KTB* and (by the results of section 4.3) all the systems in figure 4.1 of which it is an extension (i.e. those reached along the lines going in a leftward direction from *KTB* on the diagram). These systems are therefore distinct from all the others in figure 4.1, viz. those containing 4 or 5 as a theorem.

Re (2). This is a reflexive transitive model that falsifies instances of B and 5 ($\mathbb{P}_0 \to \Box\Diamond\mathbb{P}_0$ and $\Diamond\mathbb{P}_0 \to \Box\Diamond\mathbb{P}_0$ at α). So it is a model of *KT4* and all the systems in figure 4.1 of which it is an extension. This means that these systems are distinct from all the others in the figure – those containing B or 5 as a theorem.

Re (3). This is a serial transitive euclidean model that falsifies instances of T and B ($\Box\mathbb{P}_0 \to \mathbb{P}_0$ and $\neg\mathbb{P}_0 \to \Box\Diamond\neg\mathbb{P}_0$ at α). Thus this is a model of *KD45* and all the systems in figure 4.1 that this system includes. So these systems are distinct from all the others in the figure – those containing T or B as a theorem.

Re (4). This model is symmetric and transitive (hence also euclidean; see exercise 3.58(*e*)), and it falsifies an instance of D ($\Box\mathbb{P}_0 \to \Diamond\mathbb{P}_0$). Thus it is a model of *KB4* and its inclusions in figure 4.1. These systems are therefore distinct from the rest on the diagram, since these all contain D.

Re (5). This model is serial and euclidean, and it falsifies an instance of 4 ($\Box\mathbb{P}_0 \to \Box\Box\mathbb{P}_0$ at α). Thus it models *KD5* and the systems in figure 4.1 that *KD5* extends. So these systems are all distinct from the others in the figure, since these all contain 4.

Re (6). This is a serial symmetric model that falsifies an instance of T ($\Box\mathbb{P}_0 \to \mathbb{P}_0$ at α). So it is a model of *KDB* and all the systems in figure 4.1 that this system extends. Hence these systems are distinct from the rest on the diagram, since these systems all contain T.

The foregoing remarks suffice to establish the distinctness of all the systems in figure 4.1, though there is much redundancy. It is left for the reader (as an exercise) to work through the details to check that the six models have the properties alleged and to check that these properties ensure the advertised distinctness.

EXERCISES

5.1. Check that models (1)–(6) in the proof for theorem 5.2 (*a*) have the properties stated, (*b*) falsify the sentences mentioned, and (thus) (*c*) ensure the distinctness of the systems in figure 4.1. (Compare exercise 3.22.)

5.2. Using results from sections 3.2 and 3.3, including the exercises, describe classes of standard models with respect to which the following systems are sound.

(*a*) *KG*
(*b*) KD_c
(*c*) KT_c
(*d*) KB_c
(*e*) $K4_c$
(*f*) $K5_c$
(*g*) *KP*
(*h*) $K\bar{P}$
(*i*) $KG^{k,l,m,n}$
(*j*) *KU*
(*k*) *K* plus $\Diamond\top \to (\Box A \to A)$
(*l*) *K* plus $\Box^j(\Diamond^k \Box^l A \to \Box^m \Diamond^n A)$
(*m*) *K* plus $\Diamond^j \top \to (\Diamond^k \Box^l A \to \Box^m \Diamond^n A)$

5.3. Consider the following conditions on a standard model $\mathcal{M} = \langle W, R, P \rangle$.

(h^{++}) if $\alpha R \beta$ and $\alpha R \gamma$ and $\beta R \delta$ and $\gamma R \epsilon$, then $\beta = \gamma$ or $\beta R \epsilon$ or $\gamma R \delta$ or $\delta = \epsilon$

(h^{+}) if $\alpha R \beta$ and $\alpha R \gamma$, then $\beta R \gamma$ or $\gamma R \beta$

(h) if $\alpha R \beta$ and $\alpha R \gamma$, then $\beta = \gamma$ or $\beta R \gamma$ or $\gamma R \beta$

Prove that the systems KH^{++}, KH^{+}, and *KH* are sound, respectively, relative to classes of standard models satisfying (h^{++}), (h^{+}), and (h) (see exercise 3.47). Using results in exercise 4.56, show that KL^{++}, KL^{+}, and *KL* are likewise sound with respect to classes of models satisfying (h^{++}), (h^{+}), and (h).

Prove also: (1) if $\mathcal{M}$ is reflexive, then it satisfies one of (h^{++}), (h^{+}), and (h) just in case it satisfies the others; (2) if $\mathcal{M}$ is euclidean, then it satisfies all three conditions.

5.4. Consider the following conditions on a standard model $\mathcal{M} = \langle W, R, P \rangle$.

convergence: if $\alpha \neq \beta$, then for some γ in $\mathcal{M}$, $\alpha R \gamma$ and $\beta R \gamma$

strong convergence: for some γ in $\mathcal{M}$, $\alpha R \gamma$ and $\beta R \gamma$

connectedness: $\alpha = \beta$ or $\alpha R \beta$ or $\beta R \alpha$

strong connectedness: $\alpha R \beta$ or $\beta R \alpha$

Prove:

(*a*) A reflexive relation is strongly convergent if and only if it is convergent.

(*b*) The system *S4.2* (see exercise 4.55) is sound with respect to classes of reflexive transitive standard models that are convergent (and hence strongly convergent).

(*c*) A reflexive relation is strongly connected if and only if it is connected.

(*d*) The system *S4.3* (see exercise 4.56) is sound with respect to classes of reflexive transitive standard models that are connected (and hence strongly connected).

5.5. Consider the following conditions on a standard model $\mathcal{M} = \langle W, R, P \rangle$.

functionality: (for every α in $\mathcal{M}$) there is exactly one β in $\mathcal{M}$ such that $\alpha R \beta$

identity: $\alpha R \beta$ if and only if $\alpha = \beta$

Prove that *KD!* is sound with respect to classes of standard models in which the relation is functional, and that *KT!* is sound with respect to classes of standard models in which the relation is identity (see exercise 3.54).

5.6. Consider the following condition on a standard model $\mathcal{M} = \langle W, R, P \rangle$.

(for every α in $\mathcal{M}$) there is a β in $\mathcal{M}$ such that both $\alpha R \beta$ and for every γ and δ in $\mathcal{M}$, if $\beta R \gamma$ and $\beta R \delta$, then $\gamma = \delta$

Prove that the system *S4.1* of exercise 4.58 is sound with respect to classes of reflexive transitive standard models that satisfy this condition.

5.7. Identify classes of standard models with respect to which the following systems are sound.

(*a*) *KD4!* (*b*) *KD4U* (*c*) $KD4H^+$ (*d*) $K4G_c$

5.8. Prove the distinctness of the modal logics *KG*, KD_c, KT_c, $K4_c$, KB_c, $K5_c$, *KP*, $K\bar{P}$, *KU*, *K* plus $\Diamond\top \to (\Box A \to A)$, KH^{++}, KH^+, *KH*, *S4.2*, *S4.3*, *KD!*, *KT!*, *S4.1*, *KD4!*, *KD4U*, $KD4H^+$, and $K4G_c$ (mentioned in exercises 5.2–5.7), and prove that each is distinct from all the systems in figure 4.1.

5.9. Prove, for any $k > 1$:

(*a*) $KT \neq KT^k$.

(*b*) $KB \neq KB^k$.

(*c*) $K4 \neq K4^k$.

(*d*) $K5 \neq K5^k$.

(See the penultimate paragraph of section 4.3 for the point of this.)

5.10. Consider again the following rule of inference.

$$\frac{\Box A}{A}$$

Using exercise 3.63 and the soundness results in figure 5.1, show that this rule does not hold in the systems *KB*, *K5*, *KD5*, *K45*, *KD45*, and *KB4*. (In this connection the reader may wish to prove that $\Box(A \rightarrow \Diamond\Diamond A)$ is a theorem of any normal *KB*-system.)

5.11. Using theorem 5.2 and exercise 4.32, prove that the systems *K*, *KB*, *K4*, *K5*, *K45*, and *KB4* have no theorems of the form $\Diamond A$.

5.12. Consider again the following rule of inference.

$$\frac{\Diamond A}{A}$$

By the results of exercises 4.13 and 4.61 this rule holds in *K* and *KD*, and by the results of the preceding exercise it holds also in *KB*, *K4*, *K5*, *K45*, and *KB4* (vacuously, since none of these systems has theorems of the form $\Diamond A$). The remaining systems in figure 4.1 – *KT*, *KDB*, *KD4*, *KD5*, *KD45*, *KTB*, *KT4*, and *KT5* – do not have this rule. This is a consequence (as the reader should argue) of the soundness results in figure 5.1 and the following, which are for the reader to prove.

(*a*) $\Diamond(A \rightarrow \Box A)$ is a theorem of any normal *KT*-system.

(*b*) $A \rightarrow \Box A$ is not a theorem of *KT*, *KTB*, *KT4*, or *KT5*.

(*c*) $\Diamond(\Box A \rightarrow A)$ is a theorem of any normal *KD4*- or normal *KD5*-system.

(*d*) $\Box A \rightarrow A$ is not a theorem of *KD4*, *KD5*, or *KD45*.

(*e*) $\Diamond\Diamond(A \rightarrow (\Box A \rightarrow \Box\Box A))$ is a theorem of any normal *KDB*-system.

(*f*) $\Diamond(A \rightarrow (\Box A \rightarrow \Box\Box A))$ is not a theorem of *KDB*.

5.13. Prove:

(*a*) A relation is functional if and only if it is serial and partially functional.

(*b*) A relation is identity if and only if it is reflexive and vacuous.

(*c*) If a relation is functional, then it satisfies one of transitivity, density, euclideanness, and condition (*b*) in exercise 3.39 if and only if it satisfies the others.

(See exercise 3.58.)

5.2. Postscript on modalities

In section 4.4 we established maximums for the numbers of distinct modalities in normal extensions of the systems *KT4* and *K5* (fourteen each), *KD5*, *K45*, and *KB4* (ten each), and *KD45* and *KT5* (six each). Figures 4.2–4.8 show the identities of these modalities and the implications among them. Now we can prove that for the smallest of each of the seven types of systems these are exactly the distinct modalities. Formally:

THEOREM 5.3. *The systems KT4, K5, KD5, K45, KB4, KD45, and KT5 have exactly the distinct modalities diagramed in figures 4.2–4.8, respectively.*

Proof. Theorems 4.22–4.28 affirm that the systems in question have at most the advertised distinct modalities. For minimality it needs to be demonstrated that none of the modalities in a diagram for a system is equivalent to any of the others, i.e. that the logic has no reduction laws $\phi A \leftrightarrow \psi A$, where ϕ and ψ are (different) modalities in the diagram. To prove this it suffices to show that $\phi A \leftrightarrow \psi A$ is not valid in some class of models with respect to which the logic is sound. Thus, for example, to prove the theorem for *KT4* it must be shown, inter alia, that the modalities $\Diamond\Box$ and $\Box\Diamond$ are distinct, and for this it will do (by theorems 3.5 and 5.1) to exhibit a reflexive transitive countermodel for an instance of $\Diamond\Box A \leftrightarrow \Box\Diamond A$.

We invite the reader – as an exercise – to work out the details of the requisite proofs. With a little ingenuity and organization the number of cases can be reduced. For example, a reflexive transitive countermodel for $\Diamond\Box A \leftrightarrow \Box\Diamond\Box A$ also falsifies an instance of $A \leftrightarrow \Box A$ and so at one stroke establishes distinctness in *KT4* for the pairs $\cdot$, $\Box$ and $\Diamond\Box$, $\Box\Diamond\Box$.

Theorem 5.3 tells us about modalities in seven of the fifteen systems in figure 4.1. What about modalities in the remaining eight?

THEOREM 5.4. *Each of the systems K, KD, KT, KB, K4, KDB, KD4, and KTB has infinitely many distinct modalities.*

Proof. It is enough to show that for any $n > m \geqslant 0$ the schema

$$4^{m,n}!. \quad \Box^m A \leftrightarrow \Box^n A$$

is not a theorem of any of the systems in question. For this will mean that none of these systems contains any reduction laws for modalities of the form $\Box^n$, and so each such modality is distinct from all the others.

Let m and n be natural numbers such that $m < n$, and consider a standard model $\mathcal{M} = \langle W, R, P \rangle$ in which $W = \{\alpha_0, \ldots, \alpha_m, \ldots, \alpha_n\}$ (all distinct), $P_0 = \{\alpha_0, \ldots, \alpha_m\}$, and R relates worlds that are identical or adjacent (i.e. $\alpha_i R \alpha_j$ if and only if $i = j$ or $i = j+1$ or $j = i+1$). Then, as the reader should check, $\mathcal{M}$ is a reflexive symmetric countermodel for $\Box^m A \rightarrow \Box^n A$ (consider at α_0 the instance in which $A = \mathbb{P}_0$). Since this countermodel can be constructed for any m and n, it follows that for no $n > m \geqslant 0$ is the schema $4^{m,n}!$ valid in the class of reflexive symmetric models. Thus by theorems 3.5 and 5.1 this schema is never a theorem of *KTB*, nor of any of the systems in figure 4.1 that it (properly) extends – *K*, *KD*, *KT*, *KB*, and *KDB*. Therefore, each of these modal logics has infinitely many distinct modalities.

We leave it an exercise for the reader to prove that for every $n > m \geqslant 0$ there is a serial transitive countermodel for an instance of $\Box^n A \rightarrow \Box^m A$, which is all that is needed to show that the number of distinct modalities is infinite in *KD4* and *K4* (as well as *K* and *KD* again). With this the proof of the theorem will be complete.

EXERCISES

5.14. Give the details of the proof for theorem 5.3.

5.15. Verify that the model $\mathcal{M}$ in the proof of theorem 5.4 is reflexive and symmetric and falsifies $\Box^m \mathbb{P}_0 \rightarrow \Box^n \mathbb{P}_0$. Then complete the proof of the theorem by showing that for every $n > m \geqslant 0$ there is a serial transitive standard model that rejects an instance of $\Box^n A \rightarrow \Box^m A$.

5.16. Prove:

(*a*) *KD4!* and *KD4U* have exactly the distinct modalities diagramed in figure 4.9 (see exercise 4.65).

(*b*) *S4.2* (*KT4G*) and *S4.3* (*KT4H*$^+$) have exactly the distinct modalities diagramed in figure 4.10 (see exercise 4.66).

(*c*) *KD4H*$^+$ has exactly the distinct modalities diagramed in figure 4.11 (see exercise 4.67).

(*d*) *S4.1* (*KT4G*$_c$) has exactly the distinct modalities diagramed in figure 4.12 (see exercise 4.68).

(*e*) *K4G*$_c$ has exactly the distinct modalities diagramed in figure 4.13 (see exercise 4.69).

(*f*) *K5!* has exactly the distinct modalities diagramed in figure 4.7 (see exercise 4.70).

(*g*) *KD!B*, *KD!B*$_c$, *KD!4*, *KD!4*$_c$, *KD!5*, and *KD*$_c$*5*$_c$ have exactly the distinct modalities stated in exercise 4.71.

(*h*) *KT!*, *KDT*$_c$, and *KD*$_c$*T* have exactly the distinct modalities stated in exercise 4.72.

5.17. Using exercise 4.34 prove that the system *KG* has infinitely many distinct modalities.

5.18. Prove that *KD4H* has infinitely many distinct modalities (see exercise 5.3(*c*) and, perhaps, the countermodel to $\Box^n A \to \Box^m A$ in the solution to exercise 5.15). Then, using exercise 4.56(*n*, *q*), argue that there are infinitely many distinct modalities in *K* and all the systems that result from adding D, 4, G, H^{++}, and H in various combinations to *K*.

5.19. Prove that the systems *KU* and *K* plus $\Diamond\top \to (\Box A \to A)$ have infinitely many distinct modalities. Then argue that this is true of *K4*$_c$.

5.20. Prove that the systems *KD!* and *K4!* have infinitely many distinct modalities.

5.21. For each of the modal logics *KH*$^+$, *KDH*$^+$, *K4H*$^+$, *K4U*, *K4UG*, and *KG!* (where G! = $\Diamond\Box A \leftrightarrow \Box\Diamond A$) prove or disprove that it has infinitely many distinct modalities.

5.3. Completeness: basic theorems

We begin with the idea of a canonical standard model for a normal modal logic.

DEFINITION 5.5. Let $\mathcal{M} = \langle W, R, P\rangle$ be a standard model, and let Σ be a normal system of modal logic. $\mathcal{M}$ is a *canonical standard model* for Σ iff:

(1) $W = \{\Gamma : \text{Max}_\Sigma \Gamma\}$.

(2) For every α in $\mathcal{M}$,

$\Box A \in \alpha$ iff for every β in $\mathcal{M}$ such that $\alpha R \beta$, $A \in \beta$.

(3) $P_n = |\mathbb{P}_n|_\Sigma$, for $n = 0, 1, 2, \ldots$.

Thus, as in section 2.7, W is the set of Σ-maximal sets of sentences, and for each natural number n, P_n collects just those of such sets as contain the atomic sentences $\mathbb{P}_n$.

It is not immediately obvious, however, that there *are* any such models, i.e. that there are relations R satisfying the condition in clause (2) of the definition. We defer to theorem 5.11 the proof that indeed there are. Meanwhile, the following theorem gives an alternative way of characterizing canonical standard models.

THEOREM 5.6. *$\mathcal{M} = \langle W, R, P \rangle$ is a canonical standard model for a normal system Σ iff W and P are as in definition 5.2, and for every α in $\mathcal{M}$,*

$\Diamond \mathrm{A} \in \alpha$ iff for some β in $\mathcal{M}$ such that $\alpha R \beta$, $\mathrm{A} \in \beta$.

Proof. For left-to-right, suppose that $\mathcal{M} = \langle W, R, P \rangle$ is a canonical standard model for a normal system Σ. Then W and P are of course as specified, and for any world (Σ-maximal set of sentences) α in $\mathcal{M}$:

$\Diamond \mathrm{A} \in \alpha$ iff $\neg\Box\neg \mathrm{A} \in \alpha$
– Df$\Diamond$ and the Σ-maximality of α;
iff $\Box\neg \mathrm{A} \notin \alpha$
– theorem 2.18 (5);
iff not every β in $\mathcal{M}$ is such that if $\alpha R \beta$, then $\neg \mathrm{A} \in \beta$
– definition 5.5;
iff for some β in $\mathcal{M}$ such that $\alpha R \beta$, $\neg \mathrm{A} \notin \beta$;
iff for some β in $\mathcal{M}$ such that $\alpha R \beta$, $\mathrm{A} \in \beta$
– theorem 2.18 (5).

It is left for the reader to show the reverse, i.e. that $\mathcal{M}$ is a canonical standard model for Σ if W, R, and P satisfy the specified conditions.

Now we show that in a canonical standard model the worlds verify exactly the sentences they contain. This is the fundamental theorem for the completeness of normal systems of modal logic.

THEOREM 5.7. *Let $\mathcal{M}$ be a canonical standard model for a normal system Σ. Then for every α in $\mathcal{M}$:*

$\models^{\mathcal{M}}_{\alpha} \mathrm{A}$ iff $\mathrm{A} \in \alpha$.

In other words, $\|\mathrm{A}\|^{\mathcal{M}} = |\mathrm{A}|_{\Sigma}$.

Proof. The proof is by induction on the complexity of A, and the non-modal cases were discussed in section 2.7 (see exercise 2.53). Of the modal cases, we treat only that in which A is a necessitation, $\Box$B.

As an inductive hypothesis we assume that the theorem holds for sentences shorter than A, so that, in particular, for every α in $\mathcal{M}$,

$$\models^{\mathcal{M}}_{\alpha} \text{B iff B} \in \alpha.$$

Then:

$\models^{\mathcal{M}}_{\alpha} \Box$B iff for every β in $\mathcal{M}$ such that $\alpha R \beta$, $\models^{\mathcal{M}}_{\beta}$ B
– definition 3.2(1);

iff for every β in $\mathcal{M}$ such that $\alpha R \beta$, B $\in \beta$
– inductive hypothesis;

iff $\Box$B $\in \alpha$
– definition 5.5.

So the theorem holds when A is a necessitation.

As a corollary to theorem 5.7 it follows that a normal system of modal logic is determined by each of its canonical standard models; i.e. that the theorems of the system are just the sentences true in any such model. Formally:

THEOREM 5.8. *Let $\mathcal{M}$ be a canonical standard model for a normal system* Σ. *Then:*

$$\models^{\mathcal{M}} \text{A } \textit{iff} \vdash_{\Sigma} \text{A}.$$

Proof. See the remarks in section 2.7.

Nothing proved so far presupposes the existence of canonical standard models. In definition 5.9 we introduce what we call the proper canonical standard model for a normal modal logic; theorem 5.10 gives an alternative characterization. Then we prove that proper canonical standard models are, indeed, canonical standard models (theorem 5.11), which shows that such models exist.

DEFINITION 5.9. Let $\mathcal{M} = \langle W, R, P \rangle$ be a standard model, and let Σ be a normal system of modal logic. $\mathcal{M}$ is the *proper canonical standard model* for Σ iff:

(1) $W = \{\Gamma : \text{Max}_{\Sigma}\, \Gamma\}$.

(2) For every α and β in $\mathcal{M}$, $\alpha R \beta$ iff $\{\text{A} : \Box\text{A} \in \alpha\} \subseteq \beta$.

(3) $P_n = |\mathbb{P}_n|_{\Sigma}$, for $n = 0, 1, 2, \ldots$.

In other words, proper canonical standard models are like canonical standard models with respect to W and P, and R is defined so that the alternatives to a world α are just those worlds that contain the necessitates of necessitations in α. Note that such models always exist.

THEOREM 5.10. *$\mathcal{M} = \langle W, R, P \rangle$ is the proper canonical standard model for a normal system of modal logic Σ iff W and P are as in definition 5.9, and for every α and β in $\mathcal{M}$,*

$$\alpha R \beta \text{ iff } \{\Diamond A: A \in \beta\} \subseteq \alpha.$$

Proof. By definition 5.9, $\mathcal{M} = \langle W, R, P \rangle$ is the proper canonical standard model for a normal modal logic Σ just in case W and P are as specified, and for all worlds (Σ-maximal sets of sentences) α and β in $\mathcal{M}$,

$$\{A: \Box A \in \alpha\} \subseteq \beta \text{ iff } \{\Diamond A: A \in \beta\} \subseteq \alpha.$$

This is precisely theorem 4.29 (1).

Thus proper canonical standard models are like canonical standard models with respect to W and P, and R is defined so that a world collects all the possibilitations of sentences occurring in its alternatives.

THEOREM 5.11. *Proper canonical standard models are canonical standard models.*

Proof. Let $\mathcal{M}$ be the proper canonical standard model for a normal system Σ. We need only to show that for every world (Σ-maximal set of sentences) α in $\mathcal{M}$,

$$\Box A \in \alpha \text{ iff for every } \beta \text{ in } \mathcal{M} - \text{i.e. } \mathrm{Max}_{\Sigma}\beta - \text{such that } \{A: \Box A \in \alpha\} \subseteq \beta,\ A \in \beta.$$

This is theorem 4.30 (1).

EXERCISES

5.22. Give the proof of theorem 5.6, right-to-left.

5.23. Give the proof of theorem 5.7 for the case in which $A = \Diamond B$.

5.24. Let $\mathcal{M} = \langle W, R, P \rangle$ be the proper canonical standard model for a normal system Σ. Using the results of exercise 4.77, prove:

(*a*) $\mathcal{M}$ is serial if Σ contains D.

(*b*) $\mathcal{M}$ is reflexive if Σ contains T.

(*c*) $\mathcal{M}$ is symmetric if Σ contains B.

(*d*) $\mathcal{M}$ is transitive if Σ contains 4.

(*e*) $\mathcal{M}$ is euclidean if Σ contains 5.

Also prove:

(*f*) $\mathcal{M}$ is incestual if Σ contains G.

5.4. Determination

We are ready now to prove our first determination results. We begin with the system *K*.

THEOREM 5.12. *K is determined by the class of standard models.*

Proof. Soundness comes immediately from theorem 5.1. Completeness follows from the fact of the existence of canonical standard models, theorem 5.11: any sentence valid in the class of standard models is true in the proper canonical standard model for *K* and hence, by theorem 5.8, is a theorem of the system.

In general, to prove the completeness of a normal modal logic with respect to a class of models it is sufficient to show that the proper canonical standard model for the system is contained in the class. For the logic is determined by this model alone (theorem 5.8) and so is complete with respect to any class that contains it. Compare the remarks in section 2.7. This should serve to motivate the next theorem, which paves the way for the determination results in theorem 5.14.

THEOREM 5.13. *Let $\mathcal{M}$ be the proper canonical standard model for a normal system Σ. Then:*

(1) *$\mathcal{M}$ is serial if Σ contains* D.

(2) *$\mathcal{M}$ is reflexive if Σ contains* T.

(3) *$\mathcal{M}$ is symmetric if Σ contains* B.

(4) *$\mathcal{M}$ is transitive if Σ contains* 4.

(5) *$\mathcal{M}$ is euclidean if Σ contains* 5.

Proof. Let $\mathcal{M} = \langle W, R, P \rangle$ be the proper canonical standard model for a normal system Σ.

For (1). Assume that D is a theorem of Σ (so that Σ is a *KD*-system). To show that $\mathcal{M}$ is serial – that for every α in $\mathcal{M}$,

there is a β in $\mathcal{M}$ such that $\alpha R \beta$

– is to show that for every Σ-maximal set of sentences α,

$$\text{there is a } \mathrm{Max}_\Sigma\, \beta \text{ such that } \{A : \Box A \in \alpha\} \subseteq \beta.$$

And by Lindenbaum's lemma (theorem 2.19) it is enough, for this, to establish the Σ-consistency of the set $\{A : \Box A \in \alpha\}$. So suppose, to reach a contradiction, that this set is Σ-inconsistent, i.e. that $\bot$ is Σ-deducible from it. Then the set contains sentences $A_1, \ldots, A_n$ ($n \geqslant 0$) such that

$$\vdash_\Sigma (A_1 \wedge \ldots \wedge A_n) \rightarrow \bot.$$

Hence by the rule RD,

$$\vdash_\Sigma (\Box A_1 \wedge \ldots \wedge \Box A_n) \rightarrow \Diamond \bot.$$

Since α contains $\Box A_1, \ldots, \Box A_n$, this means that $\Diamond \bot$ is Σ-deducible from α, and hence (theorem 2.18 (1)) that $\Diamond \bot$ is in α. But α also contains $\neg \Diamond \bot$ (N$\Diamond$), since Σ is normal (theorems 2.18 (2) and 4.4). This cannot be (theorem 2.18 (5)).

There is a simpler proof of the seriality of $\mathcal{M}$. To wit, as a consequence of theorem 5.6 it holds of every α in $\mathcal{M}$ that

$$\Diamond \top \in \alpha \text{ iff for some } \beta \text{ in } \mathcal{M} \text{ such that } \alpha R \beta,\ \top \in \beta.$$

But $\Diamond \top$ (P) is a theorem of Σ (theorem 4.12), so every α in $\mathcal{M}$ contains it. Hence for every α in $\mathcal{M}$ there is a β in $\mathcal{M}$ such that $\alpha R \beta$, i.e. $\mathcal{M}$ is serial. Indeed, it should be noted, this proof is good for *any* canonical standard model for a normal *KD*-system.

For (2). Suppose that Σ contains the theorem T. We wish to show that $\mathcal{M}$ is reflexive, i.e. that for every α in $\mathcal{M}$,

$$\alpha R \alpha.$$

This means that for every Σ-maximal set of sentences α,

$$\{A : \Box A \in \alpha\} \subseteq \alpha.$$

So assume that $A \in \{A : \Box A \in \alpha\}$, to show that $A \in \alpha$. The assumption just means that α contains $\Box A$. Since α is Σ-maximal, $\Box A \rightarrow A$ is in α, too. But α is closed under MP, so it contains A.

For (3). Assume B to be a theorem of Σ. Let us prove that for all Σ-maximal sets of sentences α and β,

$$\text{if } \{A : \Box A \in \alpha\} \subseteq \beta, \text{ then } \{\Diamond A : A \in \alpha\} \subseteq \beta.$$

For by definition 5.9 and theorem 5.10 this means that $\mathcal{M}$ is symmetric, i.e. that for every α and β in $\mathcal{M}$,

$$\text{if } \alpha R \beta, \text{ then } \beta R \alpha.$$

We assume that $\{A : \Box A \in \alpha\} \subseteq \beta$, and also that $A \in \alpha$. It remains only to be shown that $\Diamond A \in \beta$. But if α contains A and the theorem $A \to \Box\Diamond A$ too, then $\Box\Diamond A$ is in α. Hence $\Diamond A \in \beta$.

For (4). Assume that Σ contains the theorem 4, $\Box A \to \Box\Box A$. The transitivity of $\mathcal{M}$ is expressed by saying that for every Σ-maximal set of sentences α, β, and γ,

$$\text{if } \{A : \Box A \in \alpha\} \subseteq \beta \text{ and } \{A : \Box A \in \beta\} \subseteq \gamma,$$
$$\text{then } \{A : \Box A \in \alpha\} \subseteq \gamma.$$

We suppose that $\{A : \Box A \in \alpha\} \subseteq \beta$, that $\{A : \Box A \in \beta\} \subseteq \gamma$, and that $\Box A \in \alpha$, to show that $A \in \gamma$. The presence of 4 in α and the last assumption imply that $\Box\Box A$ is in α. By the first assumption, then, $\Box A$ is in β, and by the second, A is in γ.

For (5). We leave this for the reader (compare exercise 5.24(*e*)), with the remark that the euclideanness of $\mathcal{M}$ means that for every Σ-maximal set of sentences α, β, and γ,

$$\text{if } \{A : \Box A \in \alpha\} \subseteq \beta \text{ and } \{\Diamond A : A \in \gamma\} \subseteq \alpha,$$
$$\text{then } \{\Diamond A : A \in \gamma\} \subseteq \beta.$$

This ends the proof of theorem 5.13.

We are in a position at last to prove determination theorems for the fourteen normal systems obtained by taking the schemas D, T, B, 4, and 5 as theorems in various combinations, i.e. for the systems beyond *K* in figure 4.1. We state these results formally in the following theorem.

THEOREM 5.14. *The fourteen normal systems beyond K in figure 4.1 are determined by the classes of standard models indicated in figure 5.1.*

Proof. Except perhaps for *KB4* and *KT5* the results are easily understood. The soundness parts follow from theorems 3.5 and 5.1, as we remarked earlier. For completeness it is enough to observe that the proper canonical standard models for each system are in the appropriate classes of models, by theorem 5.13. For example, by parts (2) and (4) of theorem 5.13 the proper canonical standard model for *KT4* is both reflexive and transitive. Thus *KT4* (*S4*) is determined by the class of quasi-ordered standard models.

In the case of *KB4*, note that it is identical with *KB5* (theorem 4.18), or that transitivity and euclideanness are equivalent properties in the presence of symmetry (exercise 3.58(*e*)). For *KT5* (*S5*) recall that it is the same as *KTB4*, *KDB4*, and *KDB5* (theorem 4.21) – or that anyway the four sets of properties indicated in figure 5.1 indifferently characterize an equivalence relation (exercise 3.31).

We can cover the content of theorems 5.12 and 5.14 together by putting the matter as follows. Let C_{D}, C_{T}, C_{B}, C_4, and C_5 be respectively the classes of serial, reflexive, symmetric, transitive, and euclidean standard models, and let $\mathrm{S}_1, \ldots, \mathrm{S}_n$ be any selection (possibly empty) from the schemas D, T, B, 4, and 5 (so that $\mathsf{C}_{\mathrm{S}_1}, \ldots, \mathsf{C}_{\mathrm{S}_n}$ are the corresponding classes of standard models). Then: the system $KS_1 \ldots S_n$ is determined by the class $\mathsf{C}_{\mathrm{S}_1} \cap \ldots \cap \mathsf{C}_{\mathrm{S}_n}$.

Of course this formulation does not really go beyond theorems 5.12 and 5.14, though it may seem to. It shows, for example, that the system *KDT4* is determined by the class of serial reflexive transitive standard models. But D is redundant here as an axiom (theorem 4.14), just as seriality is as a property (exercise 3.58(*a*)).

In chapter 3 we saw that systems of modal logic can be determined by more than one class of models. For by theorem 3.12 a system of modal logic determined by a class C of standard models is also determined by the class $\mathscr{G}(\mathsf{C})$ of models generated from those in C, and these classes are in general distinct. As a corollary we proved (theorem 3.13) that the modal logic determined by the class of standard models in which the relation is an equivalence is the same as the modal logic determined by the class of standard models in which the relation is universal (like those in chapter 1). By theorem 5.14 we see, at last, that this is indeed the modal logic *S5* (*KT5*). This is worth recording.

THEOREM 5.15. *KT5 is determined by the class of universal standard models.*

In section 5.6 we use filtrations to obtain more determination results for *K* and its normal extensions.

Let us close this section by recalling the schema

$$\text{G.} \quad \Diamond\Box A \to \Box\Diamond A.$$

As we saw in chapter 3, G is valid in any class of incestual models, i.e. standard models $\mathscr{M} = \langle W, R, P \rangle$ such that for every α, β, and γ in $\mathscr{M}$,

$$\text{if } \alpha R \beta \text{ and } \alpha R \gamma, \text{ then for some } \delta \text{ in } \mathscr{M}, \beta R \delta \text{ and } \gamma R \delta.$$

This means that the system *KG* is sound with respect to any class of incestual models. Let us show now that the proper canonical standard model for any normal modal logic containing the schema G is incestual – and thus that every normal *KG*-system is complete with respect to the class of incestual models – as follows.

Let $\mathcal{M} = \langle W, R, P \rangle$ be the proper canonical standard model for a normal system containing the schema G, and suppose, for α, β, and γ in $\mathcal{M}$, that $\alpha R \beta$ and $\alpha R \gamma$. By definition 5.9 and theorem 5.10, this means that

$$\{\Diamond A : A \in \beta\} \subseteq \alpha$$

and that

$$\{A : \Box A \in \alpha\} \subseteq \gamma.$$

The problem now is to show that for some δ in $\mathcal{M}$, $\beta R \delta$ and $\gamma R \delta$, i.e.

$$\{A : \Box A \in \beta\} \subseteq \delta \text{ and } \{A : \Box A \in \gamma\} \subseteq \delta$$

– in other words, that $\mathcal{M}$ has a Σ-maximal set of sentences δ that includes the set

$$\{A : \Box A \in \beta\} \cup \{A : \Box A \in \gamma\}.$$

By Lindenbaum's lemma (theorem 2.19) it is sufficient to show that this set is Σ-consistent, for then such an extension must exist. So suppose, to reach a contradiction, that the set is Σ-inconsistent, i.e. that $\perp$ is Σ-deducible from it. Then for some $i, j \geqslant 0$ there are sentences $B_1, \ldots, B_i$ in $\{A : \Box A \in \beta\}$ and sentences $C_1, \ldots, C_j$ in $\{A : \Box A \in \gamma\}$ such that

$$\vdash_\Sigma (B_1 \wedge \ldots \wedge B_i \wedge C_1 \wedge \ldots \wedge C_j) \rightarrow \perp.$$

Equivalently, by *PL*,

$$\vdash_\Sigma (B_1 \wedge \ldots \wedge B_i) \rightarrow \neg(C_1 \wedge \ldots \wedge C_j).$$

By RK it follows that

$$\vdash_\Sigma (\Box B_1 \wedge \ldots \wedge \Box B_i) \rightarrow \Box\neg(C_1 \wedge \ldots \wedge C_j).$$

Because each of $\Box B_1, \ldots, \Box B_i$ belongs to β, $\Box\neg(C_1 \wedge \ldots \wedge C_j)$ is Σ-deducible from β, and so belongs to β. By the first inclusion above, then, $\Diamond\Box\neg(C_1 \wedge \ldots \wedge C_j)$ is a member of α. But α also contains this instance of the schema G:

$$\Diamond\Box\neg(C_1 \wedge \ldots \wedge C_j) \rightarrow \Box\Diamond\neg(C_1 \wedge \ldots \wedge C_j).$$

So the consequent $\Box\Diamond\neg(C_1 \wedge \ldots \wedge C_j)$ is also in α. By the second inclusion above, then, $\Diamond\neg(C_1 \wedge \ldots \wedge C_j)$ belongs to γ. Because γ contains

Df$\Box$, we may infer further that $\neg\Box(C_1 \wedge \ldots \wedge C_j)$ is in γ. But so is $\Box(C_1 \wedge \ldots \wedge C_j)$. For by RK on a tautology,

$$\vdash_{\Sigma}(\Box C_1 \wedge \ldots \wedge \Box C_j) \to \Box(C_1 \wedge \ldots \wedge C_j);$$

and γ contains each of $\Box C_1, \ldots, \Box C_j$, and so also $\Box(C_1 \wedge \ldots \wedge C_j)$. Contradiction.

It follows from this and the soundness result that the modal logic *KG* is determined by the class of incestual standard models. In the next section we generalize these proofs to obtain similar results about *k,l,m,n*-incestuality and normal $KG^{k,l,m,n}$-systems.

EXERCISES

5.25. Verify the determination claims for the systems beyond *K* in figure 4.1 not mentioned explicitly in the proof of theorem 5.14.

5.26. Let $\mathcal{M} = \langle W, R, P \rangle$ be the proper canonical standard model for a normal system Σ. Prove:

(*a*) $\mathcal{M}$ is partially functional if Σ contains D_c.

(*b*) $\mathcal{M}$ is functional if Σ contains D!.

(*c*) $\mathcal{M}$ is vacuous if Σ contains T_c.

(*d*) $\mathcal{M}$ is identity if Σ contains T!.

(*e*) $\mathcal{M}$ is dense if Σ contains 4_c.

(*f*) $\mathcal{M}$ is empty if Σ contains $\overline{P}$.

Using these results, theorem 5.13, and the result about G and incestuality at the end of section 5.4, formulate and prove more determination theorems like those in theorem 5.14.

5.27. Prove that the proper canonical standard model for any normal extension of $KG^{k,l,m,n}$ is *k,l,m,n*-incestual. Use this result to prove theorem 5.13 and parts (*a*)–(*e*) of the preceding exercise.

5.28. Using exercise 3.57, prove that *K* is determined by each of the classes of irreflexive, asymmetric, antisymmetric, and intransitive standard models.

5.29. With reference to results in section 3.4 and exercise 5.4, prove:

(*a*) Every generated reflexive transitive incestual standard model is convergent (and hence strongly convergent).

(*b*) The system *S4.2* is determined by the class of reflexive transitive standard models that are convergent (and hence strongly convergent).

5.30. Prove that the system *KD4G* is determined by the class of serial transitive standard models that are convergent or strongly convergent.

5.31. Consider the rule of inference

$$\frac{\Box \mathrm{A}}{\mathrm{A}}$$

and the fifteen systems in figure 4.1. By the results of exercises 4.13, 4.61, 4.62, and 5.10, *K*, *KD*, *KT*, *KDB*, *KTB*, *KT4*, and *KT5* have the rule, whereas *KB*, *K5*, *KD5*, *K45*, *KD45*, and *KB4* do not.

Use theorem 5.14 and results in exercise 3.63 to prove that *K4* and *KD4* also have the rule. (Indeed, all these conclusions are immediate consequences of theorem 5.14 and results in exercise 3.63, as the reader should verify.)

5.32. Let Σ be a normal modal logic, and consider the following principle.

If $\vdash_\Sigma \Box \mathrm{A}_1 \vee \ldots \vee \Box \mathrm{A}_n$, then
$\vdash_\Sigma \mathrm{A}_i$ for some $i = 1, \ldots, n \; (n > 0)$.

This 'rule of disjunction' holds for the following systems in figure 4.1:

K, *KD*, *KT*, *K4*, *KD4*, *KT4*

We can use theorems 5.8, 5.13, and 5.14 and the safe extension theorem from exercise 3.62 to prove this. For example, let us show that the principle holds for *K*.

We argue contrapositively. Suppose that not $\vdash_K \mathrm{A}_i$ for each $i = 1, \ldots, n$. Then by theorem 5.8 each A_i is false at some world α_i in the proper canonical standard model $\mathcal{M} = \langle W, R, P \rangle$ for *K*. Let α be a world not in $\mathcal{M}$, and define the standard model $\mathcal{M}^\# = \langle W^\#, R^\#, P^\# \rangle$ as follows.

(1) $W^\# = W \cup \{\alpha\}$.
(2) $R^\# = R \cup \{\langle \alpha, \alpha_i \rangle : i = 1, \ldots, n\}$.
(3) $P^\# = P$.

Then $\mathcal{M}^\#$ is a safe extension of $\mathcal{M}$. So by the safe extension theorem each A_i is false at α_i in $\mathcal{M}^\#$ as well. But $\alpha R \alpha_i$ for each α_i. So each $\Box \mathrm{A}_i$ fails at α, which means that $\Box \mathrm{A}_1 \vee \ldots \vee \Box \mathrm{A}_n$ is false at α. By theorem 5.14, not $\vdash_K \Box \mathrm{A}_1 \vee \ldots \vee \Box \mathrm{A}_n$, which is what we wished to prove.

The arguments for *KD*, *KT*, *K4*, *KD4*, and *KT4* are left to the reader.

The idea in each case is to define for the system's proper canonical standard model a safe extension that has the desired properties. (Indeed, a single safe extension can be defined that will do the job for all the cases, including *K*.)

The remaining systems in figure 4.1,

KB, K5, KDB, KD5, K45, KD45, KB4, KTB, KT5,

do not have the rule of disjunction. For *KB*, *K5*, *KD5*, *K45*, *KD45*, and *KB4* this was shown in exercise 5.10 (let $n = 1$ in the rule). The cases of *KDB*, *KTB*, and *KT5* are for the reader; it may be helpful to recall that G is a theorem of each of these logics.

5.33. Consider again the following rules of inference.

$$\frac{\Box A \to \Box B}{A \to B} \qquad \frac{\Diamond A \to \Diamond B}{A \to B}$$

By the results of exercises 4.13 and 4.16, these rules are present in *K* and *KD*.

(*a*) Prove that *KB*, *K4*, *K5*, *KDB*, *KD4*, *KD5*, *K45*, *KD45*, *KB4*, *KTB*, *KT4*, and *KT5* do not have these rules.

(*b*) Prove or disprove that *KT* has these rules.

5.34. Consider again the following rules of inference.

$$\frac{\Box A \leftrightarrow \Box B}{A \leftrightarrow B} \qquad \frac{\Diamond A \leftrightarrow \Diamond B}{A \leftrightarrow B}$$

By the results of exercises 4.13 and 4.61, these rules hold in *K* and *KD*. In which of the other systems in figure 4.1 do these rules hold?

5.35. Referring to the discussion in section 2.8, use theorems 3.18 and 5.15 to prove that the system *S5* (*KT5*) is decidable.

5.5. $KG^{k,l,m,n}$

In section 3.3 we introduced the schema

$$G^{k,l,m,n}. \quad \Diamond^k \Box^l A \to \Box^m \Diamond^n A$$

and the property of *k,l,m,n*-incestuality for a standard model $\mathcal{M} = \langle W, R, P\rangle$: for every α, β, and γ in $\mathcal{M}$,

if $\alpha R^k \beta$ and $\alpha R^m \gamma$, then for some δ in $\mathcal{M}$, $\beta R^l \delta$ and $\gamma R^n \delta$, all where k, l, m, $n \geqslant 0$. In virtue of theorem 3.8, the modal logic $KG^{k,l,m,n}$ is sound with respect to any class of k,l,m,n-incestual models. In this section we show that any normal $KG^{k,l,m,n}$-system is complete with respect to the class of k,l,m,n-incestual models, by proving that the proper canonical standard model for such a system is k,l,m,n-incestual. These results yield determination theorems for an infinite class of normal modal logics, including the systems covered in the preceding section. We begin with the following theorem.

THEOREM 5.16. *Let* $\mathcal{M} = \langle W, R, P\rangle$ *be the proper canonical standard model for a normal system* Σ. *Then for every* α *and* β *in* $\mathcal{M}$ *and every* $k \geqslant 0$*:*

(1) $\alpha R^k \beta$ *iff* $\{A : \Box^k A \in \alpha\} \subseteq \beta$.

(2) $\alpha R^k \beta$ *iff* $\{\Diamond^k A : A \in \beta\} \subseteq \alpha$.

Proof. Assume that $\mathcal{M}$ is the proper canonical standard model for a normal system Σ. We give the proof for (1) only; (2) follows from (1) by theorem 4.29 (2). Where $k = 0$, (1) means that

$$\alpha R^0 \beta \text{ iff } \{A : \Box^0 A \in \alpha\} \subseteq \beta,$$

which just amounts to

$$\alpha = \beta \text{ iff } \alpha \subseteq \beta$$

(by definitions 3.6 (1) and 2.3 (1)). And this is obvious: for if $\alpha \subseteq \beta$, then β cannot properly include α without being inconsistent; so $\alpha = \beta$ (compare exercise 2.46). This establishes the basis of an induction. Next we make the inductive hypothesis that (1) holds for every number up to and including some number k – so that for every α and β in $\mathcal{M}$,

$$\alpha R^k \beta \text{ iff } \{A : \Box^k A \in \alpha\} \subseteq \beta$$

– and show from this that

$$\alpha R^{k+1} \beta \text{ iff } \{A : \Box^{k+1} A \in \alpha\} \subseteq \beta.$$

Thus:

$\alpha R^{k+1} \beta$ iff for some γ in $\mathcal{M}$, $\alpha R \gamma$ and $\gamma R^k \beta$
– definition 3.6 (2);

iff for some γ in $\mathcal{M}$, $\{A : \Box A \in \alpha\} \subseteq \gamma$ and $\{A : \Box^k A \in \gamma\} \subseteq \beta$
– definition 5.9 and the inductive hypothesis;

iff $\{A : \Box^{k+1} A \in \alpha\} \subseteq \beta$
– theorem 4.31.

This completes the inductive part of the proof, and hence the proof itself.

THEOREM 5.17. *The proper canonical standard model for a normal $KG^{k,l,m,n}$-system is k,l,m,n-incestual, for every $k, l, m, n \geq 0$.*

Proof. Let Σ be a normal $KG^{k,l,m,n}$-system. By theorem 5.16 and Lindenbaum's lemma the property of k,l,m,n-incestuality for the proper canonical standard model for Σ can be expressed as: for every Σ-maximal set of sentences α, β, and γ,

$$\text{if } \{\Diamond^k A : A \in \beta\} \subseteq \alpha \text{ and } \{A : \Box^m A \in \alpha\} \subseteq \gamma,$$
$$\text{then } \mathrm{Con}_\Sigma\{A : \Box^l A \in \beta\} \cup \{A : \Box^n A \in \gamma\}.$$

The proof is a simple generalization of that at the end of the preceding section for the case in which $k = l = m = n = 1$. Briefly, we assume that

$$\{\Diamond^k A : A \in \beta\} \subseteq \alpha \text{ and } \{A : \Box^m A \in \alpha\} \subseteq \gamma,$$

but also that

$$\mathrm{C\emptyset n}_\Sigma\{A : \Box^l A \in \beta\} \cup \{A : \Box^n A \in \gamma\}$$

(and argue to a contradiction). From the latter assumption it follows that

$$\vdash_\Sigma (B_1 \wedge \ldots \wedge B_i) \rightarrow \neg(C_1 \wedge \ldots \wedge C_j),$$

and from this by RK^k (theorem 4.6) that

$$\vdash_\Sigma (\Box^l B_1 \wedge \ldots \wedge \Box^l B_i) \rightarrow \Box^l \neg(C_1 \wedge \ldots \wedge C_j),$$

where $\Box^l B_1, \ldots, \Box^l B_i$ are in β, and $\Box^n C_1, \ldots, \Box^n C_j$ are in γ. Hence β contains $\Box^l \neg(C_1 \wedge \ldots \wedge C_j)$, and so (by the initial assumption) $\Diamond^k \Box^l \neg(C_1 \wedge \ldots \wedge C_j)$ belongs to α. But α contains the instance

$$\Diamond^k \Box^l \neg(C_1 \wedge \ldots \wedge C_j) \rightarrow \Box^m \Diamond^n \neg(C_1 \wedge \ldots \wedge C_j)$$

of $G^{k,l,m,n}$. So $\Box^m \Diamond^n \neg(C_1 \wedge \ldots \wedge C_j)$ is a member of α, and hence (by the initial assumption) $\Diamond^n \neg(C_1 \wedge \ldots \wedge C_j)$ is in γ. Using $\mathrm{Df}\Box^k$ (theorem 4.6) we see that γ contains $\neg\Box^n(C_1 \wedge \ldots \wedge C_j)$. However, note that by RK^k on a tautology,

$$\vdash_\Sigma (\Box^n C_1 \wedge \ldots \wedge \Box^n C_j) \rightarrow \Box^n(C_1 \wedge \ldots \wedge C_j).$$

So since γ contains each conjunct of the antecedent, γ also contains $\Box^n(C_1 \wedge \ldots \wedge C_j)$. Contradiction.

EXERCISES

5.36. Formulate and prove determination theorems for some normal modal logics not in figure 4.1 that are covered by $KG^{k,l,m,n}$.

5.37. Let $\mathcal{M} = \langle W, R, P\rangle$ be the proper canonical standard model for a normal system Σ. Prove:

(*a*) $\mathcal{M}$ is secondarily reflexive if Σ contains U.

(*b*) $\mathcal{M}$ is reverse secondarily reflexive if Σ contains $\Diamond\top \to (\Box A \to A)$.

(See exercise 3.51.)

5.38. Let $\mathcal{M} = \langle W, R, P\rangle$ be the proper canonical standard model for a normal system Σ, and consider the conditions in exercise 3.52. Prove:

(*a*) $\mathcal{M}$ satisfies condition (*a*) if Σ contains $\Box^j(\Diamond^k\Box^l A \to \Box^m\Diamond^n A)$.

(*b*) $\mathcal{M}$ satisfies condition (*b*) if Σ contains $\Diamond^j\top \to (\Diamond^k\Box^l A \to \Box^m\Diamond^n A)$.

Note that the results in exercises 5.27 and 5.37 are corollaries of these.

5.39. Prove that for each $k > 0$ the system KP^k is determined by the class serial standard models. (See exercise 3.49 and theorem 4.13.)

5.40. For $k > 0$ formulate and prove a determination theorem for the system $K\bar{P}^k$ ($\bar{P}^k = \neg\Diamond^k\top$).

5.41. Let $\mathcal{M} = \langle W, R, P\rangle$ be the proper canonical standard model for a normal system Σ, and consider conditions (h^{++}), (h^{+}), and (h) in exercise 5.3. Prove:

(*a*) $\mathcal{M}$ satisfies (h^{++}) if Σ contains H^{++}.

(*b*) $\mathcal{M}$ satisfies (h^{+}) if Σ contains H^{+}.

(*c*) $\mathcal{M}$ satisfies (h) if Σ contains H.

(See exercise 4.56 for some alternatives to H^{++}, H^{+}, and H.)

5.42. With reference to results in section 3.4 and exercises 5.3 and 5.4, prove:

(*a*) Every generated reflexive transitive standard model that satisfies (h^{++}), (h^{+}), or (h) is connected (and hence strongly connected).

(*b*) The system *S4.3* is determined by the class of reflexive transitive standard models that are connected (and hence strongly connected).

5.43. Prove that the system *KD4H* is determined by the class of serial transitive connected standard models.

5.44. Prove that the system *S4.1* of exercise 4.58 is determined by the class of reflexive transitive standard models that satisfy the condition given in exercise 5.6. To do this it may be helpful first to prove that for every $n > 0$ the schema

$$\Diamond((\Diamond A_1 \rightarrow \Box A_1) \wedge \ldots \wedge (\Diamond A_n \rightarrow \Box A_n))$$

is a theorem of *S4.1*. Then show that where Σ is any normal extension of *S4.1* the proper canonical standard model $\mathcal{M} = \langle W, R, P \rangle$ for Σ satisfies the condition in exercise 5.6. (For this, begin by arguing that for any α in $\mathcal{M}$ the set

$$\{A : \Box A \in \alpha\} \cup \{\Diamond A \rightarrow \Box A : A \text{ is a sentence}\}$$

is Σ-consistent.)

Prove also:

(*a*) $K4G_c$ is determined by the class of transitive standard models that satisfy the condition in exercise 5.6.

(*b*) The normal extension of *K* obtained by taking the schema above as a theorem for every $n > 0$ is determined by the class of standard models that satisfy the condition in exercise 5.6.

Indeed, (*a*) and (*b*) may be used as lemmas in the proof of the determination theorem for *S4.1*.

5.45. Prove:

(*a*) KB_c is determined by the class of standard models that satisfy condition (*a*) in exercise 3.39.

(*b*) $K5_c$ is determined by the class of standard models that satisfy condition (*b*) in exercise 3.39.

5.46. Investigate the status of the rules (*a*)–(*f*) in exercise 4.13 and the rule of disjunction in exercise 5.32 with respect to normal modal logics other than those in figure 4.1. (For example, compare *KU* and *K* plus $\Diamond\top \rightarrow (\Box A \rightarrow A)$.)

5.47. Consider the modal logics determined by the classes of models described in exercises 3.14 and 3.15. Describe axiomatizations for these logics (they are non-normal), and adapt the techniques of the present chapter to prove determination theorems for them.

5.48. Investigate the construction of non-proper canonical standard models for normal systems (consider extensions of *K4*).

5.6. Decidability

A system of modal logic is decidable, as we explained in section 2.8, if it is axiomatizable by a finite number of schemas and has the finite model property. In this section we show that the fifteen systems in figure 4.1 are decidable, by proving that each is axiomatizable by finitely many schemas and has the f.m.p.

THEOREM 5.18. *Each of the fifteen normal systems in figure 4.1 is axiomatizable, indeed by a finite number of schemas.*

Proof. It is sufficient to observe that in each case the logic can be axiomatized by finitely many schemas together with the reasonable rules RPL and RK.

We obtain f.m.p. results for the logics in question as a corollary to the next two theorems.

THEOREM 5.19. *K is determined by the class of finite standard models.*

Proof. Soundness follows from theorem 5.12. For completeness, suppose that A is true in every finite standard model, and let Γ be the set of subsentences of A. Then A is true in every Γ-filtration of any standard model, since these are all finite. By theorem 3.17 it follows that A is true in every standard model, and by theorem 5.12 this means that A is a theorem of *K*. To put the argument again, contrapositively, suppose A is not a theorem of *K*, so that by theorem 5.12 A is false in some standard model $\mathscr{M}$ (for example, the proper canonical standard model for *K*). Let $\mathscr{M}^*$ be a Γ-filtration of $\mathscr{M}$, where Γ is the set of subsentences of A. Then $\mathscr{M}^*$ is a finite standard model, and by theorem 3.16 A is false in $\mathscr{M}^*$. Stated in this way it is readily seen that *K* has the f.m.p.: every non-theorem of *K* is false in a finite model of *K*.

THEOREM 5.20. *The fourteen normal systems beyond K in figure 4.1 are determined by the classes of finite standard models indicated in figure 5.1.*

Proof. Soundness in each case is a consequence of theorem 5.14. The completeness of all the systems except *K5* and *KD5* may be proved using theorems 3.19 and 3.20. For example, let us show that *KTB* (the *Brouwersche* system) is complete with respect to the class of finite standard models that are reflexive and symmetric. Suppose A is a non-theorem of *KTB*. Then by theorem 5.14 A is false in a reflexive symmetric standard model $\mathcal{M}$. Let Γ be the set of subsentences of A, and let $\mathcal{M}^*$ be a symmetric Γ-filtration of $\mathcal{M}$ defined as in part (1) of theorem 3.20. Then $\mathcal{M}^*$ is finite, and by theorem 3.19 it is reflexive. So by theorem 3.16 $\mathcal{M}^*$ is a reflexive symmetric finite standard model in which A is false.

For *K5* and *KD5* we use theorems 3.19 and 3.21. Let us give the proof for *K5*, by way of illustration. Suppose A is not a theorem of *K5*, so that by theorem 5.14 it is false in some euclidean model $\mathcal{M}$. Let Γ be the modal closure of the set of subsentences of A, and let $\mathcal{M}^*$ be a coarsest Γ-filtration of $\mathcal{M}$. Now observe that Γ is logically finite relative to $\mathcal{M}$. For by theorem 4.23 *K5* has only finitely many modalities, and so since the set of subsentences of A is finite and $\mathcal{M}$ is a model of *K5*, every sentence in Γ is $\mathcal{M}$-equivalent to one or another of finitely many sentences in Γ. Thus $\mathcal{M}^*$ is finite, and by theorems 3.16 and 3.21 it is a euclidean standard model in which A is false.

This concludes the proof. We leave as exercises the details of the arguments for the remaining systems.

The corollary to theorems 5.19 and 5.20 is worth stating formally.

THEOREM 5.21. *Each of the fifteen normal systems in figure 4.1 has the finite model property.*

And as a corollary to theorems 5.18 and 5.21 we have our final theorem.

THEOREM 5.22. *Each of the fifteen normal systems in figure 4.1 is decidable.*

In closing, let us point out a certain limitation to the methods used in the proof of theorem 5.20. As we showed in section 3.6, a filtration of a euclidean model through an arbitrary set of sentences (closed under subsentences) need not yield a euclidean model. So to prove theorem 5.20 for *K5* we had to pick a more special set of sentences to get a finite euclidean filtration. Because *K5*-systems have only finitely many distinct modalities, we had recourse to filtrations through modally closed sets of sentences. (Indeed, this device is apt for all the systems in figure 4.1 in which the number of distinct modalities is finite.)

But consider the system *KG*, which is determined by the class of incestual standard models (section 5.4). There is no easy access, along the lines of the proof of theorem 5.20, to a proof that *KG* is determined by the class of finite incestual models. For a filtration of an incestual model through an arbitrary set of sentences closed under subsentences is not always an incestual model (exercise 3.75); and *KG* has infinitely many distinct modalities (exercise 5.17). We leave it to the reader to try to show the decidability of *KG* by finding appropriate filtrations for incestual models.

EXERCISES

5.49. Give the details of the proof of theorem 5.20 for the systems other than *K*, *KTB*, and *K5* in figure 4.1 (compare exercise 3.74).

5.50. Prove the decidability of the normal extensions of *K*, *KD*, *KB*, and *K4* obtained by adding U or $\Diamond\top \to (\Box A \to A)$ as theorems.

5.51. Try to prove the decidability of *KG* by defining suitable filtrations for incestual models.

5.52. Prove the decidability of some normal systems other than those mentioned in theorem 5.22 and the preceding two exercises.

6

DEONTIC LOGIC

In this chapter we introduce an operator $\bigcirc$ to represent the deontic concept of obligation. (In order not to beg any important questions, we do not – except for a single exercise – consider a correspondingly dual operator P for the deontic concept of permissibility.) In section 6.1 we present what we call standard deontic logic, and in section 6.2 we examine some further principles that have been suggested. In section 6.3 we discuss the role of time in the determination of obligations, and we introduce temporal concepts into the language and into the models for it. Section 6.4 contains a theorem about past tense obligations. Finally, in section 6.5 we point out some shortcomings with respect to the adequacy and correctness of the analysis of deontic logic in terms of normal systems and standard models.

The purpose of this chapter is illustrative: we wish to show how standard models and normal systems can be employed in the analysis of philosophical questions. The reader must judge the merit of the endeavor, as well as the extent of its success.

6.1. Standard deontic logic

Into the language of propositional logic we introduce sentences of the form $\bigcirc$A, meant to express propositions of the form *it ought to be the case that* A, or *it is obligatory that* A. Thus the operator $\bigcirc$ represents a concept of deontic necessity.

By *standard deontic logic* we mean the system *D** based on propositional logic and axiomatized by the rule of inference

$$\text{ROK.} \quad \frac{(\mathrm{A}_1 \wedge \ldots \wedge \mathrm{A}_n) \rightarrow \mathrm{A}}{(\bigcirc \mathrm{A}_1 \wedge \ldots \wedge \bigcirc \mathrm{A}_n) \rightarrow \bigcirc \mathrm{A}} \quad (n \geqslant 0)$$

and the single schema

$$\text{OD*.} \quad \neg(\bigcirc \mathrm{A} \wedge \bigcirc \neg \mathrm{A}).$$

Thus D^* is the smallest normal KD-system for $\bigcirc$ (the axiom OD* being the counterpart of O$\square$ in theorem 4.12).

The import of ROK is clear: obligation is closed under consequence, in the sense that a proposition is obligatory if it is a consequence of obligatory propositions. The import of OD* is simply that conflicts of obligation are impossible, that there are no propositions that are jointly impossible but both obligatory.

A review of theorems in chapter 4 will reveal a number of theorems and rules of inference of D^*. In particular, it can be seen that the system is equivalently axiomatized by the rule

$$\text{ROM.} \quad \frac{A \rightarrow B}{\bigcirc A \rightarrow \bigcirc B}$$

together with the following schemas.

$$\text{OC.} \quad (\bigcirc A \wedge \bigcirc B) \rightarrow \bigcirc(A \wedge B)$$
$$\text{ON.} \quad \bigcirc\top$$
$$\text{OD.} \quad \neg\bigcirc\bot$$

On the basis of propositional logic, ROM, OC, and ON are collectively equivalent to the rule ROK (compare theorem 4.3 (3)). The import of OD (the counterpart of P$\square$ in theorem 4.12) is that nothing impossible is obligatory. This is a version of the principle that *ought* implies *can*, and it may be distinguished from OD*, which more generally rules out pairs of obligations the contents of which are logically incompatible. We return to this point in section 6.5.

From the determination theorem for KD in chapter 5 it can be seen that D^* is determined by a class of standard models in which the relation – of 'deontic alternativeness' – between possible worlds is serial. Thus standard deontic logic is determined by the following account of the meaning of $\bigcirc$.

Relative to each possible world, including our own, there is a non-empty class of possible worlds that are deontic alternatives to the given world. A sentence of the form $\bigcirc A$ is true at a possible world just in case A is true at each of the world's deontic alternatives. Alternatively, one may picture the set of deontic alternatives to a world functioning collectively as a proposition that represents a standard of obligation: the proposition expressed by $\bigcirc A$ holds at a world if and only if the proposition expressed by A is entailed by the standard of obligation for the world. (Compare exercises 3.13 and 3.29.)

It is the axiom OD* (equivalently OD) that guarantees the existence of deontic alternatives – of a non-vacuous standard of obligation – for every possible world. Next we examine some further schemas that have been suggested as theorems of deontic logic, and assess them in terms of their implications for the structure of the relation of deontic alternativeness.

EXERCISES

6.1. Referring to theorem 4.12, verify that D^* is the smallest normal *KD*-system for $\bigcirc$ (ignore the lack of a counterpart to $\Diamond$ in the language).

6.2. Check that the two axiomatizations of D^* given in section 6.1 are equivalent, i.e. that they generate the same set of theorems. (See especially theorems 4.3 and 4.12.)

6.3. Formalize the idea of 'standards of obligation' in terms of models of the kind described in exercises 3.13 and 3.29.

6.4. The schema OD* is described as expressing the principle that there are no propositions that are jointly impossible but both obligatory. This suggests the following rule of inference in D^*.

$$\frac{A \to \neg B}{\bigcirc A \to \neg \bigcirc B}$$

Derive this rule. Then use it to derive OD*.

6.5. The sentence OD is described as expressing the principle that nothing impossible is obligatory. This suggests the following rule of inference in D^*.

$$\frac{\neg A}{\neg \bigcirc A}$$

Derive this rule using only propositional logic, ROM, and OD. Then use propositional logic and the rule to derive OD.

6.6. Introduce a permissibility operator P into the language and a theorem

$$\mathsf{P}A \leftrightarrow \neg \bigcirc \neg A$$

like Df$\Diamond$ into D^*. How plausible are the theorems that result ($\bigcirc A \to \mathsf{P}A$, for example)?

6.2. Further principles

By *deontic S5* is meant the normal extension of standard deontic logic obtained by adding as theorems the counterparts for $\bigcirc$ of the schemas 4 and 5, which we may put as follows.

$$\text{O4.} \quad \bigcirc A \to \bigcirc\bigcirc A$$

$$\text{O5.} \quad \neg\bigcirc A \to \bigcirc\neg\bigcirc A$$

Technically, deontic *S5* is the system *KD45* for $\bigcirc$, determined by the class of standard models in which deontic alternativeness is a serial transitive euclidean relation. So the import of deontic *S5* is that for each world there is a (non-empty) collection of 'best of all possible worlds' that form the world's standard of obligation – best in the sense that this standard holds as well for all the worlds within it.

Deontic *S5* seems too strong to capture the idea of obligation in the moral sense, though it may be appropriate to weaker notions of obligation such as that expressed by sentences in the imperative mood. So let us examine the import of O4 and O5 individually.

The axiom O4 is not altogether implausible. It is this principle that makes deontic alternativeness transitive, and thus makes possible the interpretation of the relation as leading to worlds that are in some way better from the standpoint of obligation. O4 means that what is obligatory at a world continues to be so at the world's deontic alternatives. So it rules out the possibility, for example, that some of the deontic alternatives to our world should have for themselves standards of obligation that are unrealistically high, perhaps utopian, from our point of view.

O5, on the other hand, does appear to be unreasonable. Reading $\neg\bigcirc\neg$ as expressing permissibility, for example, O5 implies that what is permissible ought to be permissible. Hence we should look for some system weaker than deontic *S5*.

One direction in which the system can be weakened is by replacing O5 by the deontic counterpart of U:

$$\text{OU.} \quad \bigcirc(\bigcirc A \to A)$$

Note that OU is a theorem of deontic *S5*, indeed follows from O5 given ROK and propositional logic (compare exercise 4.38). The schema expresses the thesis that it ought to be the case that whatever ought to be the case be the case. It is a much discussed principle in deontic logic, because it is one of the few plausible cases of a theorem of the form $\bigcirc A$ in which A is non-trivial (compare ON for example).

Semantically, OU means that the relation of deontic alternativeness is secondarily reflexive – deontic alternatives to a world are always deontic alternatives to themselves. But this condition has the following perhaps untoward consequence: if there are any unfulfilled obligations, then ours is one of the worst of all possible worlds. For suppose that for some A both $\bigcirc A$ and $\neg A$ are true. Then our world cannot be a deontic alternative to – cannot contribute to the standard of obligation of – any possible world (for at all deontic alternatives $\bigcirc A \rightarrow A$ is true). Thus ours is not among the better worlds relative to any world; no world is worse than our own.

The significance of OU, then, is that since what is not obligatory is not the case in some deontic alternative, it cannot be obligatory in *that* deontic alternative. That is, if $\bigcirc A$ is false, then, since A is false in some deontic alternative, $\bigcirc A$ cannot be true there.

Given the pessimism implied by this construal of deontic alternativeness, a more reasonable contention might be that since what is not obligatory is not the case in some deontic alternative, it cannot be obligatory in *every* deontic alternative. That is, one might argue the weaker conclusion that if $\bigcirc A$ is false, then, since A is false in at least one deontic alternative, $\bigcirc A$ cannot be true in all.

This weaker assumption is equivalent to the condition that the relation of deontic alternativeness be dense, i.e. that every deontic alternative is a deontic alternative to a deontic alternative (possibly, but not necessarily, itself). This at least has the merit that our world can have unfulfilled obligations without being at the bottom of the scale of standards of obligation.

Density as a condition on deontic alternativeness validates the schema

$$\text{O4}_c. \quad \bigcirc\bigcirc A \rightarrow \bigcirc A,$$

which is implied (given ROK and *PL*) by OU, just as secondary reflexivity implies density (compare exercises 3.58(*c*) and 4.33(*b*)).

Thus it appears that standard deontic logic might be strengthened by the addition of a reduction law,

$$\text{O4!}. \quad \bigcirc A \leftrightarrow \bigcirc\bigcirc A,$$

for the operator $\bigcirc$. (Note in this connection that from the standpoint of modalities the normal extension of *D** obtained by adding O4 and O4_c is equivalent to that obtained by adding O4 and OU; compare exercise 4.65.) We leave it for the reader to judge.

EXERCISES

6.7. Describe the conditions on the 'standards of obligation' models $\mathcal{M} = \langle W, f, P \rangle$ of exercise 6.3 that are required for the validity of the schemas O4 and O5 (as well as OD or OD*).

6.8. What is the effect on the set of valid sentences if 'best of all possible worlds' is construed as singular, i.e. if $f(\alpha)$ is required always to be a singleton set in models $\mathcal{M} = \langle W, f, P \rangle$ for deontic logic? That is, what is the significance of this constraint for the idea of a standard of obligation? (Exercises 4.52, 5.5, and 5.26 may be helpful.)

6.9. Let D_1^*, D_2^*, and D_3^* be the smallest normal extensions of D^* containing, respectively, O4 and O4$_c$, O4 and OU, and O4 and O5 (thus D_3^* is deontic *S5*). Prove that

$$D^* \subset D_1^* \subset D_2^* \subset D_3^*.$$

Each of the systems D_1^*, D_2^*, and D_3^* has only finitely many distinct modalities (constructed from $\cdot$, $\neg$, and $\bigcirc$). Identify these modalities and the implications among them. (There are relevant results in sections 4.4 and 5.2.)

6.10. Prove that D^* has the rule of inference

$$\frac{\bigcirc \mathrm{A}}{\mathrm{A}},$$

which means, roughly speaking, that there are no logically true statements of obligation with non-trivial content. Which of the systems D_1^*, D_2^*, and D_3^* in the preceding exercise have this rule? (Exercise 5.32 is relevant here.)

6.11. Which of the deontic logics discussed in sections 6.1 and 6.2 are decidable?

6.12. Examine some further schemas containing $\bigcirc$ with a view to their plausibility as theorems of deontic logic.

6.3. Obligation and time

Obligations arise and pass away as a function of circumstances. In particular, what is obligatory at a possible world varies from time to time. Some obligations endure, of course, but others are merely transient.

To this point we have analyzed the logic of obligation in terms of standard models $\mathcal{M} = \langle W, R, P \rangle$ in which R is a serial relation of deontic alternativeness. But these models do not reflect the dependence of obligation on time. To remedy this, we shall modify the models so as to recognize explicitly the role of temporal parameters in the determination of obligation. What follows is but one way of accomplishing this – there are others – but it gives a striking result about what we call 'past tense obligations'.

Let us think of time as a set of discrete moments, ordered in a linear way, without end in past or future. Then time can be represented by the set of integers,

$$Z = \{..., -1, 0, +1, ...\},$$

and the relations of earlier and later are represented by $<$ and $>$.

Since we wish to regard possible worlds as time-stretched, let us construe a possible world as a function on Z into an otherwise unspecified set of momentary world-states. Assuming this much, we can define a relation of historical identity: worlds α and β have the same history at a time t – written $\alpha \sim_t \beta$ – just in case they are identical up to t. Formally:

$$\alpha \sim_t \beta \text{ iff } \alpha(t') = \beta(t') \text{ for every } t' < t.$$

The relation of deontic alternativeness between worlds is of course relativized to times, and it is to be constrained by the relation of historical identity. Specifically, we insist that β is a deontic alternative to α at a time t only if α and β are historically identical at t. Formally:

$$(\mathrm{R}_t) \quad \text{if } \alpha R_t \beta, \text{ then } \alpha \sim_t \beta$$

This constraint gives voice to the view that obligations accrue to a world as a function of events and actions that have occurred there; a world and its deontic alternatives are merely different outcomes of their yesterdays' events.

The evaluation of atomic sentences is also indexed by the set of times. So $P_t(n)$ is the set of worlds for which $\mathbb{P}_n$ is true at t. In addition we put the following condition on P.

$$(\mathrm{P}_t) \quad \text{if } \alpha(t) = \beta(t), \text{ then } \alpha \in P_t(n) \text{ iff } \beta \in P_t(n)$$

This realizes our intention that the value of $\mathbb{P}_n$ at a time t in a world α be a function solely of the world-state $\alpha(t)$.

With models $\mathcal{M} = \langle W, R, P \rangle$ thus revised so as to account for time, sentences are evaluated with respect to pairs $\langle \alpha, t \rangle$ of worlds and times. In other words, we write $\models^{\mathcal{M}}_{\langle \alpha, t \rangle} \mathrm{A}$ to mean that A is true in α at t. The following are the interesting clauses of the definition of truth.

(1) $\models^{\mathscr{M}}_{\langle\alpha,t\rangle} \mathbb{P}_n$ iff $\alpha \in P_t(n)$, for $n = 0, 1, 2, \ldots$.

(2) $\models^{\mathscr{M}}_{\langle\alpha,t\rangle} \bigcirc$A iff for every β in $\mathscr{M}$ such that $\alpha R_t \beta$, $\models^{\mathscr{M}}_{\langle\beta,t\rangle}$ A.

Despite the temporal relativization of the models, nothing changes in the logic of $\bigcirc$: the valid sentences are just as they were before (valid now means true at every point $\langle\alpha, t\rangle$ in every model). So, on the assumption that deontic alternativeness is serial at every moment, we still have standard deontic logic, D^*.

The effect of the new modeling emerges more clearly, however, if we introduce into the language two further operators, $\boxdot$ and $\diamond$, to express notions of historical necessity and possibility. Then we add to the truth definition the following clauses.

(3) $\models^{\mathscr{M}}_{\langle\alpha,t\rangle} \boxdot$A iff for every β in $\mathscr{M}$ such that $\alpha \sim_t \beta$, $\models^{\mathscr{M}}_{\langle\beta,t\rangle}$ A.

(4) $\models^{\mathscr{M}}_{\langle\alpha,t\rangle} \diamond$A iff for some β in $\mathscr{M}$ such that $\alpha \sim_t \beta$, $\models^{\mathscr{M}}_{\langle\beta,t\rangle}$ A.

In other words, $\boxdot$A holds at a world and time if and only if A holds at all worlds having the same history as the given world at that time, and $\diamond$A holds just in case there is some such world at which A is true.

Since $\sim_t$ is an equivalence relation for each time t, the logic of $\boxdot$ and $\diamond$ is at least *S5* (i.e. *KT5*; compare theorem 5.14). The following validities are representative of the interaction of the operators $\boxdot$, $\diamond$, and $\bigcirc$.

(*a*) $\boxdot$A $\to \bigcirc$A

(*b*) $\bigcirc$A $\to \diamond$A

(*c*) $\boxdot$A $\leftrightarrow \bigcirc\boxdot$A

(*d*) $\diamond$A $\leftrightarrow \bigcirc\diamond$A

The first of these corresponds to condition (R_t). The second, it should be noted, is another, stronger version of the thesis that *ought* implies *can*; it is a consequence of (R_t) and the timewise seriality of deontic alternativeness. According to the last two schemas, statements of obligation are deontically vacuous when they concern what is historically necessary or possible – vacuous in the sense that the deontic sentences are equivalent to their contents.

Let us define the operator $\odot$ as follows.

$$\odot A = \boxdot A \vee \boxdot \neg A.$$

This operator expresses a notion of historical determinacy. Given a possible world and time, $\odot$A means that the proposition expressed by A either holds at every world having the same history or fails at all such.

The point of introducing $\odot$ is to give succinct expression to the fact that historically determinate propositions are deontically vacuous. That is, the following schema is valid.

$$(e)\ \odot A \rightarrow (A \leftrightarrow \bigcirc A)$$

This raises the question of whether there are any valid sentences of the form $\odot$A. We turn to this in the next section.

EXERCISES

6.13. Verify that, for each t in Z, $\sim_t$ is an equivalence relation. Then argue that the logic of $\boxdot$ and $\Diamond$ is at least a normal *KT5*-system.

6.14. Prove the validity of the following schemas in section 6.3.

$$(a)\ \boxdot A \rightarrow \bigcirc A$$
$$(b)\ \bigcirc A \rightarrow \Diamond A$$
$$(c)\ \boxdot A \leftrightarrow \bigcirc \boxdot A$$
$$(d)\ \Diamond A \leftrightarrow \bigcirc \Diamond A$$

6.15. Derive ON ($\bigcirc\top$) and OD ($\neg\bigcirc\bot$) from (*a*) and (*b*) in the preceding exercise together with appropriate principles for $\boxdot$ and $\Diamond$.

6.16. State formally the truth conditions for sentences of the form $\odot$A.

6.17. Prove the validity of the following schemas.

$$\odot A \leftrightarrow \odot \neg A \qquad \odot A \rightarrow (\neg A \leftrightarrow \bigcirc \neg A)$$

6.18. Using the schema

$$(e)\ \odot A \rightarrow (A \leftrightarrow \bigcirc A)$$

together with principles for $\boxdot$ and $\Diamond$, derive schemas (*c*) and (*d*) in exercise 6.14.

6.4. Past tense obligation

Let us add to the language operators [P], $(-)$, and $\langle P \rangle$. The members of this trio correspond to simple past tense constructions, with readings, respectively, *it always was the case that, at the moment just past it was the case that*, and *it* (*at least*) *once was the case that*. Formally, these operators are evaluated as follows.

(1) $\models^{\mathcal{M}}_{\langle \alpha, t \rangle}$ [P]A iff for every t' in Z such that $t' < t$, $\models^{\mathcal{M}}_{\langle \alpha, t' \rangle}$ A.

(2) $\models^{\mathcal{M}}_{\langle \alpha, t \rangle}$ (−)A iff $\models^{\mathcal{M}}_{\langle \alpha, t-1 \rangle}$ A.

(3) $\models^{\mathcal{M}}_{\langle \alpha, t \rangle}$ ⟨P⟩A iff for some t' in Z such that $t' < t$, $\models^{\mathcal{M}}_{\langle \alpha, t' \rangle}$ A.

At the end of the last section we asked whether there are any valid sentences of the form ⊙A, i.e. whether there are any valid historically determinate propositions. With the introduction of the operators [P], (−), and ⟨P⟩, the answer is affirmative.

By Ⓟ let us understand any genuine past tense modality, i.e. any finite, non-empty sequence of the operators [P], (−), and ⟨P⟩. Then where A contains no occurrences of ⊡, ◇, and ○, the sentence

⊙ⓅA

is valid. That is to say, pure past tense sentences are historically determinate.

Without going into the details of a full proof, we can illustrate this claim by showing the validity of the simple sentence

⊙(−)$\mathbb{P}_0$.

For this it is enough to prove that, for any pair $\langle \alpha, t \rangle$ in a model $\mathcal{M} = \langle W, R, P \rangle$, if

(*a*) $\models^{\mathcal{M}}_{\langle \beta, t \rangle}$ (−)$\mathbb{P}_0$ for some β in $\mathcal{M}$ such that $\alpha \sim_t \beta$,

then

(*b*) $\models^{\mathcal{M}}_{\langle \gamma, t \rangle}$ (−)$\mathbb{P}_0$ for every γ in $\mathcal{M}$ such that $\alpha \sim_t \gamma$.

So assume (*a*). Then

$\models^{\mathcal{M}}_{\langle \beta, t-1 \rangle}$ $\mathbb{P}_0$ for some β in $\mathcal{M}$ such that $\alpha \sim_t \beta$,

which means that for some β in $\mathcal{M}$ such that $\alpha \sim_t \beta$,

$\beta \in P_{t-1}(0)$.

To show (*b*), let γ be a world in $\mathcal{M}$ such that $\alpha \sim_t \gamma$; it is sufficient to argue that $\gamma \in P_{t-1}(0)$. But since $t-1 < t$, $\beta(t-1) = \gamma(t-1)$, and so by condition (P_t) in the preceding section,

$\beta \in P_{t-1}(0)$ iff $\gamma \in P_{t-1}(0)$.

From this the conclusion follows at once.

Given the historical determinacy of pure past tense propositions and the deontic vacuity of historically determinate propositions ((*e*) in the preceding section), we see that pure past tense propositions are always deontically vacuous. In other words, where A is devoid of ⊡, ◇, and ○, the sentence

$$Ⓟ\mathrm{A} \leftrightarrow ○Ⓟ\mathrm{A}$$

is valid. According to this, present obligations concerning past events cannot have any deontic force: 'You ought (now) to go to the zoo yesterday' means no more than 'You went to the zoo yesterday'.

EXERCISES

6.19. Describe some principles about past tenses, i.e. some valid sentences involving the operators [P], $(-)$, and $\langle P \rangle$.

6.20. Prove the validity of the following 'induction schema' for past tenses.

$$((-)\mathrm{A} \wedge [\mathrm{P}](\mathrm{A} \rightarrow (-)\mathrm{A})) \rightarrow [\mathrm{P}]\mathrm{A}$$

6.21. Examine some more instances of ⊙ⓅA (where A does not contain ⊡, ◇, or ○) and show their validity. Explain how a general proof of the validity of this schema might proceed.

6.22. Introduce into the language operators [F], $(+)$, and $\langle F \rangle$ with evaluations as follows.

(1) $\models^{\mathscr{M}}_{\langle \alpha, t\rangle} [\mathrm{F}]\mathrm{A}$ iff for every t' in Z such that $t' > t$, $\models^{\mathscr{M}}_{\langle \alpha, t'\rangle} \mathrm{A}$.

(2) $\models^{\mathscr{M}}_{\langle \alpha, t\rangle} (+)\mathrm{A}$ iff $\models^{\mathscr{M}}_{\langle \alpha, t+1\rangle} \mathrm{A}$.

(3) $\models^{\mathscr{M}}_{\langle \alpha, t\rangle} \langle \mathrm{F}\rangle\mathrm{A}$ iff for some t' in Z such that $t' > t$, $\models^{\mathscr{M}}_{\langle \alpha, t'\rangle} \mathrm{A}$.

Identify some validities involving these 'future tense' operators. What is valid when these are mixed with [P], $(-)$, and $\langle P \rangle$?

6.23. Using the operators introduced in the preceding exercise, as well as the rest, formulate a principle of fatalism – 'What will be will be'.

6.5. Shortcomings

There are two important criticisms of standard deontic logic and its account of obligation. One raises doubt about the adequacy of the analysis; the other calls into question its correctness.

The doubt about adequacy concerns the expression of conditional obligation. Some obligations seem to be unconditional. For example, you ought not to cough during the concert. But if you do, you ought to apologize. The obligation to apologize is conditional on having coughed.

Let us represent the conditional obligation of B given that A by a sentence of the form

$$\bigcirc(B/A).$$

The question then is whether this form is definable in terms of $\bigcirc$ and other familiar operators. The obvious suggestions for a definition are

$$A \rightarrow \bigcirc B,$$

$$\bigcirc(A \rightarrow B),$$

and

$$\Box(A \rightarrow \bigcirc B),$$

where $\Box$ expresses some suitable notion of necessity.

But all these fail. The first makes $\bigcirc(B/A)$ true whenever A is false or $\bigcirc B$ is true. The second makes $\bigcirc(B/A)$ true whenever $\bigcirc\neg A$ or $\bigcirc B$ is. And all three have the unwanted consequence that if $\bigcirc(B/A)$ is true so is $\bigcirc(B/A \wedge A')$, where A' is any additional condition whatsoever.

So it seems that the operator $\bigcirc(\ /\ )$ is either primitive – i.e. genuinely novel – or definable only by means at present beyond ours. We return to this matter in chapter 10.

The criticism touching on the correctness of standard deontic logic has two parts.

First, by ROK, $\bigcirc A$ is a theorem if A is. So the logic is committed to the view that obligations always exist, however trivial they may be (compare ON). But it seems reasonable to assume that there exist possible worlds (presumably very unlike our own) at which nothing at all is obligatory.

Second, and more serious, is the question of the correctness of the theorem

$$\text{OD*}. \quad \neg(\bigcirc A \wedge \bigcirc\neg A).$$

It is a matter of controversy whether deontic logic should thus rule on the question whether contrary obligations $\bigcirc A$ and $\bigcirc\neg A$ are always inconsistent. Indeed, it is arguable that the possibility of such conflict is a main feature of some concepts of obligation, that it is often this, for example, that gives moral dilemmas their poignancy.

The difficulty is that OD* is a theorem of any normal system for $\bigcirc$ in which the sentence

$$\text{OD}. \quad \neg\bigcirc\bot$$

is a theorem. But OD is relatively uncontroversial, since it merely denies the existence of impossible obligations.

There is a more persuasive point to be made here. In any normal system for $\bigcirc$ the deontic theses OD and OD* are equivalent – in the strong sense that their biconditional,

$$\neg\bigcirc\bot \leftrightarrow \neg(\bigcirc A \wedge \bigcirc\neg A),$$

is a theorem. Semantically, this means that in any standard model the propositions expressed by OD and OD* are identical: there is no way to distinguish the principle that *ought* implies *can* from the principle that conflicts of obligation cannot exist.

To the extent that the propositions expressed by OD and OD* are indistinguishable, so that OD cannot be a theorem without OD*, the analysis of deontic logic in terms of normal systems and standard models is a failure.

If we examine the axiomatization of *D** in terms of ROM, OC, ON, and OD, in section 6.1, we can extract a weaker, more plausible system of deontic logic – to wit, the system formed on the basis of propositional logic by ROM and OD. Let us call this *minimal deontic logic*, or simply *D*.

By adopting ROM we accept the principle that obligation is closed under implication – that a proposition is obligatory if it is implied by an obligatory proposition. And, of course, OD rules out obligations that are impossible simpliciter.

By abandoning ON we give up the view that obligations are present at every possible world. By rejecting OC, on the other hand, the assumption of the uniqueness of a world's standard of obligation falls away. And without this there is a no implication from OD to OD*.

The system *D* is not normal, and models for it cannot be standard. So we must seek more subtlety in our semantic and proof-theoretic analyses of modality generally. This is the point of the chapters that follow.

EXERCISES

6.24. With regard to the suggested definientia for $\bigcirc(B/A)$,

$$A \rightarrow \bigcirc B,\ \bigcirc(A \rightarrow B),\ \Box(A \rightarrow \bigcirc B),$$

prove:

(*a*) The first makes $\bigcirc(B/A)$ true if A is false or $\bigcirc B$ is true.

(*b*) The second makes $\bigcirc(B/A)$ true if either $\bigcirc\neg A$ or $\bigcirc B$ is.

(*c*) For all three, if $\bigcirc(B/A)$ is true then so is $\bigcirc(B/A \wedge A')$. (Assume the logic of $\Box$ is normal.)

6.25. Prove that the schema

$$\neg \bigcirc \bot \leftrightarrow \neg(\bigcirc A \wedge \bigcirc \neg A)$$

is a theorem of any normal system for $\bigcirc$. (Compare exercise 4.7.)

6.26. Using the result in the preceding exercise, explain why the propositions expressed by OD and OD* are the same in any standard model.

6.27. By retaining the rule ROM in the system *D* we embrace the theorem

$$\bigcirc A \rightarrow \bigcirc(A \vee B).$$

(Prove this.) This means that the inference from 'You should post this letter' to 'You should post or burn this letter' is correct. Is this defensible?

6.28. Describe a class of models of the kind in exercise 3.14 with respect to which the system *D* is sound. Describe countermodels in this class for ON and instances of OC and OD*.

PART III

7

MINIMAL MODELS FOR MODAL LOGICS

The truth conditions for modal sentences in minimal models are a generalization of those in chapter 3. Possible worlds continue to figure in the semantic analysis of necessity and possibility, but the meanings of modal sentences are given a much simpler account. A necessitation $\Box$A is said to be true at a possible world just in case the proposition expressed by A is in a certain but very general sense necessary with respect to the world; and $\Diamond$A is true at a world if and only if the proposition expressed by A is, in a corresponding sense, possible. The resulting notion of validity is such that far fewer principles hold generally on this account than did in chapter 3.

In section 7.1 we set out the definition of a minimal model, state the truth conditions of modal sentences, and prove the basic theorem about validity in classes of minimal models. In section 7.2 we examine M, C, and N from the standpoint of minimal models and define some key concepts for the treatment of certain logics involving these schemas. Section 7.3 contains a theorem to the effect that standard models can be identified with minimal models of a special kind. Section 7.4 briefly describes conditions on minimal models sufficient for the validation of the schemas D, T, B, 4, and 5. In section 7.5 we introduce the idea of filtration for minimal models and state some theorems that we use at the end of chapter 9 to prove determination theorems.

7.1. Minimal models

A *minimal model* is a structure

$$\mathcal{M} = \langle W, N, P \rangle$$

in which, as before, W is a set of possible worlds and P gives a truth value to each atomic sentence at each world. The new component, N, is a function that associates with each possible world a collection of sets of possible worlds. Formally:

DEFINITION 7.1. $\mathcal{M} = \langle W, N, P \rangle$ is a *minimal model* iff:

(1) W is a set.

(2) N is a mapping from W to sets of subsets of W (i.e. $N_\alpha \subseteq \mathscr{P}(W)$, for each world α in W).

(3) P is a mapping from natural numbers to subsets of W (i.e. $P_n \subseteq W$, for each natural number n).

Once again, as we indicated in the introduction, the idea of this modeling is that each possible world α in a minimal model $\mathcal{M} = \langle W, N, P \rangle$ has associated with it a set N_α of propositions that are in some sense necessary at α. Since we shall identify a proposition in $\mathcal{M}$ with a set of possible worlds in $\mathcal{M}$, N_α becomes a collection of subsets of W. It must be emphasized that N_α may be any collection of propositions, including the empty collection; we make no assumptions about the nature of N except that it be a function from W to $\mathscr{P}(\mathscr{P}(W))$. This point will be better appreciated when we describe some minimal countermodels to familiar schemas in the next section.

The interpretation of the necessity operator in minimal models is thus quite simple and natural: we shall say that a sentence of the form $\Box$A is true at α in $\mathcal{M}$ just in case the proposition expressed by A – the truth set $\|A\|^{\mathcal{M}}$ – is among those necessary at α, i.e. is among those in N_α.

DEFINITION 7.2. Let α be a world in a minimal model $\mathcal{M} = \langle W, N, P \rangle$.

(1) $\models^{\mathcal{M}}_{\alpha} \Box A$ iff $\|A\|^{\mathcal{M}} \in N_\alpha$.
(2) $\models^{\mathcal{M}}_{\alpha} \Diamond A$ iff $-\|A\|^{\mathcal{M}} \notin N_\alpha$.

Clause (2) above deserves comment. This treatment of the meaning of the possibility operator in minimal models simply reflects our intention that the notions of necessity and possibility be dual, in particular that $\Diamond$ have the meaning $\neg\Box\neg$ (as we see in theorem 7.3 below). For clause (2) stipulates that $\Diamond$A is true at α in $\mathcal{M}$ just in case the denial of the proposition expressed by A, i.e. $-\|A\|^{\mathcal{M}}$, is not necessary at α, i.e. is not in N_α. (By $-\|A\|^{\mathcal{M}}$ we mean of course the set $W - \|A\|^{\mathcal{M}}$; wherever possible we shall use the shorter form of expression.)

By way of an example of the behavior of the function N in minimal models, let us return to the interpretation of $\Box$ as expressing a notion of obligation. We suggested in chapter 6 that P$\Box$, $\neg\Box\bot$, may be a law of deontic logic while O$\Box$, $\neg(\Box A \wedge \Box\neg A)$, is not; but we have observed that the biconditional of these is true in every standard model, so that one is valid in a class of standard models if and only if the other is. Within the framework of minimal models it is possible to distinguish the propositions expressed by P$\Box$ and O$\Box$ and so have one valid without the other. Specifically, consider the class of minimal models $\mathcal{M} = \langle W, N, P \rangle$ such that for no α in $\mathcal{M}$ does N_α contain $\emptyset$. Then the proposition $\|\bot\|^{\mathcal{M}}$ is contained in no N_α, and so P$\Box$ is true at every world in every such model. None the less, there are countermodels to instances of O$\Box$ within this class of models; we leave it as an exercise for the reader to discover some.

For another illustration of the role of N in minimal models, suppose $\Box$ to be a present continuous tense operator, with a reading like 'it is (being) the case that'. We think of the possible worlds in a minimal model as points in time (on the real line, let us say), and for each moment α we take N_α to be the set of open intervals around α. In this way $\Box$A is true at a moment α if and only if the proposition expressed by A is an open interval around α, i.e. if and only if A is continuously true throughout some (open) interval that contains the (present) moment α.

Let us close this section with the following theorem, which states that the schema Df$\Diamond$ is true in every minimal model, and that validity in a class of minimal models is preserved by the rule RE.

THEOREM 7.3. *Let* C *be a class of minimal models. Then:*

(1) $\models_{\mathsf{C}} \Diamond A \leftrightarrow \neg\Box\neg A$.

(2) *If* $\models_{\mathsf{C}} A \leftrightarrow B$, *then* $\models_{\mathsf{C}} \Box A \leftrightarrow \Box B$.

Proof. The proof for (1) rehearses our remarks earlier in the section. Let α be a world in a minimal model $\mathcal{M} = \langle W, N, P \rangle$. Then:

$\models_\alpha^{\mathcal{M}} \Diamond A$ iff $-\|A\|^{\mathcal{M}} \notin N_\alpha$
– definition 7.2 (2);

iff $\|\neg A\|^{\mathcal{M}} \notin N_\alpha$
– theorem 2.10;

iff not $\models_\alpha^{\mathcal{M}} \Box\neg A$
– definition 7.2 (1);

iff $\models_\alpha^{\mathcal{M}} \neg\Box\neg A$
– definition 2.5 (4).

Therefore, $\models_\alpha^{\mathcal{M}} \Diamond A \leftrightarrow \neg\Box\neg A$, for every world α in every minimal model $\mathcal{M}$.

For (2). Suppose that C is a class of minimal models such that $\models_{\mathsf{C}} A \leftrightarrow B$, so that $\|A\|^{\mathcal{M}} = \|B\|^{\mathcal{M}}$ for every $\mathcal{M}$ in C. From this it follows that for any world α in any model $\mathcal{M} = \langle W, N, P\rangle$ in C, $\|A\|^{\mathcal{M}} \in N_\alpha$ if and only if $\|B\|^{\mathcal{M}} \in N_\alpha$. So foɹ any α in any $\mathcal{M}$ in C, $\models^{\mathcal{M}}_{\alpha} \Box A$ if and only if $\models^{\mathcal{M}}_{\alpha} \Box B$, which means that $\models_{\mathsf{C}} \Box A \leftrightarrow \Box B$.

Theorem 7.3 provides (as the discerning reader no doubt suspects) the basis of the soundness results for the classical modal logics introduced in the next chapter. Indeed, as we shall see, the theorem means that every class of minimal models determines a classical modal logic.

EXERCISES

7.1. Show that each of the following can be falsified in minimal models.

M. $\Box(A \wedge B) \to (\Box A \wedge \Box B)$

C. $(\Box A \wedge \Box B) \to \Box(A \wedge B)$

N. $\Box\top$

7.2. Falsify an instance of O$\Box$ – $\neg(\Box A \wedge \Box\neg A)$ – in a minimal model $\mathcal{M} = \langle W, N, P\rangle$ for which it holds that $\varnothing \notin N_\alpha$ for every α in $\mathcal{M}$.

7.3. Let C be any class of minimal models. Prove:

(*a*) $\models_{\mathsf{C}} \Box A \leftrightarrow \neg\Diamond\neg A$.

(*b*) If $\models_{\mathsf{C}} A \leftrightarrow B$, then $\models_{\mathsf{C}} \Diamond A \leftrightarrow \Diamond B$.

7.4. Prove that the following are valid in any class of minimal models.

(*a*) $\Diamond\top \leftrightarrow \neg\Box\bot$ (*b*) $\Box\top \leftrightarrow \neg\Diamond\bot$

7.5. We say that minimal models $\mathcal{M} = \langle W, N, P\rangle$ and $\mathcal{M}' = \langle W', N', P'\rangle$ agree on the atoms of A just in case (i) $W = W'$, (ii) $N = N'$, and (iii) $P_n = P'_n$ whenever $\mathbb{P}_n$ is an atomic subsentence of A. Prove that if $\mathcal{M}$ and $\mathcal{M}'$ agree on the atoms of A then they agree on A, in the sense that $\models^{\mathcal{M}}_{\alpha} A$ if and only if $\models^{\mathcal{M}'}_{\alpha} A$ for every α in $\mathcal{M}$. (The proof is by induction on the complexity of A. Give it at least for the cases in which A is atomic, the falsum, a conditional, and a necessitation. Compare exercise 3.4.)

7.6. Consider the following conditions on a minimal model $\mathcal{M} = \langle W, N, P\rangle$, for every world α and proposition (i.e. set of worlds) X and Y in $\mathcal{M}$:

(m) if $X \cap Y \in N_\alpha$, then $X \in N_\alpha$ and $Y \in N_\alpha$

(c) if $X \in N_\alpha$ and $Y \in N_\alpha$, then $X \cap Y \in N_\alpha$

(n) $W \in N_\alpha$

Prove that the schemas M, C, and N are valid in classes of minimal models satisfying, respectively, conditions (m), (c), and (n).

7.7. For a minimal model $\mathcal{M} = \langle W, N, P \rangle$ inductively define N^n so that for every α in $\mathcal{M}$ and every natural number n:

(*a*) $\models_\alpha^{\mathcal{M}} \Box^n A$ iff $\|A\|^{\mathcal{M}} \in N_\alpha^n$.
(*b*) $\models_\alpha^{\mathcal{M}} \Diamond^n A$ iff $-\|A\|^{\mathcal{M}} \notin N_\alpha^n$.

Then prove (*a*) and (*b*) (by induction on n).

7.8. Let $\mathcal{M} = \langle W, N, P \rangle$ be a minimal model, but suppose the truth conditions of modal sentences are given like this:

(*a*) $\models_\alpha^{\mathcal{M}} \Diamond A$ iff $\|A\|^{\mathcal{M}} \in N_\alpha$.
(*b*) $\models_\alpha^{\mathcal{M}} \Box A$ iff $-\|A\|^{\mathcal{M}} \notin N_\alpha$.

Prove theorem 7.3 and (*a*) and (*b*) in exercise 7.3, using these truth conditions. What difference, if any, is there between the systems of modal logic determined by the class of all minimal models under the different ways of evaluating modal sentences?

7.9. Let $\mathcal{M} = \langle W, N, P \rangle$ be a minimal model, but suppose truth conditions for necessitations are given by:

$$\models_\alpha^{\mathcal{M}} \Box A \text{ iff for some } X \in N_\alpha,\ X \subseteq \|A\|^{\mathcal{M}}.$$

So to speak, $\Box A$ is true at α in $\mathcal{M}$ if and only if there is a proposition necessary at α that entails the proposition expressed by A in $\mathcal{M}$.

(*a*) State truth conditions for possibilitations so that Df$\Diamond$ is valid in any class of minimal models.
(*b*) Let C be any class of minimal models. Prove that if $\models_{\mathsf{C}} A \rightarrow B$, then $\models_{\mathsf{C}} \Box A \rightarrow \Box B$.
(*c*) Prove theorem 7.3, relative to the revised truth conditions for modal sentences.
(*d*) Which, if any, of schemas M, C, and N are valid in the class of minimal models, given the revised truth conditions?

7.10. Consider a model $\mathcal{M} = \langle W, f, P \rangle$ in which W and P are as usual and f is a mapping from propositions to propositions (i.e. for any set X of worlds in $\mathcal{M}$, $f(X)$ is a set of worlds in $\mathcal{M}$). Relative to a world α in $\mathcal{M}$ we define the truth conditions for necessitations by:

$$\models_\alpha^{\mathcal{M}} \Box A \text{ iff } \alpha \in f(\|A\|^{\mathcal{M}}).$$

Equivalently we may say:

$$\|\Box A\|^{\mathcal{M}} = f(\|A\|^{\mathcal{M}}).$$

(*a*) Prove the equivalence of these formulations.

(*b*) Prove part (2) of theorem 7.3, where C is any class of models of this sort.

(*c*) State truth conditions for possibilitations so that Df$\Diamond$ is valid in any class of such models.

(*d*) Given a minimal model $\mathcal{M} = \langle W, N, P\rangle$ we can define an equivalent model $\mathcal{M}' = \langle W, f, P\rangle$ by stipulating that for any α and X in $\mathcal{M}$, $\alpha \in f(X)$ if and only if $X \in N_\alpha$. Prove that $\mathcal{M}$ and $\mathcal{M}'$ are pointwise equivalent.

(*e*) Given a model $\mathcal{M} = \langle W, f, P\rangle$, define an equivalent minimal model $\mathcal{M}' = \langle W, N, P\rangle$, and prove their pointwise equivalence.

(*f*) Conclude from (*d*) and (*e*) that models of this new sort are essentially the same as minimal models.

7.11. (This exercise presupposes a knowledge of boolean algebra.) A boolean algebra is a structure

$$\mathscr{B} = \langle B, 1, 0, -, \cap, \cup\rangle$$

in which B is a set containing 1 (the unit element) and 0 (the zero element) and closed under the unary operation $-$ (boolean complementation) and the binary operations $\cap$ (boolean intersection or meet) and $\cup$ (boolean union or join). We assume the relation $\leqslant$ of boolean inclusion to be defined – for example, $a \leqslant b$ if and only if any one of the following:

$$a = a \cap b, \quad a \cup b = b, \quad a \cap -b = 0, \quad -a \cup b = 1.$$

By a *modal algebra* we mean a structure $\langle \mathscr{B}, *\rangle$ in which $\mathscr{B}$ is a boolean algebra and $*$ is a unary operation in $\mathscr{B}$, an algebraic counterpart of necessitation.

An *algebraic model* is a structure $\mathcal{M} = \langle \mathscr{B}, *, P\rangle$ in which $\mathscr{B}$ and $*$ form a modal algebra and P is a mapping from the set of natural numbers to elements of $\mathscr{B}$. $\mathcal{M}$ is said to be *finite* just in case $\mathscr{B}$ has a finite number of elements.

Intuitively, the points in $\mathcal{M}$ (i.e. in $\mathscr{B}$) may be thought of as propositions – including 'truth' and 'falsity' (1 and 0) – closed under propositional analogues of negation, conjunction, disjunction, and necessitation ($-$, $\cap$, $\cup$, and $*$). Then P is in effect an assignment of propositions in $\mathcal{M}$ to the atomic sentences.

It should be emphasized in this connection that the points in an algebraic model are not possible worlds, nor even, necessarily, sets of possible worlds. Though sometimes definable, possible worlds are not in general a feature of algebraic models.

In an algebraic model $\mathcal{M} = \langle \mathcal{B}, *, P \rangle$ sentences are evaluated by a mapping $\| \ \|^{\mathcal{M}}$ to points (propositions) in $\mathcal{M}$. Here is the definition:

(1) $\|\mathbb{P}_n\|^{\mathcal{M}} = P_n$, for $n = 0, 1, 2, \ldots$.
(2) $\|\top\|^{\mathcal{M}} = 1$.
(3) $\|\bot\|^{\mathcal{M}} = 0$.
(4) $\|\neg A\|^{\mathcal{M}} = -\|A\|^{\mathcal{M}}$.
(5) $\|A \wedge B\|^{\mathcal{M}} = \|A\|^{\mathcal{M}} \cap \|B\|^{\mathcal{M}}$.
(6) $\|A \vee B\|^{\mathcal{M}} = \|A\|^{\mathcal{M}} \cup \|B\|^{\mathcal{M}}$.
(7) $\|A \rightarrow B\|^{\mathcal{M}} = -\|A\|^{\mathcal{M}} \cup \|B\|^{\mathcal{M}}$.
(8) $\|A \leftrightarrow B\|^{\mathcal{M}} = (-\|A\|^{\mathcal{M}} \cup \|B\|^{\mathcal{M}}) \cap (-\|B\|^{\mathcal{M}} \cup \|A\|^{\mathcal{M}})$.
(9) $\|\Box A\|^{\mathcal{M}} = *\|A\|^{\mathcal{M}}$.
(10) $\|\Diamond A\|^{\mathcal{M}} = -*-\|A\|^{\mathcal{M}}$.

A sentence A is said to be true in an algebraic model $\mathcal{M}$ – written $\models^{\mathcal{M}} A$ – if and only if $\|A\|^{\mathcal{M}} = 1$. And A is valid in a class C of algebraic models – $\models_{C} A$ – just in case $\models^{\mathcal{M}} A$ for every $\mathcal{M}$ in C.

Prove, for any algebraic model $\mathcal{M}$:

(*a*) $\models^{\mathcal{M}} A \rightarrow B$ iff $\|A\|^{\mathcal{M}} \leqslant \|B\|^{\mathcal{M}}$.
(*b*) $\models^{\mathcal{M}} A \leftrightarrow B$ iff $\|A\|^{\mathcal{M}} = \|B\|^{\mathcal{M}}$.

Prove, for any class C of algebraic models:

(*c*) $\models_{C} \Diamond A \leftrightarrow \neg\Box\neg A$.
(*d*) If $\models_{C} A \leftrightarrow B$, then $\models_{C} \Box A \leftrightarrow \Box B$.

Prove:

(*e*) None of the schemas M, C, and N is valid in the class of all algebraic models.
(*f*) For every minimal model there is an equivalent algebraic model. (Compare exercise 7.10(*d*).)

We should note that the set of sentences valid in a class C of algebraic models is closed under tautological consequence (RPL); i.e. if A is a tautological consequence of $A_1, \ldots, A_n$, each of which is valid in C, then A is valid in C too. This is perhaps obvious, but we shall not prove it.

7.2. The schemas M, C, and N

Let us consider the following schemas.

M. $\Box(A \wedge B) \rightarrow (\Box A \wedge \Box B)$

C. $(\Box A \wedge \Box B) \rightarrow \Box(A \wedge B)$

N. $\Box\top$

Though these are all valid in any class of standard models, each has a counterexample in a minimal model. This is the content of the following theorem.

THEOREM 7.4. *None of the schemas* M, C, *and* N *is valid in the class of all minimal models.*

Proof. For each schema it is sufficient to describe an instance and a minimal model that falsifies it. We begin with N, since it is the simplest.

For N. Here the instance is just $\Box\top$. Let $\mathcal{M} = \langle W, N, P\rangle$ be a minimal model such that $W = \{\alpha\}$ and $N_\alpha = \varnothing$ (it does not matter about P). Thus $\mathcal{M}$ contains just one world, α, and at that world there are no necessary propositions. In particular, the proposition W – i.e. $\|\top\|^{\mathcal{M}}$ – is not necessary at α, so that $\Box\top$ is false at α. So N has a minimal countermodel.

Notice that the model above continues to falsify N if $N_\alpha = \{\varnothing\}$. In this variation the proposition $\|\bot\|^{\mathcal{M}}$ is necessary at α even though $\|\top\|^{\mathcal{M}}$ is not.

For M. Consider the instance $\Box(\mathbb{P}_0 \wedge \mathbb{P}_1) \rightarrow (\Box\mathbb{P}_0 \wedge \Box\mathbb{P}_1)$, and let $\mathcal{M} = \langle W, N, P\rangle$ be a minimal model in which $W = \{\alpha, \beta\}$ (distinct), $N_\alpha = \{\varnothing\}$, $P_0 = \{\alpha\}$, and $P_1 = \{\beta\}$. Then neither the proposition $\{\alpha\}$ expressed by $\mathbb{P}_0$ nor the proposition $\{\beta\}$ expressed by $\mathbb{P}_1$ is necessary at α; but the proposition expressed by $\mathbb{P}_0 \wedge \mathbb{P}_1$ – viz. $\{\alpha\} \cap \{\beta\} = \varnothing$ – is necessary at α. So $\Box(\mathbb{P}_0 \wedge \mathbb{P}_1)$ is true at α, whereas $\Box\mathbb{P}_0$ and $\Box\mathbb{P}_1$ are both false at α. Hence $\Box(\mathbb{P}_0 \wedge \mathbb{P}_1) \rightarrow (\Box\mathbb{P}_0 \wedge \Box\mathbb{P}_1)$ is false at α. Thus M has a minimal countermodel.

For C. Consider the instance $(\Box\mathbb{P}_0 \wedge \Box\mathbb{P}_1) \rightarrow \Box(\mathbb{P}_0 \wedge \mathbb{P}_1)$, and let $\mathcal{M}$ be a minimal model like that above for M except that $N_\alpha = \{\{\alpha\}, \{\beta\}\}$. Then the situation above is reversed: the propositions expressed by $\mathbb{P}_0$ and $\mathbb{P}_1$ are both necessary at α, whereas the proposition expressed by their conjunction is not necessary at α. Thus $\Box\mathbb{P}_0$ and $\Box\mathbb{P}_1$ are true at α, and $\Box(\mathbb{P}_0 \wedge \mathbb{P}_1)$ is false at α. From this it follows that $(\Box\mathbb{P}_0 \wedge \Box\mathbb{P}_1) \rightarrow \Box(\mathbb{P}_0 \wedge \mathbb{P}_1)$ is false at α, so that C has a minimal countermodel.

This concludes the proof.

It follows from theorem 7.4 that the set of sentences valid in a class of minimal models is not in general closed under any of the rules of inference RM, RR, RK, and RN, and also that the schemas R and K are not generally valid. We set it as an exercise for the reader to prove these things.

Let us turn now to some positive results about minimal models.

We consider the following conditions on the function N in a minimal model $\mathcal{M} = \langle W, N, P \rangle$, for every world α in $\mathcal{M}$ and every proposition (i.e. set of worlds) X and Y in $\mathcal{M}$:

(m) if $X \cap Y \in N_\alpha$, then $X \in N_\alpha$ and $Y \in N_\alpha$

(c) if $X \in N_\alpha$ and $Y \in N_\alpha$, then $X \cap Y \in N_\alpha$

(n) $W \in N_\alpha$

It is important to observe that condition (m) is equivalently expressed in terms of closure under supersets:

(m′) if $X \subseteq Y$ and $X \in N_\alpha$, then $Y \in N_\alpha$

For suppose that (m) holds, and that X and Y are propositions in $\mathcal{M}$ such that $X \subseteq Y$ and $X \in N_\alpha$. Then $X = X \cap Y$, so that N_α contains $X \cap Y$ and hence, by (m), Y. Conversely, suppose that (m′) holds, and that X and Y are propositions in $\mathcal{M}$ such that $X \cap Y \in N_\alpha$. Then $Y \in N_\alpha$, by (m′), since $X \cap Y \subseteq Y$.

According as the function in a minimal model satisfies conditions (m), (c), or (n), we shall say that the model is *supplemented*, is *closed under intersections*, or *contains the unit*. When a model satisfies the first two conditions, i.e. when it is supplemented and closed under intersections, we shall say that it is a *quasi-filter*. When all three conditions are met, i.e. in the case of a quasi-filter that contains the unit, we call the model a *filter*. Thus every filter is a quasi-filter, and every quasi-filter (and hence every filter) is supplemented. Note that filters are equally well characterized as non-empty quasi-filters – non-empty in the sense that $N_\alpha \neq \varnothing$ for every α in a filter. This follows from the fact that in any supplemented model N_α is non-empty just in case it contains W; see exercise 7.13. This terminology facilitates the statement of results about classes of minimal models, for example in the next theorem.

THEOREM 7.5. *The following schemas are valid respectively in the indicated classes of minimal models.*

(1) M: *supplemented*

(2) C: *closed under intersections*

(3) N: *contains the unit*

Proof

For (1). Let α be a world in a supplemented minimal model $\mathcal{M} = \langle W, N, P \rangle$, and suppose that $\models_\alpha^\mathcal{M} \Box(A \wedge B)$. This means that $\|A \wedge B\|^\mathcal{M} \in N_\alpha$, i.e. that $\|A\|^\mathcal{M} \cap \|B\|^\mathcal{M} \in N_\alpha$. Because $\mathcal{M}$ is supplemented, N_α contains $\|A\|^\mathcal{M}$ and $\|B\|^\mathcal{M}$, which means that $\models_\alpha^\mathcal{M} \Box A$ and $\models_\alpha^\mathcal{M} \Box B$. This is enough to establish the contention that M is valid in the class of supplemented minimal models.

For (2). Let α be a world in a minimal model $\mathcal{M} = \langle W, N, P \rangle$ that is closed under intersections, and suppose that $\Box A \wedge \Box B$ is true at α. We leave it as an exercise for the reader to argue that $\Box(A \wedge B)$ is true at α, which is all that is required now to show that C is valid in the class of minimal models closed under intersections.

For (3). Let $\mathcal{M} = \langle W, N, P \rangle$ be a minimal model that contains the unit. Then N_α contains $\|\top\|^\mathcal{M}$, for every α in $\mathcal{M}$, which means that $\Box\top$ is true in $\mathcal{M}$. Therefore N is valid in the class of minimal models containing the unit.

The reader should observe that the countermodels for M, C, and N in the proof of theorem 7.4 fail to satisfy the respective conditions in theorem 7.5 for the validity of these schemas.

The classes of minimal models selected for consideration here determine some important systems of modal logic. For the most part these are weaker than the smallest normal system, *K*, in the sense of being properly included in *K*. As we shall see, however, *K* is determined by the class of filters, i.e. by the class of supplemented minimal models closed under intersections and containing the unit. Indeed, *K* is determined by a special class of filters, the augmented minimal models introduced in the next section.

We close the section with definitions of three types of construction on minimal models.

DEFINITION 7.6. Let $\mathcal{M} = \langle W, N, P \rangle$ be a minimal model. The *supplementation* of $\mathcal{M}$ is the minimal model $\mathcal{M}^+ = \langle W, N^+, P \rangle$ in which N_α^+ is the superset closure of N_α, for each α in $\mathcal{M}$. That is, for every α and X in $\mathcal{M}$,

$$X \in N_\alpha^+ \text{ iff } Y \subseteq X \text{ for some } Y \in N_\alpha.$$

Thus $\mathcal{M}^+$ differs from $\mathcal{M}$ only in that N_α^+ contains every proposition in $\mathcal{M}$ that includes any proposition in N_α. Notice that $N_\alpha \subseteq N_\alpha^+$, since every proposition includes itself. Thus we may characterize a minimal model as supplemented just in case it is its own supplementation, i.e. when $N_\alpha^+ = N_\alpha$ for every α.

DEFINITION 7.7. Let $\mathcal{M} = \langle W, N, P \rangle$ be a minimal model. The *intersection closure* of $\mathcal{M}$ is the minimal model $\mathcal{M}^- = \langle W, N^-, P \rangle$ in which, for every α and X in $\mathcal{M}$,

$$X \in N_\alpha^- \text{ iff } X = X_1 \cap \ldots \cap X_n \text{ for some } n > 0 \text{ and } X_1, \ldots, X_n \in N_\alpha.$$

Of course $N_\alpha \subseteq N_\alpha^-$, since a proposition is identical with the intersection of it with itself. So a minimal model is closed under intersections if and only if it is its own intersection closure.

The models $\mathcal{M}^{+-}$ and $\mathcal{M}^{-+}$ are always the same, which is to say that it is a matter of indifference whether we first form the supplementation of a minimal model and then take the intersection closure of the result, or vice versa. For suppose that $X \in N_\alpha^{+-}$, so that $X = X_1 \cap \ldots \cap X_n$, where each X_i is a superset of some $Y_i \in N_\alpha$. Then N_α^- contains $Y_1 \cap \ldots \cap Y_n$, which is a subset of $X_1 \cap \ldots \cap X_n$ and hence of X, so that $X \in N_\alpha^{-+}$. Thus $N_\alpha^{+-} \subseteq N_\alpha^{-+}$ (and we leave the argument for the reverse as an exercise). This result makes possible the following definition.

DEFINITION 7.8. Let $\mathcal{M} = \langle W, N, P \rangle$ be a minimal model. The *quasi-filtering* of $\mathcal{M}$ is the minimal model $\mathcal{M}^\pm = \langle W, N^\pm, P \rangle$, where $\mathcal{M}^\pm = \mathcal{M}^{+-} = \mathcal{M}^{-+}$.

So a quasi-filter is a minimal model identical with its own quasi-filtering.

The supplementation, intersection closure, or quasi-filtering of a minimal model does not in general produce an equivalent model. But these constructions are valuable in connection with filtrations and canonical models, as we shall see.

EXERCISES

7.12. Using theorem 7.4, prove that R and K are not generally valid in classes of minimal models, and that validity in a class of minimal models is not always preserved by the rules of inference RM, RR, RK, and RN.

7.13. Prove that for any α in any supplemented minimal model $\mathcal{M} = \langle W, N, P \rangle$, $N_\alpha \neq \varnothing$ if and only if $W \in N_\alpha$.

7.14. Complete the proof of part (2) of theorem 7.5 (compare exercise 7.6).

7.15. Complete the proof that $\mathcal{M}^{+-} = \mathcal{M}^{-+}$ by showing that $N_\alpha^{-+} \subseteq N_\alpha^{+-}$ for every α (see before theorem 7.8).

7.16. Let $\mathcal{M}$ be a minimal model. Prove:

(*a*) $\mathcal{M}^+$ is closed under intersections if $\mathcal{M}$ is.

(*b*) $\mathcal{M}^-$ is supplemented if $\mathcal{M}$ is.

7.17. Give examples to show that the supplementation, intersection closure, or quasi-filtering of a minimal model does not always yield an equivalent model.

7.18. Consider the duals of M, C, and N:

$$\text{M}\Diamond.\quad (\Diamond \text{A} \vee \Diamond \text{B}) \rightarrow \Diamond(\text{A} \vee \text{B})$$
$$\text{C}\Diamond.\quad \Diamond(\text{A} \vee \text{B}) \rightarrow (\Diamond \text{A} \vee \Diamond \text{B})$$
$$\text{N}\Diamond.\quad \neg\Diamond\bot$$

Show that these schemas are valid in classes of minimal models that are respectively supplemented, closed under intersections, and contain the unit.

7.19. Give examples of minimal models satisfying the following conditions.

(*a*) supplemented, closed under intersections, and falsifying N

(*b*) supplemented, containing the unit, and falsifying an instance of C

(*c*) closed under intersections, containing the unit, and falsifying an instance of M

These results aid in the proof of the distinctness of the modal logics listed in figure 8.1; see theorem 9.2.

7.20. Identify conditions on minimal models to validate the schemas R and K.

7.21. Identify a condition on minimal models to validate the following schemas.

$$\Box(\text{A} \wedge \text{B}) \rightarrow \Box \text{A} \qquad \Diamond \text{A} \rightarrow \Diamond(\text{A} \vee \text{B})$$

7.22. Describe minimal countermodels for instances of each of the following schemas.

D. $\Box A \rightarrow \Diamond A$

T. $\Box A \rightarrow A$

B. $A \rightarrow \Box \Diamond B$

4. $\Box A \rightarrow \Box \Box A$

5. $\Diamond A \rightarrow \Box \Diamond A$

7.23. Consider the truth conditions for modal sentences given in exercise 7.8. Identify conditions on minimal models to validate the schemas M$\Diamond$, C$\Diamond$, and N$\Diamond$. Compare the logics determined by classes of minimal models satisfying these conditions with the logics determined (under the usual interpretation of $\Box$ and $\Diamond$) by the classes of minimal models that are supplemented, closed under intersections, or possessed of the unit.

7.24. Let $\mathcal{M} = \langle W, N, P \rangle$ be a supplemented minimal model. Prove, for every α in $\mathcal{M}$:

$$\models_{\alpha}^{\mathcal{M}} \Box A \text{ iff for some } X \in N_{\alpha},\ X \subseteq \|A\|^{\mathcal{M}}.$$

7.25. Consider the truth conditions for modal sentences in exercise 7.9. Prove:

(*a*) The schema M is valid in any class of minimal models, relative to these truth conditions.

(*b*) For every minimal model with truth conditions of this sort there is an equivalent supplemented minimal model with truth conditions of the usual kind; and vice versa.

7.26. Consider the models $\mathcal{M} = \langle W, f, P \rangle$ in exercise 7.10. Define classes of models of this sort equivalent to the classes of minimal models that are supplemented, closed under intersection, and possessed of the unit.

7.27. Consider the following conditions on the algebraic models $\mathcal{M} = \langle \mathcal{B}, *, P \rangle$ described in exercise 7.11.

(m*) $*(a \cap b) \leqslant *a \cap *b$

(c*) $*a \cap *b \leqslant *(a \cap b)$

(n*) $*1 = 1$

Prove that the schemas M, C, and N are valid in classes of algebraic models satisfying, respectively, conditions (m*), (c*), and (n*).

7.3. Augmentation

What is the relationship between standard models and minimal models? To answer this question precisely we introduce the idea of an augmented model.

A minimal model $\mathscr{M} = \langle W, N, P \rangle$ is *augmented* if and only if it is supplemented and, for every world α in it,

$$\bigcap N_\alpha \in N_\alpha.$$

Thus in an augmented model each N_α contains a smallest proposition, the set comprising just those worlds that are members of every proposition in N_α. Models of this kind are equally well characterized by the condition that N_α always contains $\bigcap N_\alpha$ and every superset thereof. In other words, $\mathscr{M}$ is augmented just in case for every α and X in $\mathscr{M}$,

$$\text{(a)} \quad X \in N_\alpha \text{ iff } \bigcap N_\alpha \subseteq X.$$

(In one direction this condition is of course trivial: $\bigcap N_\alpha$ is a subset of every set in N_α.) The equivalence of this characterization is readily seen. If $\mathscr{M}$ is augmented and $\bigcap N_\alpha \subseteq X$, then by supplementation $X \in N_\alpha$. On the other hand, suppose $\mathscr{M}$ satisfies the condition. Then if $X \subseteq Y$ and $X \in N_\alpha$ it follows that $\bigcap N_\alpha \subseteq Y$, which means that $Y \in N_\alpha$. So $\mathscr{M}$ is supplemented. Moreover, $\bigcap N_\alpha \in N_\alpha$, since $\bigcap N_\alpha \subseteq \bigcap N_\alpha$. Hence the model is augmented.

Notice, too, that an augmented model contains the unit, i.e. N_α always contains W (since W always includes $\bigcap N_\alpha$).

The condition that $\bigcap N_\alpha \in N_\alpha$ may be described as closure under intersection, as distinguished from closure under finite intersections (condition (c) of the preceding section). In the present context, closure under intersection is equivalent to saying that the model is closed under arbitrary intersections, i.e. that each N_α contains the intersection of any collection of its members. For this implies that, in particular, $\bigcap N_\alpha \in N_\alpha$; and, conversely, if $\mathscr{X}$ is a subset of N_α, then $\bigcap N_\alpha \subseteq \bigcap \mathscr{X}$, whence $\bigcap \mathscr{X} \in N_\alpha$ by supplementation. It follows as a corollary to this that augmented models are closed under (finite) intersections.

Thus every augmented model is a filter: supplemented, closed under intersections, and possessed of the unit. Moreover, when a filter contains only finitely many worlds it is augmented. For then each N_α contains only finitely many propositions, and so closure under arbitrary intersections reduces to closure under intersections. In short, every finite filter is augmented.

Not every filter is augmented, however. For example, consider a minimal model $\mathcal{M} = \langle W, N, P\rangle$ in which W is the set of real numbers and, for each real number α,

$$N_\alpha = \{X \subseteq W : (\alpha, \beta) \subseteq X \text{ for some } \beta \in W \text{ such that } \alpha < \beta\}.$$

That is, N_α contains every set of real numbers that includes some open interval (α, β) where β is larger than α. It is left as an exercise for the reader to show that $\mathcal{M}$ is supplemented and contains the unit; we complete the proof that $\mathcal{M}$ is a filter by arguing, as follows, that it is closed under intersections. Suppose for α in $\mathcal{M}$ that N_α contains sets X and Y. Then there exist intervals (α, β_1) and (α, β_2) such that $(\alpha, \beta_1) \subseteq X$ and $(\alpha, \beta_2) \subseteq Y$. We may assume without loss of generality that $\beta_1 \leqslant \beta_2$, so that $(\alpha, \beta_1) \subseteq (\alpha, \beta_2)$. Then both X and Y include (α, β_1), from which it follows that $X \cap Y$ includes this interval. By supplementation, then, $X \cap Y$ is in N_α. So $\mathcal{M}$ is closed under intersections. But it is not closed under intersection; indeed, $\bigcap N_\alpha$ is not a member of N_α for any α in $\mathcal{M}$. For $\bigcap N_\alpha$ is always the empty set (since there is no smallest interval (α, β) where $\beta > \alpha$), whereas none of the sets in N_α is empty (since for no $\beta > \alpha$ is (α, β) empty). In short, $\bigcap N_\alpha = \varnothing$, but $\varnothing \notin N_\alpha$, for every α in $\mathcal{M}$.

The relationship between standard models and minimal models can now be stated: a standard model is essentially an augmented minimal model. We put this more precisely as a theorem.

THEOREM 7.9. *For every standard model $\mathcal{M}^s = \langle W, R, P\rangle$ there is a pointwise equivalent augmented minimal model $\mathcal{M}^m = \langle W, N, P\rangle$, and vice versa.*

Proof. Let $\mathcal{M}^s$ be a standard model and define the minimal model $\mathcal{M}^m$ by stipulating that

$$X \in N_\alpha \text{ iff } \{\beta \in W : \alpha R \beta\} \subseteq X,$$

for every $\alpha \in W$ and every $X \subseteq W$. Then $\bigcap N_\alpha = \{\beta \in W : \alpha R \beta\}$ for each $\alpha \in W$. So $\mathcal{M}^m$ satisfies condition (a), which means that it is augmented. The proof that $\mathcal{M}^s$ and $\mathcal{M}^m$ are pointwise equivalent, i.e. that a world verifies the same sentences in each model, is by induction on the complexity of a sentence A. The only case of interest is that in which A is a necessitation, $\Box$B, where the argument goes like this:

$$\models_\alpha^{\mathcal{M}^s} \Box \text{B iff for every } \beta \in W \text{ such that } \alpha R \beta, \models_\beta^{\mathcal{M}^s} \text{B}$$
– definition 3.2(1);

$$\text{iff } \{\beta \in W : \alpha R \beta\} \subseteq \|\text{B}\|^{\mathcal{M}^s}$$
– compare exercise 3.13;

iff $\|B\|^{\mathcal{M}^m} \in N_\alpha$
– definition of N and the inductive hypothesis;
iff $\models_\alpha^{\mathcal{M}^m} \Box B$
– definition 7.2(1).

For the other half of the theorem, let $\mathcal{M}^m$ be an augmented minimal model and define the standard model $\mathcal{M}^s$ by:

$$\alpha R \beta \text{ iff } \beta \in \bigcap N_\alpha,$$

for every α and β in W. As before, necessitation is the only case of interest in the inductive proof that the models are pointwise equivalent. The argument proceeds as follows.

$\models_\alpha^{\mathcal{M}^m} \Box B$ iff $\|B\|^{\mathcal{M}^m} \in N_\alpha$
– definition 7.2(1);
iff $\bigcap N_\alpha \subseteq \|B\|^{\mathcal{M}^m}$
– because $\mathcal{M}^m$ is augmented;
iff for every $\beta \in W$ such that $\alpha R \beta$, $\models_\beta^{\mathcal{M}^s} B$
– definition of R and the inductive hypothesis;
iff $\models_\alpha^{\mathcal{M}^s} \Box B$
– definition 3.2(1).

This completes the proof of the theorem.

Thus we see that the standard models can be paired one-to-one with the augmented minimal models in such a way that paired models are pointwise equivalent. It is in this sense that we identify the two classes of models.

In view of this we have already a completeness theorem for classes of minimal models. In particular, every normal system is complete with respect to the class of augmented models, since by theorems 5.8 and 7.9 any non-theorem of such a logic is false in some such model.

With an eye to proving this result directly in chapter 9 we define the operation of augmentation, which turns a minimal model (of any sort) into an augmented model.

DEFINITION 7.10. Let $\mathcal{M} = \langle W, N, P \rangle$ be a minimal model. The *augmentation* of $\mathcal{M}$ is the minimal model $\mathcal{M}^! = \langle W, N^!, P \rangle$ in which, for each $\alpha \in$ W,

$$N_\alpha^! = \{X \subseteq W : \bigcap N_\alpha \subseteq X\}.$$

That is, $\mathcal{M}^!$ is the supplementation of $\mathcal{M}$ closed under intersection. An augmented model is thus a minimal model identical with its own augmentation.

EXERCISES

7.28. Show that the model defined two paragraphs before theorem 7.9 is supplemented and contains the unit.

7.29. For both halves of theorem 7.9 give the proofs of pointwise equivalence for the cases in which A is atomic, the falsum, a conditional, and a possibilitation.

7.30. Consider the following conditions on a minimal model $\mathcal{M} = \langle W, N, P \rangle$, for every α, X, and Y in $\mathcal{M}$:

(r) $X \cap Y \in N_\alpha$ iff $X \in N_\alpha$ and $Y \in N_\alpha$

(k) if $-X \cup Y \in N_\alpha$, then if $X \in N_\alpha$ then $Y \in N_\alpha$

Prove that the schemas R and K are valid in classes of minimal models satisfying respectively conditions (r) and (k).

7.31. Prove that a supplemented minimal model is closed under intersections just in case it satisfies condition (k) in the preceding exercise.

7.32. Describe conditions on minimal models to validate the schemas D, T, B, 4, and 5. (Compare exercise 7.22.)

7.4. The schemas D, T, B, 4, and 5

We consider once again the following schemas.

D. $\Box A \rightarrow \Diamond A$

T. $\Box A \rightarrow A$

B. $A \rightarrow \Box \Diamond A$

4. $\Box A \rightarrow \Box \Box A$

5. $\Diamond A \rightarrow \Box \Diamond A$

None of these is valid in the class of all minimal models; see exercise 7.22. But for each of these schemas we can identify a class of minimal models that validates it. We wish to consider the following conditions on a minimal model $\mathcal{M} = \langle W, N, P \rangle$, for every world α and proposition X in $\mathcal{M}$:

(d) if $X \in N_\alpha$, then $-X \notin N_\alpha$

(t) if $X \in N_\alpha$, then $\alpha \in X$

(b) if $\alpha \in X$, then $\{\beta \text{ in } \mathcal{M} : -X \notin N_\beta\} \in N_\alpha$

(iv) if $X \in N_\alpha$, then $\{\beta \text{ in } \mathcal{M} : X \in N_\beta\} \in N_\alpha$

(v) if $X \notin N_\alpha$, then $\{\beta \text{ in } \mathcal{M} : X \notin N_\beta\} \in N_\alpha$

THEOREM 7.11. *The following schemas are valid respectively in the indicated classes of minimal models.*

(1) D: *condition* (d)

(2) T: *condition* (t)

(3) B: *condition* (b)

(4) 4: *condition* (iv)

(5) 5: *condition* (v)

Proof. Let α be a world in a minimal model $\mathcal{M} = \langle W, N, P \rangle$.

For (1). Suppose $\mathcal{M}$ satisfies (d), and that $\Box A$ is true at α. Then $\|A\|^{\mathcal{M}} \in N_\alpha$, and so by (d), $-\|A\|^{\mathcal{M}} \notin N_\alpha$, which means that $\Diamond A$ is true at α. It follows that D is valid in the class of minimal models that satisfy condition (d).

For (2). Assume that $\mathcal{M}$ satisfies condition (t), and suppose $\Box A$ to be true at α. Then $\|A\|^{\mathcal{M}} \in N_\alpha$, and so by (t), $\alpha \in \|A\|^{\mathcal{M}}$, which means that A is true at α. This suffices to show that T is valid in the class of minimal models that satisfy condition (t).

For (3). Here we suppose that $\mathcal{M}$ satisfies condition (b), and that A is true at α. In other words, $\alpha \in \|A\|^{\mathcal{M}}$, from which it follows by (b) that $\{\beta \text{ in } \mathcal{M}: -\|A\|^{\mathcal{M}} \notin N_\beta\} \in N_\alpha$. This means that $\{\beta \text{ in } \mathcal{M}: \models^{\mathcal{M}}_\beta \Diamond A\} \in N_\alpha$, i.e. that $\|\Diamond A\|^{\mathcal{M}} \in N_\alpha$. But this last just means that $\Box\Diamond A$ is true at α, which is what we needed to show. Thus the schema B is valid in the class of minimal models satisfying condition (b).

For (4). Exercise.

For (5). Exercise.

As the reader has probably noticed, it is often possible to 'read off' from a schema a constraint on the class of minimal models that yields a class of models that validates the schema. We close this section by remarking that the schema

G. $\Diamond\Box A \to \Box\Diamond A$

is valid in the class of minimal models $\mathcal{M} = \langle W, N, P \rangle$ that satisfy the condition that for every world α and proposition X in $\mathcal{M}$:

(g) if $-\{\beta \text{ in } \mathcal{M}: X \in N_\beta\} \notin N_\alpha$,
then $\{\beta \text{ in } \mathcal{M}: -X \notin N_\beta\} \in N_\alpha$

The proof is not difficult and is left as an exercise. In the exercises we invite the reader to generalize this result for the schema $G^{k,l,m,n}$, as well as for its converse.

EXERCISES

7.33. Prove parts (4) and (5) of theorem 7.11.

7.34. Prove that the schema G is valid in any class of minimal models that satisfy the condition (g) (in the last paragraph of section 7.4).

7.35. Consider the duals of T, B, 4, and 5:

T$\Diamond$. $A \rightarrow \Diamond A$

B$\Diamond$. $\Diamond\Box A \rightarrow A$

4$\Diamond$. $\Diamond\Diamond A \rightarrow \Diamond A$

5$\Diamond$. $\Diamond\Box A \rightarrow \Box A$

Show that these schemas are valid in classes of minimal models satisfying respectively conditions (t), (b), (iv), and (v).

7.36. Prove that a minimal model satisfies conditions (d), (b), (iv), and (n) (contains the unit) if it satisfies (t) and (v).

7.37. Identify conditions on minimal models to validate the following sentences.

P. $\Diamond\top$ $\quad$ $\bar{\text{P}}$. $\neg\Diamond\top$

7.38. For α and X in a minimal model $\mathcal{M} = \langle W, N, P \rangle$ we define N^n inductively as follows.

(1) $X \in N^0_\alpha$ iff $\alpha \in X$.

(2) $X \in N^n_\alpha$ iff $\{\beta \text{ in } \mathcal{M}: X \in N^{n-1}_\beta\} \in N_\alpha$, for $n > 0$.

Given this definition one can prove (*a*) and (*b*) in exercise 7.7.

Prove that the schema

$G^{k,l,m,n}$. $\Diamond^k\Box^l A \rightarrow \Box^m\Diamond^n A$

is valid in any class of minimal models $\mathcal{M} = \langle W, N, P\rangle$ that satisfy the condition,

$$(\mathrm{g}^{k,l,m,n}) \quad \text{if } -\{\beta \text{ in } \mathcal{M}: X \in N^l_\beta\} \notin N^k_\alpha,$$
$$\text{then } \{\beta \text{ in } \mathcal{M}: -X \notin N^n_\beta\} \in N^m_\alpha.$$

Identify a condition $(\mathrm{g}_c^{k,l,m,n})$ on minimal models to validate the schema

$$\mathrm{G}_c^{k,l,m,n}. \quad \Box^k \Diamond^l A \to \Diamond^m \Box^n A.$$

Derive validating conditions on minimal models for schemas like D, T, B, 4, and 5 from conditions $(\mathrm{g}^{k,l,m,n})$ or $(\mathrm{g}_c^{k,l,m,n})$.

7.39. Identify conditions on minimal models to validate schemas (like U, $\Box(\Box A \to A)$) not covered by $\mathrm{G}^{k,l,m,n}$ or $\mathrm{G}_c^{k,l,m,n}$.

7.40. Define conditions on the models $\mathcal{M} = \langle W, f, P\rangle$ of exercise 7.10 equivalent to (d), (t), (b), (iv), (v), and (g).

7.41. For a model $\mathcal{M} = \langle W, f, P\rangle$ in exercise 7.10 inductively define f^n so that, for every n, $\|\Box^n A\|^{\mathcal{M}} = f^n(\|A\|^{\mathcal{M}})$.

7.42. Consider the following conditions on the algebraic models $\mathcal{M} = \langle \mathcal{B}, *, P\rangle$ of exercise 7.11.

$$(\mathrm{d}^*) \quad *a \leqslant -*-a$$
$$(\mathrm{t}^*) \quad *a \leqslant a$$
$$(\mathrm{b}^*) \quad a \leqslant *-*-a$$
$$(\mathrm{iv}^*) \quad *a \leqslant **a$$
$$(\mathrm{v}^*) \quad -*-a \leqslant *-*-a$$
$$(\mathrm{g}^*) \quad -*-*a \leqslant *-*-a$$

Prove that the schemas D, T, B, 4, 5, and G are valid in classes of algebraic models satisfying, respectively, conditions (d*), (t*), (b*), (iv*), (v*), and (g*).

7.43. Let $\mathcal{M} = \langle \mathcal{B}, *, P\rangle$ be an algebraic model, as in exercise 7.11, and define:

(1) $*^0 a = a$.

(2) $*^n a = **^{n-1} a$, for $n > 0$.

Prove, for any $n \geqslant 0$:

(*a*) $\|\Box^n A\|^{\mathcal{M}} = *^n \|A\|^{\mathcal{M}}$. (*b*) $\|\Diamond^n A\|^{\mathcal{M}} = -*^n - \|A\|^{\mathcal{M}}$.

Identify conditions on algebraic models to validate the schemas $\mathrm{G}^{k,l,m,n}$ and $\mathrm{G}_c^{k,l,m,n}$ and other schemas not covered by these.

7.5. Filtrations

Filtrations of minimal models are very simply defined (recall the meanings of $\equiv$, $[\alpha]$, and $[X]$ from section 2.3).

DEFINITION 7.12. Let $\mathcal{M} = \langle W, N, P \rangle$ be a minimal model, and let Γ be a set of sentences closed under subsentences. Then a *filtration of $\mathcal{M}$ through* Γ is any minimal model $\mathcal{M}^* = \langle W^*, N^*, P^* \rangle$ such that:

(1) $W^* = [W]$.

(2) For every α in $\mathcal{M}$:

(*a*) for every sentence $\Box A \in \Gamma$,
$\|A\|^{\mathcal{M}} \in N_\alpha$ iff $[\|A\|^{\mathcal{M}}] \in N^*_{[\alpha]}$;

(*b*) for every sentence $\Diamond A \in \Gamma$,
$-\|A\|^{\mathcal{M}} \in N_\alpha$ iff $-[\|A\|^{\mathcal{M}}] \in N^*_{[\alpha]}$.

(3) $P^*_n = [P_n]$, for each n such that $\mathbb{P}_n \in \Gamma$.

By a *finest* Γ-filtration of $\mathcal{M}$ we mean one in which each $N^*_{[\alpha]}$ contains just (*a*) the sets $[\|A\|^{\mathcal{M}}]$ such that $\Box A \in \Gamma$ and $\|A\|^{\mathcal{M}} \in N_\alpha$ and (*b*) the sets $-[\|A\|^{\mathcal{M}}]$ such that $\Diamond A \in \Gamma$ and $-\|A\|^{\mathcal{M}} \in N_\alpha$. This is perhaps the simplest example of a minimal filtration. We leave it for the reader to explain what is meant by a *coarsest* filtration of this sort.

The definition is designed precisely to make possible the proof of the following basic theorem.

THEOREM 7.13. *Let $\mathcal{M}^* = \langle W^*, N^*, P^* \rangle$ be a Γ-filtration of a minimal model $\mathcal{M} = \langle W, N, P \rangle$. Then for every* $A \in \Gamma$ *and every α in $\mathcal{M}$:*

$\models^{\mathcal{M}}_\alpha A$ *iff* $\models^{\mathcal{M}^*}_{[\alpha]} A$.

Equivalently: $[\|A\|^{\mathcal{M}}] = \|A\|^{\mathcal{M}^*}$, *for every* $A \in \Gamma$.

Proof. We give the proof only for the case in which A in Γ is a necessitation, $\Box B$. Note that the inductive hypothesis implies that $[\|B\|^{\mathcal{M}}] = \|B\|^{\mathcal{M}^*}$. The reasoning then is straightforward, for any α in $\mathcal{M}$:

$\models^{\mathcal{M}}_\alpha \Box B$ iff $\|B\|^{\mathcal{M}} \in N_\alpha$
– definition 7.2(1);

iff $[\|B\|^{\mathcal{M}}] \in N^*_{[\alpha]}$
– (2)(*a*) of definition 7.12;

iff $\|B\|^{\mathcal{M}^*} \in N^*_{[\alpha]}$
– inductive hypothesis;
iff $\models^{\mathcal{M}^*}_{[\alpha]} \Box B$
– definition 7.2(1).

The following two theorems are corollaries.

THEOREM 7.14. *Let $\mathcal{M}^*$ be a Γ-filtration of a minimal model $\mathcal{M}$. Then $\mathcal{M}$ and $\mathcal{M}^*$ are equivalent modulo Γ – i.e. for every $A \in \Gamma$:*
$\models^{\mathcal{M}} A$ *iff* $\models^{\mathcal{M}^*} A$.

THEOREM 7.15. *Let* C *be a class of minimal models, and let $\Gamma(\mathsf{C})$ be the class of Γ-filtrations of models in* C. *Then for every $A \in \Gamma$:*
$\models_{\mathsf{C}} A$ *iff* $\models_{\Gamma(\mathsf{C})} A$.

The next two theorems are useful when it comes to proving finite determination theorems in chapter 9.

THEOREM 7.16. *Let $\mathcal{M}^*$ be a finest Γ-filtration of a minimal model $\mathcal{M}$, and consider the supplementation $\mathcal{M}^{*+}$, the intersection closure $\mathcal{M}^{*-}$, and the quasi-filtering $\mathcal{M}^{*\pm}$ of $\mathcal{M}^*$. Then:*

(1) *$\mathcal{M}^{*+}$ is a Γ-filtration of $\mathcal{M}$ if $\mathcal{M}$ is supplemented.*

(2) *$\mathcal{M}^{*-}$ is a Γ-filtration of $\mathcal{M}$ if $\mathcal{M}$ is closed under intersections.*

(3) *$\mathcal{M}^{*\pm}$ is a Γ-filtration of $\mathcal{M}$ if $\mathcal{M}$ is a quasi-filter.*

Proof. We give the proof for part (3) only; parts (1) and (2) are left as exercises.

Let $\mathcal{M}^{*\pm} = \langle W^*, N^{*\pm}, P^* \rangle$ be the quasi-filtering of a finest Γ-filtration $\mathcal{M}^* = \langle W^*, N^*, P^* \rangle$ of a quasi-filter $\mathcal{M} = \langle W, N, P \rangle$ To show that $\mathcal{M}^{*\pm}$ is a Γ-filtration of $\mathcal{M}$ we must prove that $N^{*\pm}$ satisfies the following conditions, for every α in $\mathcal{M}$.

(*a*) for every $\Box A \in \Gamma$, $\|A\|^{\mathcal{M}} \in N_\alpha$ iff $[\|A\|^{\mathcal{M}}] \in N^{*\pm}_{[\alpha]}$

(*b*) for every $\Diamond A \in \Gamma$, $-\|A\|^{\mathcal{M}} \in N_\alpha$ iff $-[\|A\|^{\mathcal{M}}] \in N^{*\pm}_{[\alpha]}$

We argue for (*a*) only; (*b*) is left as an exercise.

Let $\Box A$ be a sentence in Γ. For left-to-right, suppose that $\|A\|^{\mathcal{M}} \in N_\alpha$. Then $[\|A\|^{\mathcal{M}}] \in N^*_{[\alpha]}$, since $\mathcal{M}^*$ is a Γ-filtration of $\mathcal{M}$, and so $[\|A\|^{\mathcal{M}}] \in N^{*\pm}_{[\alpha]}$ by quasi-filtering. For the reverse suppose $[\|A\|^{\mathcal{M}}] \in N^{*\pm}_{[\alpha]}$. Because $\mathcal{M}^*$ is a finest Γ-filtration of $\mathcal{M}$, this means that

$$[\|A_1\|^{\mathcal{M}}] \cap \ldots \cap [\|A_n\|^{\mathcal{M}}] \subseteq [\|A\|^{\mathcal{M}}],$$

for some $\Box A_1, \ldots, \Box A_n \in \Gamma$ such that $\|A_1\|^{\mathcal{M}}, \ldots, \|A_n\|^{\mathcal{M}} \in N_\alpha$. Since $A_1, \ldots, A_n$, and A are in Γ, the inclusion implies that

$$\|A_1\|^{\mathcal{M}} \cap \ldots \cap \|A_n\|^{\mathcal{M}} \subseteq \|A\|^{\mathcal{M}}$$

(exercise 7.47). But $\mathcal{M}$ is a quasi-filter. So $\|A_1\|^{\mathcal{M}} \cap \ldots \cap \|A_n\|^{\mathcal{M}} \in N_\alpha$, and hence $\|A\|^{\mathcal{M}} \in N_\alpha$, which is what we wished to show.

THEOREM 7.17. *Let $\mathcal{M}^* = \langle W^*, N^*, P^* \rangle$ be any Γ-filtration of a minimal model $\mathcal{M} = \langle W, N, P \rangle$, and suppose that $\Box\top \in \Gamma$. Then $\mathcal{M}^*$ contains the unit if $\mathcal{M}$ does.*

Proof. We have that for every α in $\mathcal{M}$,

$$\text{for every } \Box A \in \Gamma,\ \|A\|^{\mathcal{M}} \in N_\alpha \text{ iff } [\|A\|^{\mathcal{M}}] \in N^*_\alpha.$$

So $[\|\top\|^{\mathcal{M}}] \in N^*_\alpha$ if $\mathcal{M}$ contains the unit and $\Box\top \in \Gamma$. But $[\|\top\|^{\mathcal{M}}] = W^*$. So $\mathcal{M}^*$ contains the unit.

EXERCISES

7.44. Explain what is meant by a coarsest Γ-filtration of a minimal model. (See the text following definition 7.12.)

7.45. Prove that N^* in a finest or coarsest minimal filtration satisfies the following.

$$\text{If } \alpha \equiv \beta, \text{ then } N^*_{[\alpha]} = N^*_{[\beta]}.$$

7.46. Give the proof of theorem 7.13 for the case in which $A = \Diamond B$.

7.47. Prove parts (1) and (2) of theorem 7.16, and give the argument for condition (*b*) in the proof of part (3). In connection with the argument for condition (*a*) in the proof of (3) show that

$$\|A_1\|^{\mathcal{M}} \cap \ldots \cap \|A_n\|^{\mathcal{M}} \subseteq \|A\|^{\mathcal{M}},$$

given the assumption that $\Box A_1, \ldots, \Box A_n \in \Gamma$, $\|A_1\|^{\mathcal{M}}, \ldots, \|A_n\|^{\mathcal{M}} \in N_\alpha$, and

$$[\|A_1\|^{\mathcal{M}}] \cap \ldots \cap [\|A_n\|^{\mathcal{M}}] \subseteq [\|A\|^{\mathcal{M}}].$$

7.48. Let $\mathscr{B}(\Gamma)$ be the boolean closure of a set of sentences Γ closed under subsentences (see exercise 3.68), and let $\mathcal{M}^* = \langle W^*, N^*, P^* \rangle$ be a Γ-filtration of a minimal model $\mathcal{M} = \langle W, N, P \rangle$. Prove that for every $A \in \mathscr{B}(\Gamma)$ and every α in $\mathcal{M}$,

$$\models^{\mathcal{M}}_\alpha A \text{ iff } \models^{\mathcal{M}^*}_{[\alpha]} A,$$

i.e. that $[\|A\|^{\mathscr{M}}] = \|A\|^{\mathscr{M}^*}$, for every $A \in \mathscr{B}(\Gamma)$. (The proof is inductive; give it for the cases in which A is atomic, the falsum, a conditional, and a necessitation.)

7.49. Let C be a class of minimal models, and let C_{FIN} be the class of all finite models in C. Using theorems 7.14, 7.16, and 7.17 prove that, for every sentence A,

$$\models_{\mathsf{C}} A \text{ iff } \models_{\mathsf{C}_{FIN}} A,$$

where C is any one of the following classes of minimal models.

(*a*) all

(*b*) supplemented

(*c*) closed under intersections

(d) containing the unit

(*e*) quasi-filters

(*f*) supplemented, containing the unit

(*g*) closed under intersections, containing the unit

(*h*) filters

7.50. Define the idea of a Γ-filtration for the models $\mathscr{M} = \langle W, f, P\rangle$ of exercise 7.10.

7.51. Let $\mathscr{M} = \langle \mathscr{B}, *, P\rangle$ be an algebraic model (see exercise 7.11), let Γ be a set of sentences closed under subsentences, and let

$$\mathscr{M}^{\Gamma} = \langle \mathscr{B}^{\Gamma}, *^{\Gamma}, P^{\Gamma}\rangle$$

be an algebraic model satisfying the following conditions.

(1) $\mathscr{B}^{\Gamma}$ is the boolean subalgebra of $\mathscr{B}$ based on the set $\{\|A\|^{\mathscr{M}} : A \in \Gamma\}$, i.e. the boolean algebra formed from this set by adding 1 and 0 from $\mathscr{B}$ and closing under the operations $-$, $\cap$, and $\cup$ in $\mathscr{B}$.

(2) $*^{\Gamma}$ agrees with $*$ on points in $\mathscr{B}^{\Gamma}$; i.e. $*^{\Gamma}a = *a$ for all a in $\mathscr{B}^{\Gamma}$.

(3) P^{Γ} agrees with P on all atomic sentences in Γ; i.e. $P^{\Gamma}_n = P_n$ for all n such that $\mathbb{P}_n \in \Gamma$.

We may call $\mathscr{M}^{\Gamma}$ an *algebraic* Γ-*filtration* of $\mathscr{M}$. Note that $\mathscr{M}^{\Gamma}$ is finite if Γ is. Prove that for every $A \in \Gamma$,

$$\models^{\mathscr{M}} A \text{ iff } \models^{\mathscr{M}^{\Gamma}} A.$$

The proof is by induction on the complexity of A. Give it for the cases in which A is atomic, the falsum, a conditional, and a necessitation.

8

CLASSICAL SYSTEMS OF MODAL LOGIC

In this chapter we examine from a deductive point of view a class of systems of modal logic we call *classical.*

In section 8.1 we define the class of classical systems, point out some alternative characterizations, and state some theorems on replacement and duality.

The smallest classical modal logic is called E. To name classical systems we write

$$ES_1 \ldots S_n$$

for the classical modal logic that results when the schemas $S_1, \ldots, S_n$ are taken as theorems; i.e.

$$ES_1 \ldots S_n = \text{the smallest classical system of modal logic containing (every instance of) the schemas } S_1, \ldots, S_n.$$

The order of the schema names is a matter of indifference; for example, we may write either *EMC* or *ECM* for the smallest classical logic containing the schemas M and C. And note that when there are no schemas the definition leaves E as the smallest classical system.

In section 8.2 we introduce two further classes of modal logics – *monotonic* and *regular* – and set forth some alternative characterizations of them.

Section 8.3 is a brief look at classical systems containing schemas such as D, T, B, 4, and 5.

8.1. Classical systems

We define classical systems of modal logic in terms of the schema

$$\text{Df}\Diamond. \quad \Diamond A \leftrightarrow \neg\Box\neg A$$

and the rule of inference

$$\text{RE.} \quad \frac{A \leftrightarrow B}{\Box A \leftrightarrow \Box B}.$$

DEFINITION 8.1. A system of modal logic is *classical* iff it contains Df$\Diamond$ and is closed under RE.

The following theorem provides another way of characterizing classical systems.

THEOREM 8.2. *A system of modal logic is classical iff it contains the schema*

Df$\Box$. $\Box A \leftrightarrow \neg\Diamond\neg A$

and is closed under the rule of inference

$$\text{RE}\Diamond.\quad \frac{A \leftrightarrow B}{\Diamond A \leftrightarrow \Diamond B}.$$

Proof. To see that a classical logic always contains Df$\Box$ the reader should examine the proof of this principle in section 1.2 and note that it uses only *PL*, Df$\Diamond$, and RE. For RE$\Diamond$, observe that if $A \leftrightarrow B$ is a theorem of a classical system then so are $\neg A \leftrightarrow \neg B$, $\Box\neg A \leftrightarrow \Box\neg B$, and $\neg\Box\neg A \leftrightarrow \neg\Box\neg B$ (by *PL*, RE, and *PL*); but by Df$\Diamond$ the last sentence is equivalent to $\Diamond A \leftrightarrow \Diamond B$.

We leave the argument for the reverse – that a modal logic is classical if it has Df$\Box$ and RE$\Diamond$ – as an exercise.

We turn next to theorems on replacement and duality in classical systems.

THEOREM 8.3. *Every classical system of modal logic has the rule of replacement:*

$$\text{REP.}\quad \frac{B \leftrightarrow B'}{A \leftrightarrow A[B/B']}$$

(Recall the definition of A[B/B′] in section 2.1.)

Proof. The proof is exactly like that for theorem 4.7 – replacement in normal systems – since it appeals only to *PL*, RE, and RE$\Diamond$ (as the reader should confirm).

Notice that the rules RE and RE$\Diamond$ are special cases of REP. This suggests the following characterization of classical modal logics.

THEOREM 8.4. *A system of modal logic is classical iff it contains* Df$\Diamond$ *or* Df$\Box$ *and is closed under* REP.

Proof. Exercise.

For the next theorem the reader should recall the meaning of A* given in definition 2.4.

THEOREM 8.5. *Every classical system of modal logic has* DUAL, *i.e. all the following theorems and rules of inference.*

(1) $A \leftrightarrow \neg A^*$

(2) $\dfrac{A}{\neg A^*} \quad \dfrac{\neg A}{A^*}$

(3) $\dfrac{A \rightarrow B}{B^* \rightarrow A^*}$

(4) $\dfrac{A \leftrightarrow B}{A^* \leftrightarrow B^*}$

Proof. See the proof of this for normal systems, theorem 4.8, which uses only *PL*, Df$\Diamond$, Df$\Box$, and REP.

Finally, a theorem about duals of modalities.

THEOREM 8.6. *Let* Σ *be a classical system of modal logic. Then:*

(1) $\vdash_\Sigma \phi A \leftrightarrow \neg \phi^* \neg A$.

(2) $\vdash_\Sigma \phi A$ *iff* $\vdash_\Sigma \neg \phi^* \neg A$.

(3) $\vdash_\Sigma \phi A \rightarrow \psi A$, *for every* A, *iff* $\vdash_\Sigma \psi^* A \rightarrow \phi^* A$, *for every* A.

(4) $\vdash_\Sigma \phi A \leftrightarrow \psi A$, *for every* A, *iff* $\vdash_\Sigma \phi^* A \leftrightarrow \psi^* A$, *for every* A.

Proof. This is just theorem 4.9 (now for classical logics), the proof for which suffices here as well.

EXERCISES

8.1. Complete the proof of theorem 8.2 by showing that a modal logic is classical if it contains Df$\Box$ and is closed under RE$\Diamond$.

8.2. Examine the proof of theorem 4.7 to confirm that it appeals only to *PL*, RE, and RE$\Diamond$ (with regard to theorem 8.3).

8.3. Prove theorem 8.4.

8.4. Examine the proof of theorem 4.8 to confirm that it uses only *PL*, Df$\Diamond$, Df$\Box$, and REP (with regard to theorem 8.5).

8.5. Examine the proof of theorem 4.9 to see that it suffices as well for theorem 8.6.

8.6. Prove that $A \leftrightarrow A^{**}$ is a theorem of any classical modal logic.

8.7. Where S and S$\Diamond$ are the schemas $\phi A \rightarrow \psi A$ and $\psi^* A \rightarrow \phi^* A$, for affirmative modalities ϕ and ψ, show that a classical system contains S if and only if it contains S$\Diamond$ (compare theorem 4.10).

8.8. Show that a classical modal logic has one of the following theorems and rules of inference if and only if it has them all.

$$\phi A \leftrightarrow \psi A \qquad \phi^* A \leftrightarrow \psi^* A$$

$$\frac{A \leftrightarrow B}{\phi A \leftrightarrow \psi B} \qquad \frac{A \leftrightarrow B}{\phi^* A \leftrightarrow \psi^* B}$$

8.2. Monotonic and regular systems

Thus far in this book we have distinguished two main classes of modal logics: classical systems, which are closed under the rule RE; and normal systems, which are closed under the rule

$$\text{RK.} \quad \frac{(A_1 \wedge \ldots \wedge A_n) \rightarrow A}{(\Box A_1 \wedge \ldots \wedge \Box A_n) \rightarrow \Box A} \quad (n \geqslant 0).$$

In this section we shall be interested as well in classes of modal logics defined by the rules

$$\text{RM.} \quad \frac{A \rightarrow B}{\Box A \rightarrow \Box B}$$

and

$$\text{RR.} \quad \frac{(A \wedge B) \rightarrow C}{(\Box A \wedge \Box B) \rightarrow \Box C}.$$

DEFINITION 8.7. A system of modal logic is *monotonic* iff it contains Df$\Diamond$ and is closed under RM.

DEFINITION 8.8. A system of modal logic is *regular* iff it contains Df$\Diamond$ and is closed under RR.

We denote the smallest monotonic system by *M*, and the smallest regular system by *R*. As with classical and normal systems we write

$MS_1 \dots S_n$ and $RS_1 \dots S_n$, respectively, for the smallest monotonic and regular systems that have the schemas $S_1, \dots, S_n$ as theorems.

The classes of classical, monotonic, regular, and normal systems form a sequence of ever more encompassing kinds of modal logics, in the sense of the following theorem.

THEOREM 8.9

(1) *Every monotonic system of modal logic is classical.*

(2) *Every regular system of modal logic is monotonic, and hence classical.*

(3) *Every normal system of modal logic is regular, and hence monotonic and classical.*

Proof. The proofs of parts (1) and (3) are to be found in the proof of theorem 4.2 (as the reader should verify). For (2):

1. $A \to B$	hypothesis
2. $(A \wedge A) \to B$	1, *PL*
3. $(\Box A \wedge \Box A) \to \Box B$	2, RR
4. $\Box A \to \Box B$	3, *PL*

Thus a modal logic has the rule RM if it has the rule RR; so every regular system is monotonic.

In particular, then, the smallest classical, monotonic, regular, and normal logics – *E*, *M*, *R*, and *K* – are increasingly inclusive. (That the inclusions are in fact proper is proved in chapter 9.)

Let us consider now the following schemas.

M. $\Box(A \wedge B) \to (\Box A \wedge \Box B)$

C. $(\Box A \wedge \Box B) \to \Box(A \wedge B)$

R. $\Box(A \wedge B) \leftrightarrow (\Box A \wedge \Box B)$

K. $\Box(A \to B) \to (\Box A \to \Box B)$

THEOREM 8.10

(1) *Every monotonic system of modal logic contains* M.

(2) *Every regular system of modal logic contains* M, C, R, *and* K.

Proof. That monotonic (and hence regular) systems contain M is shown in the proof, for M, of theorem 4.2. Likewise, the proofs there for C, R, and K show that these are theorems of every regular system.

The schemas mentioned in theorem 8.10 can be used to provide several alternative characterizations of monotonic and regular systems of modal logic. We single out three of these, using M and C, in the next theorem; there are others in the exercises.

THEOREM 8.11. *Let Σ be a system of modal logic containing* Df$\Diamond$. *Then:*

(1) Σ *is monotonic iff it contains* M *and is closed under* RE.

(2) Σ *is regular iff it contains* C *and is closed under* RM.

(3) Σ *is regular iff it contains* C *and* M *and is closed under* RE.

Proof. We consider systems containing Df$\Diamond$. The left-to-right cases are covered by theorem 8.10, so we treat only right-to-left.

For (1):

1. $A \rightarrow B$	hypothesis
2. $A \leftrightarrow (A \wedge B)$	1, *PL*
3. $\Box A \leftrightarrow \Box(A \wedge B)$	2, RE
4. $\Box(A \wedge B) \rightarrow (\Box A \wedge \Box B)$	M
5. $\Box A \rightarrow \Box B$	3, 4, *PL*

Thus a system has RM – and so is monotonic – if it has M and RE.

For (2). The proof of theorem 4.3 (3) shows that a system has RR, and so is regular, if it has C and RM.

For (3). If a modal logic has C, M, and RE, then by part (1) it has RM, and so by part (2) it has RR, which means that it is regular.

By way of summarizing theorem 8.11: Monotonic logics are the classical logics containing M. Regular logics are the monotonic logics containing C, or the classical logics containing C and M. In particular, $M = EM$ and $R = MC = EMC$.

In theorem 4.3 (2,3,4) normal systems are characterized by using the rules and theorems mentioned so far together with

N. $\Box\top$.

We may summarize those results as follows. Normal modal logics are: (1) the regular systems containing N; (2) the monotonic systems containing N and C; (3) the classical systems containing N, C, and M. In particular, K (the smallest normal system) $= RN = MCN = EMCN$.

Other such characterizations of normal systems, involving the schema K and the rule

$$\text{RN.} \quad \frac{A}{\Box A},$$

appear in theorem 4.3 (1) and exercise 4.5.

Theorem 8.11 makes it clear that in studying monotonic, regular, and normal modal logics we may confine our attention to the classical extensions of *E* produced by adding M, C, and N as theorems. Including *E*, eight different classical logics result from taking these schemas as theorems in all possible combinations. These systems and the inclusions among them are registered on the diagram in figure 8.1 (an extension of

Figure 8.1

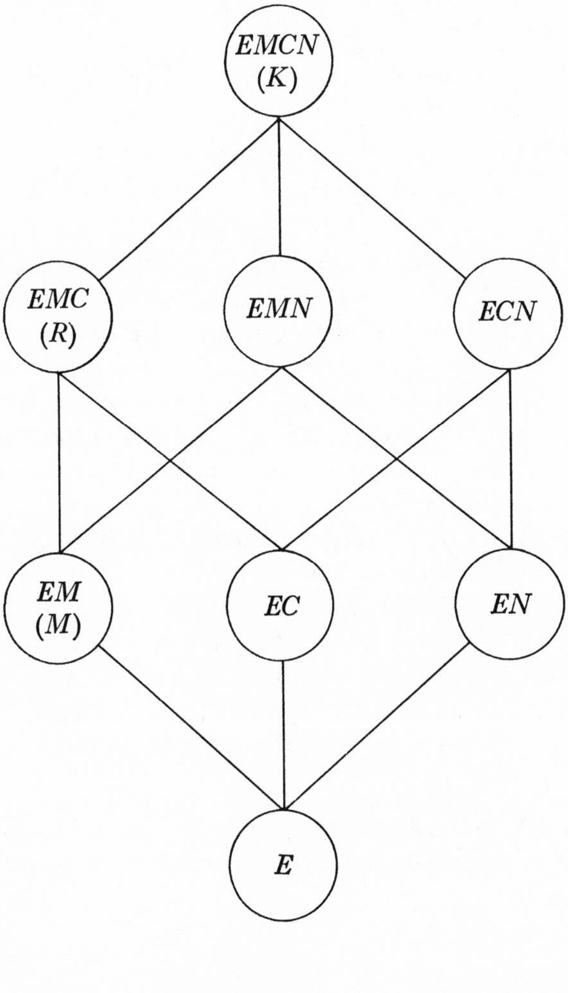

a system is reached upward along the lines). That the systems are distinct is proved in chapter 9.

Let us turn now to the following rules and schemas, in which the possibility operator is featured.

RM$\Diamond$. $\dfrac{A \to B}{\Diamond A \to \Diamond B}$

RR$\Diamond$. $\dfrac{A \to (B \vee C)}{\Diamond A \to (\Diamond B \vee \Diamond C)}$

M$\Diamond$. $(\Diamond A \vee \Diamond B) \to \Diamond(A \vee B)$

C$\Diamond$. $\Diamond(A \vee B) \to (\Diamond A \vee \Diamond B)$

R$\Diamond$. $\Diamond(A \vee B) \leftrightarrow (\Diamond A \vee \Diamond B)$

K$\Diamond$. $(\neg \Diamond A \wedge \Diamond B) \to \Diamond(\neg A \wedge B)$

THEOREM 8.12

(1) *Every monotonic system of modal logic contains* M$\Diamond$ *and is closed under* RM$\Diamond$.

(2) *Every regular system of modal logic contains* M$\Diamond$, C$\Diamond$, R$\Diamond$, *and* K$\Diamond$ *and is closed under* RM$\Diamond$ *and* RR$\Diamond$.

Proof

For (1). If $A \to B$ is a theorem of a monotonic modal logic, then so are $\neg B \to \neg A$, $\Box\neg B \to \Box\neg A$, and $\neg\Box\neg A \to \neg\Box\neg B$ (by *PL*, RM, and *PL*), and the last is equivalent to $\Diamond A \to \Diamond B$ by Df$\Diamond$. So monotonic systems are closed under RM$\Diamond$. That M$\Diamond$ is a theorem when RM$\Diamond$ is present is proved in the proof, for M$\Diamond$, of theorem 4.4. Alternatively, we may observe that by DUAL, M$\Diamond$ is a theorem of a classical logic if and only if M is (see exercise 8.11(*a*)) and then argue for RM$\Diamond$ as in the proof for theorem 4.5(5).

For (2). Because regular systems are monotonic they have RM$\Diamond$ and M$\Diamond$, by part (1). Moreover, if $A \to (B \vee C)$ is a theorem of a regular logic, then so are $(\neg B \wedge \neg C) \to \neg A$, $(\Box\neg B \wedge \Box\neg C) \to \Box\neg A$, and $\neg\Box\neg A \to (\neg\Box\neg B \vee \neg\Box\neg C)$ – by *PL*, RR, and *PL* – and hence so is $\Diamond A \to (\Diamond B \vee \Diamond C)$, by Df$\Diamond$. So regular logics are closed under RR$\Diamond$. By RR$\Diamond$ on a tautology we obtain $\Diamond(A \vee B) \to (\Diamond A \vee \Diamond B)$; so regular systems have C$\Diamond$, and hence R$\Diamond$. The argument for K$\Diamond$ can be found in the proof of theorem 4.4.

Theorem 8.12 provides the basis for some possibility-based characterizations of monotonic and regular systems. We present five in the next theorem.

THEOREM 8.13. *Let* Σ *be a system of modal logic containing* Df$\Box$. *Then:*

(1) Σ *is monotonic iff it is closed under* RM$\Diamond$.

(2) Σ *is monotonic iff it contains* M$\Diamond$ *and is closed under* RE$\Diamond$.

(3) Σ *is regular iff it is closed under* RR$\Diamond$.

(4) Σ *is regular iff it contains* C$\Diamond$ *and is closed under* RM$\Diamond$.

(5) Σ *is regular iff it contains* C$\Diamond$ *and* M$\Diamond$ *and is closed under* RE$\Diamond$.

Proof. We consider logics containing Df$\Box$. Theorem 8.12 takes care of left-to-right, so we treat only the converses.

For (1). It suffices to show that a system has RM if it has RM$\Diamond$. The argument is like that for RM$\Diamond$ (given Df$\Diamond$ and RM) in the proof of theorem 8.12. Exercise.

For (2). By part (1) we need only argue that a system has RM$\Diamond$ if it has M$\Diamond$ and RE$\Diamond$, as we did in the proof of theorem 4.5 (5).

For (3). Here it is enough to prove that a system has RR if it is closed under RR$\Diamond$. Exercise.

For (4):

1. $A \to (B \vee C)$	hypothesis
2. $\Diamond A \to \Diamond(B \vee C)$	1, RM$\Diamond$
3. $\Diamond(B \vee C) \to (\Diamond B \vee \Diamond C)$	C$\Diamond$
4. $\Diamond A \to (\Diamond B \vee \Diamond C)$	2, 3, *PL*

Thus a system has RR$\Diamond$ – and so by part (3) is regular – if it has C$\Diamond$ and RM$\Diamond$.

For (5). If a system has C$\Diamond$, M$\Diamond$, and RE$\Diamond$, then by part (2) it has RM$\Diamond$, and so by part (4) it has RR$\Diamond$, which means by part (3) that it is regular.

Further characterizations of monotonic and regular modal logics, especially using K$\Diamond$, appear in the exercises.

We also leave it as an exercise for the reader to summarize (as we did following the proof of theorem 8.11) the content of theorem 8.13. By attending to theorem 4.5 and exercise 4.6 – which involve RK$\Diamond$, RN$\Diamond$, and N$\Diamond$ as well as the principles mentioned in theorem 8.13 – one can discern a number of possibility-based characterizations of normal systems as certain kinds of regular, monotonic, and classical systems.

The principles canvassed in this section can be generalized along the modal dimension. For example, for each $k \geqslant 0$, classical, monotonic, and regular logics are closed respectively under the rules

$$\text{RE}^k. \quad \frac{A \leftrightarrow B}{\Box^k A \leftrightarrow \Box^k B},$$

$$\text{RM}^k. \quad \frac{A \rightarrow B}{\Box^k A \rightarrow \Box^k B},$$

and

$$\text{RR}^k. \quad \frac{(A \wedge B) \rightarrow C}{(\Box^k A \wedge \Box^k B) \rightarrow \Box^k C},$$

and they all contain, for each $k \geqslant 0$,

$$\text{Df}\Diamond^k. \quad \Diamond^k A \leftrightarrow \neg\Box^k \neg A.$$

(For a proof of the last see the discussion preceding theorem 4.6.) For the record we state the following theorem.

THEOREM 8.14. *Let* Σ *be a system of modal logic, and let* $k \geqslant 0$. *If* Σ *is classical it has the principles* RE^k, $\text{Df}\Diamond^k$, $\text{RE}\Diamond^k$, *and* $\text{Df}\Box^k$; *if* Σ *is monotonic it has in addition* RM^k, M^k, $\text{RM}\Diamond^k$, *and* $\text{M}\Diamond^k$; *if* Σ *is regular it has in addition* RR^k, C^k, K^k, $\text{RR}\Diamond^k$, $\text{C}\Diamond^k$, *and* $\text{K}\Diamond^k$.

Proof. Exercise. (See the discussion preceding theorem 4.6.)

EXERCISES

Where appropriate, freely make use of theorems and rules of inference established in sections 8.1 and 8.2, as well as the results of previous exercises.

8.9. Check that the proofs of theorems 8.9 (1, 3) and 8.10 are in the proof of theorem 4.2.

8.10. Check that the proof of theorem 8.11 (2) is in the proof of theorem 4.3 (3).

8.11. Let Σ be a classical system of modal logic. Prove:

(*a*) Σ contains M iff it contains M$\Diamond$.

(*b*) Σ contains C iff it contains C$\Diamond$.

(*c*) Σ contains R iff it contains R$\Diamond$.

(*d*) Σ contains K iff it contains K$\Diamond$.

(*e*) Σ contains N iff it contains N$\Diamond$.

8.12. Prove that C is a theorem of a monotonic system if and only if K is.

8.13. Extend theorem 8.11 by proving (where Σ is a system containing Df$\Diamond$):

(*a*) Σ is regular iff it contains K and is closed under RM.

(*b*) Σ is regular iff it contains K and M and is closed under RE.

(*c*) Σ is regular iff it contains R and is closed under RE.

8.14. Draw a diagram like that in figure 8.1 showing all the classical systems obtained by adding M, C, K, and N as theorems to *E* (there are ten).

8.15. Check the proof of theorem 4.4 to see that M$\Diamond$ is a theorem of any monotonic logic (theorem 8.12(1)) and that K$\Diamond$ is a theorem of any regular modal logic (theorem 8.12(2)).

8.16. With regard to theorem 8.13:

(*a*) For part (1), show that a system has RM if it has RM$\Diamond$ and Df$\Box$.

(*b*) For part (2), check the proof of theorem 4.5(5) to see that a system has RM$\Diamond$ if it has RE$\Diamond$ and M$\Diamond$.

(*c*) For part (3), show that a system has RR if it has RR$\Diamond$ and Df$\Box$.

8.17. Give a summary of theorem 8.13 like that following theorem 8.11.

8.18. Extend theorem 8.13 by proving (where Σ is a system containing Df$\Box$):

(*a*) Σ is regular iff it contains K$\Diamond$ and is closed under RM$\Diamond$.

(*b*) Σ is regular iff it contains K$\Diamond$ and M$\Diamond$ and is closed under RE$\Diamond$.

(*c*) Σ is regular iff it contains R$\Diamond$ and is closed under RE$\Diamond$.

8.19. Prove some of the parts of theorem 8.14.

8.20. Prove that the following are theorems of any classical modal logic.

(*a*) $\Diamond\top \leftrightarrow \neg\Box\bot$ (*b*) $\Box\top \leftrightarrow \neg\Diamond\bot$

8.21. Prove that the following are theorems of any monotonic modal logic.

(*a*) $\Box A \to \Box(B \to A)$

(*b*) $\Box\neg A \to \Box(A \to B)$

(*c*) $(\Box A \vee \Box B) \to \Box(A \vee B)$

(*d*) $\Diamond(A \wedge B) \rightarrow (\Diamond A \wedge \Diamond B)$

(*e*) $\Diamond(A \rightarrow B) \vee \Box(B \rightarrow A)$

(*f*) $(\Box A \rightarrow \Diamond B) \rightarrow \Diamond(A \rightarrow B)$

(*g*) $(\Box A \rightarrow \Diamond A) \rightarrow \Diamond \top$

(*h*) $(\Diamond A \rightarrow \Box B) \rightarrow \Box(A \rightarrow B)$

8.22. Prove that the following are theorems of any regular modal logic.

(*a*) $\Box(A \rightarrow B) \rightarrow (\Diamond A \rightarrow \Diamond B)$

(*b*) $\Box(A \leftrightarrow B) \rightarrow (\Box A \leftrightarrow \Box B)$

(*c*) $\Box(A \leftrightarrow B) \rightarrow (\Diamond A \leftrightarrow \Diamond B)$

(*d*) $(\Box A \wedge \Diamond B) \rightarrow \Diamond(A \wedge B)$

(*e*) $\Box(A \vee B) \rightarrow (\Diamond A \vee \Box B)$

(*f*) $\Diamond(A \rightarrow B) \leftrightarrow (\Box A \rightarrow \Diamond B)$

(*g*) $\Diamond \top \leftrightarrow (\Box A \rightarrow \Diamond A)$

(*h*) $(\Diamond A \rightarrow \Box B) \rightarrow (\Box A \rightarrow \Box B)$

(*i*) $(\Diamond A \rightarrow \Box B) \rightarrow (\Diamond A \rightarrow \Diamond B)$

8.23. Prove that the following are theorems of any monotonic modal logic (for any $n \geqslant 2$).

(*a*) $\Box(A_1 \wedge \ldots \wedge A_n) \rightarrow (\Box A_1 \wedge \ldots \wedge \Box A_n)$

(*b*) $(\Diamond A_1 \vee \ldots \vee \Diamond A_n) \rightarrow \Diamond(A_1 \vee \ldots \vee A_n)$

(*c*) $(\Box A_1 \vee \ldots \vee \Box A_n) \rightarrow \Box(A_1 \vee \ldots \vee A_n)$

(*d*) $\Diamond(A_1 \wedge \ldots \wedge A_n) \rightarrow (\Diamond A_1 \wedge \ldots \wedge \Diamond A_n)$

8.24. Prove that the following are theorems of any regular modal logic (for any $n \geqslant 2$).

(*a*) $\Box(A_1 \wedge \ldots \wedge A_n) \leftrightarrow (\Box A_1 \wedge \ldots \wedge \Box A_n)$

(*b*) $\Diamond(A_1 \vee \ldots \vee A_n) \leftrightarrow (\Diamond A_1 \vee \ldots \vee \Diamond A_n)$

(*c*) $(\Box A_1 \wedge \ldots \wedge \Box A_{n-1} \wedge \Diamond A_n) \rightarrow \Diamond(A_1 \wedge \ldots \wedge A_n)$

(*d*) $\Box(A_1 \vee \ldots \vee A_n) \rightarrow (\Diamond A_1 \vee \ldots \vee \Diamond A_{n-1} \vee \Box A_n)$

8.25. Prove that the following are theorems of any monotonic modal logic whenever $m \leqslant n$.

(*a*) $\Box^n \top \rightarrow \Box^m \top$

(*b*) $\Diamond^m \bot \rightarrow \Diamond^n \bot$

(*c*) $\Diamond^n \top \rightarrow \Diamond^m \top$

(*d*) $\Box^m \bot \rightarrow \Box^n \bot$

8.26. Prove that a system is monotonic if and only if it is classical and contains $\Box(A \wedge B) \to \Box A$ or $\Diamond A \to \Diamond(A \vee B)$.

8.27. Prove that a system containing Df$\Diamond$ is regular if and only if, for every $n \geqslant 1$, it is closed under the rule

$$\frac{(A_1 \wedge \ldots \wedge A_n) \to A}{(\Box A_1 \wedge \ldots \wedge \Box A_n) \to \Box A}.$$

8.28. Prove that a classical system contains N if and only if it is closed under the rule RN.

8.29. Let ϕ and ψ be affirmative modalities. Prove that a monotonic system has one of the following theorems and rules of inference if and only if it has them all.

$$\phi A \to \psi A \quad \psi^* A \to \phi^* A$$

$$\frac{A \to B}{\phi A \to \psi B} \quad \frac{A \to B}{\psi^* A \to \phi^* B}$$

8.30. Use the erasure transformation ϵ in exercise 1.27 to prove the consistency of the eight systems in figure 8.1.

8.31. Prove that the systems in figure 8.1 have the following rules of inference.

(a) $\dfrac{\Box A}{A}$ $\quad$ (d) $\dfrac{\Diamond A}{A}$

(b) $\dfrac{\Box A \to \Box B}{A \to B}$ $\quad$ (e) $\dfrac{\Diamond A \to \Diamond B}{A \to B}$

(c) $\dfrac{\Box A \leftrightarrow \Box B}{A \leftrightarrow B}$ $\quad$ (f) $\dfrac{\Diamond A \leftrightarrow \Diamond B}{A \leftrightarrow B}$

(Compare exercise 4.13.)

8.32. Let Γ be a maximal set of sentences in a monotonic system Σ. Prove:

$$\Box A \in \Gamma \text{ iff } |B|_\Sigma \subseteq |A|_\Sigma \text{ for some } \Box B \in \Gamma.$$

8.33. Let $\sim_\Sigma$ be the relation of Σ-equivalence (see exercise 2.37) in a system Σ. We define $|A|_\Sigma$ to be the set of all sentences Σ-equivalent to A; i.e. for every sentence A,

$$|A|_\Sigma = \{B : A \sim_\Sigma B\}.$$

(Note that here $|A|_\Sigma$ is *not* the proof set of A in Σ.) Consider the structure

$$\mathscr{B} = \langle B, 1, 0, -, \cap, \cup \rangle$$

in which B is the set of all Σ-equivalence classes of sentences – $\{|A|_\Sigma : A$ is a sentence$\}$ – and the elements 1 and 0 of B and operations $-$, $\cap$, and $\cup$ in B are defined as follows.

(1) $1 = |\top|_\Sigma$.

(2) $0 = |\bot|_\Sigma$.

(3) $-|A|_\Sigma = |\neg A|_\Sigma$.

(4) $|A|_\Sigma \cap |B|_\Sigma = |A \wedge B|_\Sigma$.

(5) $|A|_\Sigma \cup |B|_\Sigma = |A \vee B|_\Sigma$.

Then $\mathscr{B}$ is called the *lindenbaum algebra* of Σ. It is readily verified that these definitions are unambiguous and that $\mathscr{B}$ is a boolean algebra (see exercise 7.11). Prove:

(*a*) $|A \rightarrow B|_\Sigma = -|A|_\Sigma \cup |B|_\Sigma$.

(*b*) $|A \leftrightarrow B|_\Sigma = (-|A|_\Sigma \cup |B|_\Sigma) \cap (-|B|_\Sigma \cup |A|_\Sigma)$.

(*c*) $\vdash_\Sigma A$ iff $|A|_\Sigma = 1$.

(*d*) $\vdash_\Sigma A \rightarrow B$ iff $|A|_\Sigma \leqslant |B|_\Sigma$.

(*e*) $\vdash_\Sigma A \leftrightarrow B$ iff $|A|_\Sigma = |B|_\Sigma$.

Now suppose that Σ is a classical system, and define the operation $*$ on the lindenbaum algebra $\mathscr{B}$ by:

$$*|A|_\Sigma = |\Box A|_\Sigma.$$

Verify that this definition is unambiguous by proving:

(*f*) If $|A|_\Sigma = |B|_\Sigma$, then $*|A|_\Sigma = *|B|_\Sigma$.

The structure $\langle \mathscr{B}, * \rangle$ is a modal algebra in the sense of exercise 7.11; we may call it the *lindenbaum modal algebra* of Σ. Prove:

(*g*) $|\Diamond A|_\Sigma = -*-|A|_\Sigma$.

Finally, consider the following conditions on an algebraic model $\mathscr{M} = \langle \mathscr{B}, *, P \rangle$ formed from the lindenbaum modal algebra $\langle \mathscr{B}, * \rangle$ (see exercises 7.11 and 7.27).

(m*) $*(a \cap b) \leqslant *a \cap *b$

(c*) $*a \cap *b \leqslant *(a \cap b)$

(n*) $*1 = 1$

Prove:

(*h*) $\mathscr{M}$ satisfies (m*) iff Σ contains M.

(*i*) $\mathcal{M}$ satisfies (c*) iff Σ contains C.

(*j*) $\mathcal{M}$ satisfies (n*) iff Σ contains N.

8.3. Other schemas

We confine ourselves in this section to two theorems about classical modal logics containing as theorems the schemas

T. $\Box A \to A$

and

5. $\Diamond A \to \Box\Diamond A$.

THEOREM 8.15. *Every classical ET5-system has the following theorems and rules of inference.*

P. $\Diamond\top$

RP. $\dfrac{A}{\Diamond A}$

D. $\Box A \to \Diamond A$

B. $A \to \Box\Diamond A$

G. $\Diamond\Box A \to \Box\Diamond A$

4. $\Box A \to \Box\Box A$

N. $\Box\top$

RN. $\dfrac{A}{\Box A}$

Proof. Except for N the proofs are left as exercises. For N we argue as follows.

1. $\Diamond\top$	P
2. $\Diamond\top \leftrightarrow \top$	1, *PL*
3. $\Box\Diamond\top \leftrightarrow \Box\top$	2, RE
4. $\Diamond\top \to \Box\Diamond\top$	5
5. $\Box\top$	1, 3, 4, *PL*

THEOREM 8.16. *Every classical ET5-system has at most six distinct modalities, viz.* $\cdot$, $\Box$, $\Diamond$, *and their negations, with implications among the affirmative three as diagramed in figure 8.2.*

Proof. Recall from section 4.4 that modalities ϕ and ψ are equivalent in a system when $\phi A \leftrightarrow \psi A$ is a theorem; otherwise they are distinct. Thus to prove the theorem it is enough to show that any classical *ET5*-system has the following reduction laws.

$$\Box A \leftrightarrow \Box\Box A \quad \Diamond A \leftrightarrow \Diamond\Diamond A$$
$$\Diamond A \leftrightarrow \Box\Diamond A \quad \Box A \leftrightarrow \Diamond\Box A$$

This is left for the reader as an exercise, along with the demonstration that the implications are as advertised.

The reader should compare theorems 4.28 and 8.16 (and the diagrams in figures 4.8 and 8.2).

Theorems 8.15 and 8.16 are meant to be illustrative of results for classical systems containing as theorems schemas like D, T, B, 4, and 5. Some further results are suggested in the exercises.

EXERCISES

8.34. Complete the proof of theorem 8.15.

8.35. Give the proof of theorem 8.16.

8.36. Use theorems 8.9 and 8.16 to prove theorem 4.28.

8.37. Let Σ be a monotonic modal logic. Prove:

(*a*) Σ is an *MN*-system iff it has theorems of the form $\Box$A.

(*b*) Σ is an *MP*-system iff it has theorems of the form $\Diamond$A.

8.38. Recall the schemas 4! ($\Box A \leftrightarrow \Box\Box A$) and 5! ($\Diamond A \leftrightarrow \Box\Diamond A$). Prove that 4! is a theorem of any classical *E5!*-system.

8.39. Investigate the question of the number of distinct modalities in some classical systems having as theorems schemas such as the ones mentioned in theorem 8.15. For example:

(*a*) Every regular *R5*-system has at most ten distinct modalities.
(*b*) Every monotonic *MD5*-system has at most ten distinct modalities.

Figure 8.2. Modalities in classical *ET5*-systems.

(*c*) Every classical *ED45*-system has at most six distinct modalities.

Identify the modalities in these systems and describe the implications among them.

8.40. Investigate the question of consistency for some classical systems having as theorems schemas like those in theorem 8.15. Use the erasure transformation of exercise 1.27, or argue by way of results in exercises 4.12 and 4.60 and theorem 8.9.

8.41. Investigate the presence of rules (*a*)–(*f*) from exercise 8.31 in classical logics containing theorems like those in theorem 8.15. (Compare exercises 4.13, 4.61, and 4.62.)

8.42. Let $\mathscr{M} = \langle \mathscr{B}, *, P \rangle$ be an algebraic model based on the lindenbaum modal algebra of a classical system Σ (see exercise 8.33), and consider the conditions (d*), (t*), (b*), (iv*), (v*), and (g*) in exercise 7.42. Prove:

(*a*) $\mathscr{M}$ satisfies (d*) iff Σ contains D.

(*b*) $\mathscr{M}$ satisfies (t*) iff Σ contains T.

(*c*) $\mathscr{M}$ satisfies (b*) iff Σ contains B.

(*d*) $\mathscr{M}$ satisfies (iv*) iff Σ contains 4.

(*e*) $\mathscr{M}$ satisfies (v*) iff Σ contains 5.

(*f*) $\mathscr{M}$ satisfies (g*) iff Σ contains G.

Also prove that $\mathscr{M}$ satisfies the condition

$$(\text{p}^*) \quad -*-1 = 1$$

if and only if Σ contains P.

Referring to the definition of $*^n$ in exercise 7.43, prove:

(*g*) $*^n|\text{A}|_\Sigma = |\Box^n\text{A}|_\Sigma$.

(*h*) $-*^n-|\text{A}|_\Sigma = |\Diamond^n\text{A}|_\Sigma$.

Finally, consider the following conditions on $\mathscr{M}$.

$$(\text{g}^{*k,l,m,n}) \quad -*^k-*^l a \leqslant *^m-*^n-a$$

$$(\text{g}_\text{c}^{*k,l,m,n}) \quad *^k-*^l-a \leqslant -*^m-*^n a$$

Prove:

(*i*) $\mathscr{M}$ satisfies $(\text{g}^{*k,l,m,n})$ iff Σ contains $\text{G}^{k,l,m,n}$.

(*j*) $\mathscr{M}$ satisfies $(\text{g}_\text{c}^{*k,l,m,n})$ iff Σ contains $\text{G}_\text{c}^{k,l,m,n}$.

9

DETERMINATION AND DECIDABILITY FOR CLASSICAL SYSTEMS

In this chapter we connect classical modal logics and classes of minimal models by way of determination theorems. Our method is much the same as in chapter 5 for normal systems and standard models. In section 9.1 we treat questions of soundness and prove the distinctness of the eight classical systems on the diagram in figure 8.1 having M, C, and N as theorems. Section 9.2 contains the definition of canonical minimal models and the fundamental theorems for completeness. We do not single out any particular canonical minimal model as 'proper', as we did in the case of normal systems. But we indicate a uniform way of describing canonical minimal models – a way that highlights two very useful such models, which we call the smallest and the largest, for any classical logic.

In section 9.3 there are determination theorems for the systems in figure 8.1. The idea of supplementation plays an important part in obtaining completeness results for the monotonic systems, and we use augmentation to reach, again, the conclusion that normal systems are complete with respect to the class of standard models. Then in section 9.4 we treat in an abridged fashion questions of completeness for classical systems having as theorems familiar schemas such as D, T, B, 4, and 5. Finally, in section 9.5 we prove the decidability of the systems *E*, *M*, *R*, and, once again, *K*.

9.1. Soundness

The basic theorem for the soundness of classical modal logics with respect to classes of minimal models is the following.

THEOREM 9.1. *Let* $\mathrm{S}_1, \ldots, \mathrm{S}_n$ *be schemas valid respectively in classes of minimal models* $\mathsf{C}_1, \ldots, \mathsf{C}_n$. *Then the system of modal logic* $ES_1 \ldots S_n$ *is sound with respect to the class* $\mathsf{C}_1 \cap \ldots \cap \mathsf{C}_n$.

Proof. By theorems 2.8 and 7.3, Df$\Diamond$ is valid in any class of minimal models, and the rules RE and RPL preserve validity in any such class. Moreover, if $S_1, \ldots, S_n$ are valid respectively in $C_1, \ldots, C_n$, then they are valid in the intersection of these classes. Therefore, every theorem of $ES_1 \ldots S_n$ is valid in this intersection, which means that this logic is sound with respect to that class.

Thus *E*, the smallest classical system, is sound with respect to every class of minimal models, in particular the class of all such models. Moreover, by referring to theorem 7.5 we obtain soundness results for all the classical systems on the diagram in figure 8.1. In particular, the smallest monotonic, regular, and normal systems – *M*, *R*, and *K* – are sound with respect to classes of supplementations, quasi-filters, and filters, respectively. The proofs of these and other soundness theorems we leave as exercises.

Let us use theorem 9.1 to prove the distinctness of the systems in figure 8.1.

THEOREM 9.2. *The eight classical systems in figure 8.1 are all distinct.*

Proof. Consider the following three minimal models $\mathcal{M} = \langle W, N, P \rangle$.

(1) $W = \{\alpha\}$; $N_\alpha = \emptyset$; $P_n = \emptyset$ for $n \geqslant 0$.

(2) $W = \{\alpha, \beta\}$ (distinct); $N_\alpha = N_\beta = \{\{\alpha\}, \{\beta\}, W\}$; $P_0 = \{\alpha\}$ and $P_n = \{\beta\}$ for $n \geqslant 0$.

(3) $W = \{\alpha, \beta\}$ (distinct); $N_\alpha = N_\beta = \{\emptyset, W\}$; $P_0 = \{\alpha\}$ and $P_n = \{\beta\}$ for $n \geqslant 0$.

Re (1). $\mathcal{M}$ is supplemented and closed under intersections, and it falsifies N. By theorem 9.1 this is a model of *EMC* and all the systems below it on the diagram in figure 8.1. Thus these systems are distinct from all the others on the diagram, viz. those containing N as a theorem.

Re (2). Here $\mathcal{M}$ is supplemented and contains the unit, and it falsifies an instance of C: $(\Box \mathbb{P}_0 \wedge \Box \mathbb{P}_1) \to \Box(\mathbb{P}_0 \wedge \mathbb{P}_1)$. So $\mathcal{M}$ is a model of *EMN* and all the systems in figure 8.1 that it extends. These systems are therefore distinct from the others on the diagram, which all contain C.

Re (3). In this case $\mathcal{M}$ is closed under intersections and contains the unit, and it falsifies an instance of M: $\Box(\mathbb{P}_0 \wedge \mathbb{P}_1) \to (\Box \mathbb{P}_0 \wedge \Box \mathbb{P}_1)$. Hence $\mathcal{M}$ is a model of *ECN* and the systems below it in figure 8.1, which means that these systems are all distinct from the others on the diagram, wherein M is a theorem.

We leave it as an exercise for the reader to check that the foregoing considerations yield the pairwise distinctness of the eight systems in figure 8.1.

EXERCISES

9.1. Prove:

(*a*) E is sound with respect to any class of minimal models.

(*b*) M is sound with respect to any class of supplementations.

(*c*) R is sound with respect to any class of quasi-filters.

(*d*) K is sound with respect to any class of filters.

Formulate and prove similar soundness theorems for the other four systems in figure 8.1.

9.2. Verify that models (1)–(3) in the proof of theorem 9.2 (*a*) meet the conditions stated, (*b*) falsify the sentences in question, and (thus) (*c*) ensure the distinctness of the systems in figure 8.1. (Compare exercise 7.19.)

9.3. Using results from section 7.4, including the exercises, describe classes of minimal models with respect to which the following systems are sound.

(*a*) ED (*e*) $E5$

(*b*) ET (*f*) EG

(*c*) EB (*g*) $EG^{k,l,m,n}$

(*d*) $E4$

9.4. Consider the following conditions on a minimal model $\mathcal{M} = \langle W, N, P\rangle$.

$$(\mathrm{p})\ \varnothing \notin \mathrm{N}_\alpha \quad (\bar{\mathrm{p}})\ \varnothing \in \mathrm{N}_\alpha$$

Prove that EP and $E\bar{P}$ are sound respectively relative to classes of minimal models satisfying (p) and ($\bar{\mathrm{p}}$).

9.5. Prove that $EG_c^{k,l,m,n}$ is sound with respect to any class of minimal models $\mathcal{M} = \langle W, N, P\rangle$ for which the following condition holds.

$$(\mathrm{g}_c^{k,l,m,n})\ \text{if } \{\beta \text{ in } \mathcal{M}: -X \notin N_\beta^l\} \in N_\alpha^k, \text{ then } -\{\beta \text{ in } \mathcal{M}: X \in N_\beta^n\} \notin N_\alpha^m$$

(Compare exercise 7.38.)

9.6. Prove that *EU* is sound with respect to any class of minimal models $\mathcal{M} = \langle W, N, P \rangle$ satisfying the following condition.

(u) $\{\beta \text{ in } \mathcal{M}\colon \text{if } X \in N_\beta, \text{ then } \beta \in X\} \in N_\alpha$

9.7. Using exercise 8.37(*a*) and theorem 9.2, prove that *E*, *EM* (*M*), *EC*, and *EMC* (*R*) have no theorems of the form $\Box$A.

9.8. Prove that *M* is sound with respect to any class of minimal models in which modalities are evaluated as in exercise 7.9.

9.9. Prove that *ET5* has exactly the distinct modalities diagramed in figure 8.2.

9.10. Extend the results of exercise 8.39 by establishing the exact numbers of distinct modalities in the systems investigated. In particular, prove:

(*a*) *R5* has exactly ten distinct modalities, the same as *K5* (see theorems 4.23 and 5.3).

(*b*) *MD5* has exactly ten distinct modalities, the same as *KD5* (see theorems 4.24 and 5.3).

(*c*) *ED45* has exactly six distinct modalities, the same as *KD45* (see theorems 4.27 and 5.3).

9.11. Prove that each of the systems in figure 8.1 has infinitely many distinct modalities.

9.12. Consider the following conditions on a model $\mathcal{M} = \langle W, f, P \rangle$ of the kind described in exercise 7.10.

(m_f) $f(X \cap Y) \subseteq f(X) \cap f(Y)$

(c_f) $f(X) \cap f(Y) \subseteq f(X \cap Y)$

(n_f) $f(W) = W$

(d_f) $f(X) \subseteq -f(-X)$

(t_f) $f(X) \subseteq X$

(b_f) $X \subseteq f(-f(-X))$

(iv_f) $f(X) \subseteq f(f(X))$

(v_f) $-f(-X) \subseteq f(-f(-X))$

(g_f) $-f(-f(X)) \subseteq f(-f(-X))$

Prove that *E* is sound with respect to any class of such models, and that the following systems are sound relative to classes of such models satisfying the indicated conditions.

(*a*) *EM* (*M*): (m_f)	(*f*) *EB*: (b_f)
(*b*) *EC*: (c_f)	(*g*) *E4*: (iv_f)
(*c*) *EN*: (n_f)	(*h*) *E5*: (v_f)
(*d*) *ED*: (d_f)	(*i*) *EG*: (g_f)
(*e*) *ET*: (t_f)	

9.13. Prove that *E* is sound with respect to any class of algebraic models (defined in exercise 7.11). Then, referring to exercises 7.27 and 7.42, show that the following systems are sound with respect to classes of algebraic models that satisfy the indicated conditions.

(*a*) *EM* (*M*): (m*)	(*f*) *EB*: (b*)
(*b*) *EC*: (c*)	(*g*) *E4*: (iv*)
(*c*) *EN*: (n*)	(*h*) *E5*: (v*)
(*d*) *ED*: (d*)	(*i*) *EG*: (g*)
(*e*) *ET*: (t*)	

Finally, referring to exercises 7.43 and 8.42, show the soundness of the following systems relative to classes of algebraic models satisfying the indicated conditions.

(*j*) *EP*: (p*)

(*k*) $EG^{k,l,m,n}$: $(g^{*k,l,m,n})$

(*l*) $EG_c^{k,l,m,n}$: $(g_c^{*k,l,m,n})$

9.2. Completeness: basic theorems

We begin with the key idea of a canonical minimal model for a classical modal logic.

DEFINITION 9.3. Let $\mathcal{M} = \langle W, N, P\rangle$ be a minimal model, and let Σ be a classical system of modal logic. $\mathcal{M}$ is a *canonical minimal model* for Σ iff:

(1) $W = \{\Gamma\colon \mathrm{Max}_\Sigma\, \Gamma\}$.

(2) For every α in $\mathcal{M}$, $\Box A \in \alpha$ iff $|A|_\Sigma \in N_\alpha$.

(3) $P_n = |\mathbb{P}_n|_\Sigma$, for $n = 0, 1, 2, \ldots$.

Parts (1) and (3) of the definition conform to the account in section 2.7. The acute reader will have noticed that the condition on N in part (2) must be shown not to be ambiguous. That is, it must be shown that whenever A and B are sentences such that $|A|_{\Sigma} = |B|_{\Sigma}$, then $|A|_{\Sigma} \in N_{\alpha}$ just in case $|B|_{\Sigma} \in N_{\alpha}$, for every world α in $\mathscr{M}$. Only if this is true is the definition correct.

That it is true depends on the assumption that the system Σ is classical. For suppose that $|A|_{\Sigma} = |B|_{\Sigma}$. By theorem 2.22(3) this means that $\vdash_{\Sigma} A \leftrightarrow B$. Because Σ is classical we may infer by RE that $\vdash_{\Sigma} \Box A \leftrightarrow \Box B$. So by theorem 2.20(2) $\Box A \leftrightarrow \Box B$ belongs to every Σ-maximal set of sentences, i.e. to every world α in $\mathscr{M}$. By theorem 2.18(9) it follows that $\Box A \in \alpha$ if and only if $\Box B \in \alpha$, for every α in $\mathscr{M}$. Hence, by the condition on N in the definition, $|A|_{\Sigma} \in N_{\alpha}$ just in case $|B|_{\Sigma} \in N_{\alpha}$, for every α in $\mathscr{M}$.

It should be clear that functions N exist that satisfy the condition in a canonical minimal model for a classical system. We mention some specific examples at the end of the section, in connection with an important alternative characterization of the class of models of this kind. Meanwhile, the following theorem characterizes canonical minimal models in terms of the behavior of the possibility operator.

THEOREM 9.4. *$\mathscr{M} = \langle W, N, P\rangle$ is a canonical minimal model for a classical system Σ iff W and P are as in definition 9.3, and for every α in $\mathscr{M}$,*

$$\Diamond A \in \alpha \text{ iff } -|A|_{\Sigma} \notin N_{\alpha}.$$

Proof. For left-to-right, let $\mathscr{M} = \langle W, N, P\rangle$ be a canonical minimal model for a classical modal logic Σ (so that W and P are as specified). Then for any world α in $\mathscr{M}$:

$\Diamond A \in \alpha$ iff $\neg\Box\neg A \in \alpha$
– Df$\Diamond$ and the Σ-maximality of α;

iff $\Box\neg A \notin \alpha$
– theorem 2.18(5);

iff $|\neg A|_{\Sigma} \notin N_{\alpha}$
– definition 9.3;

iff $-|A|_{\Sigma} \notin N_{\alpha}$
– theorem 2.22(4, 6).

The reverse is left as an exercise for the reader.

We come now to the fundamental theorem for the completeness of classical modal logics, to wit, that a world in a canonical minimal model verifies precisely the sentences it contains.

THEOREM 9.5. *Let $\mathcal{M}$ be a canonical minimal model for a classical system Σ. Then for every α in $\mathcal{M}$:*

$\models_{\alpha}^{\mathcal{M}}$ A *iff* A $\in \alpha$.

In other words, $\|A\|^{\mathcal{M}} = |A|_{\Sigma}$.

Proof. The only case that need be examined is that in which A is a necessitation, $\Box$B. By the inductive hypothesis, $\|B\|^{\mathcal{M}} = |B|_{\Sigma}$. So for any α in $\mathcal{M}$, $\|B\|^{\mathcal{M}} \in N_{\alpha}$ if and only if $|B|_{\Sigma} \in N_{\alpha}$. By definitions 7.2(1) and 9.3 it follows that $\models_{\alpha}^{\mathcal{M}} \Box$B if and only if $\Box$B $\in \alpha$.

From theorem 9.5 it follows at once (see the remarks in section 2.7) that the theorems of a classical modal logic are exactly the sentences true in a canonical minimal model for the logic. Formally:

THEOREM 9.6. *Let $\mathcal{M}$ be a canonical minimal model for a classical system Σ. Then:*

$\models^{\mathcal{M}}$ A *iff* $\vdash_{\Sigma}$ A.

What are canonical minimal models like? The question comes to asking what the function N may be like in a canonical minimal model $\mathcal{M}$ for a classical modal logic Σ. The simplest example of such a model is that in which, for each world α in $\mathcal{M}$, N_{α} is the set $\{|A|_{\Sigma}: \Box A \in \alpha\}$; i.e. N_{α} consists just of those proof sets $|A|_{\Sigma}$ such that $\Box$A $\in \alpha$, and nothing else. In this case we have what we shall call the *smallest* canonical minimal model for Σ. At the opposite extreme there is the canonical minimal model $\mathcal{M}$ in which, for each α, N_{α} consists of $\{|A|_{\Sigma}: \Box A \in \alpha\}$ together with the set of all *non*-proof sets relative to Σ – i.e. $\{|A|_{\Sigma}: \Box A \in \alpha\}$ plus every set X of worlds in $\mathcal{M}$ that is not a proof set in Σ for any sentence. This is the *largest* canonical minimal model for Σ.

These examples should make it clear that as long as N_{α} always consists of the set $\{|A|_{\Sigma}: \Box A \in \alpha\}$ together with a collection of non-proof sets (relative to Σ) $\mathcal{M}$ will be a canonical minimal model for Σ. Thus we have an alternative and often simpler way of recognizing canonical minimal models. Because of the usefulness of this way of putting the matter, we state it formally in the following theorem (the proof of which is left as an exercise).

THEOREM 9.7. *$\mathcal{M} = \langle W, N, P \rangle$ is a canonical minimal model for a classical system of modal logic Σ iff W and P are as in definition 9.3, and for every α in $\mathcal{M}$,*

$$N_\alpha = \{|A|_\Sigma : \Box A \in \alpha\} \cup \mathcal{X},$$

where $\mathcal{X}$ is any collection of non-proof sets relative to Σ (i.e. $\mathcal{X} \subseteq \{X \subseteq W: X \neq |A|_\Sigma$ for every sentence A$\}$).

EXERCISES

9.14. Complete the proof of theorem 9.4.

9.15. Give the proof of theorem 9.5. for the case in which A = $\Diamond$B.

9.16. Prove theorem 9.7.

9.17. Let $\mathcal{M} = \langle W, N, P \rangle$ be a canonical minimal model for a classical system Σ. Use exercise 7.7 and theorem 9.5 to prove, for any $n \geqslant 0$:

(*a*) $|\Box^n A|_\Sigma = \{\alpha \text{ in } \mathcal{M}: |A|_\Sigma \in N^n_\alpha\}$.
(*b*) $|\Diamond^n A|_\Sigma = \{\alpha \text{ in } \mathcal{M}: -|A|_\Sigma \notin N^n_\alpha\}$.

9.18. Use theorem 9.6 to prove that the system E is complete with respect to the class of all minimal models, and so is determined by it, in virtue of exercise 9.1(*a*).

9.19. Let $\mathcal{M} = \langle W, N, P \rangle$ be the smallest canonical minimal model for a monotonic system Σ. Prove:

(*a*) $\mathcal{M}$ is not supplemented.
(*b*) The supplementation $\mathcal{M}^+ = \langle W, N^+, P \rangle$ of $\mathcal{M}$ is a canonical minimal model for Σ. (Exercise 8.32 may be helpful.)

9.20. Use the second result in the preceding exercise to argue that the system M is complete with respect to the class of supplemented minimal models (and so is determined by it; see exercise 9.1(*b*)).

9.21. Let $\mathcal{M} = \langle W, N, P \rangle$ be the smallest canonical minimal model for a classical system containing the schema C. Prove that $\mathcal{M}$ is closed under intersections.

9.22. Using results in exercises 9.19 and 9.21, prove that the system R is complete with respect to the class of quasi-filters (and hence is determined by it; see exercise 9.1(*c*)).

9.23. Let $\mathcal{M} = \langle W, N, P\rangle$ be any canonical minimal model for a classical system in which N is a theorem. Prove that $\mathcal{M}$ contains the unit.

9.24. Using results in exercises 9.19, 9.21, and 9.23, prove that the system K is complete with respect to the class of filters (and is thus determined by it, given exercise 9.1(d)).

9.25. Let $\mathcal{M}^! = \langle W, N^!, P\rangle$ be the augmentation of the smallest canonical minimal model $\mathcal{M} = \langle W, N, P\rangle$ for a normal modal logic. Prove that $\mathcal{M}^!$ is also a canonical minimal model for the logic. (Theorem 4.30(1) may be helpful.)

9.26. Conclude from the preceding exercise that the system K is complete with respect to the class of augmented minimal models, and use exercise 9.24 to argue that this class determines K.

9.27. Let Σ be a monotonic system, and let $\mathcal{M} = \langle W, N, P\rangle$ be a minimal model in which W and P are as in a canonical model for Σ, and N satisfies the following condition.

$$\Box\mathrm{A} \in \alpha \text{ iff for some } X \in N_\alpha,\ X \subseteq |\mathrm{A}|_\Sigma$$

Relative to the truth conditions for modalities given in exercise 7.9, prove that for every α in $\mathcal{M}$,

$$\models^{\mathcal{M}}_\alpha \mathrm{A} \text{ iff } \mathrm{A} \in \alpha,$$

i.e. that $\|\mathrm{A}\|^{\mathcal{M}} = |\mathrm{A}|_\Sigma$. The proof is by induction on the complexity of A. Give it for the cases in which A is atomic, the falsum, a conditional, and a necessitation.

Is $\mathcal{M}$ a canonical minimal model for Σ, in the sense of definition 9.3?

9.28. Define the idea of a canonical model for a classical modal logic in terms of the models $\mathcal{M} = \langle W, f, P\rangle$ of exercise 7.10.

9.29. Let $\langle \mathcal{B}, *\rangle$ be the lindenbaum modal algebra of a classical system Σ, and define the algebraic model $\mathcal{M} = \langle \mathcal{B}, *, P\rangle$ by:

$$P_n = |\mathbb{P}_n|_\Sigma, \text{ for } n = 0, 1, 2, \ldots.$$

(See exercises 7.11 and 8.33.) Then $\mathcal{M}$ is a *canonical algebraic model* (indeed, *the* such model) for Σ, in the sense that, for every sentence A,

$$\models^{\mathcal{M}} \mathrm{A} \text{ iff } \vdash_\Sigma \mathrm{A}.$$

Prove this by induction on the complexity of A, at least for the cases in which A is atomic, the falsum, a conditional, and a necessitation.

9.3. Determination

We wish to prove determination theorems for the eight classical modal logics obtained by taking the schemas M, C, and N as theorems in all possible combinations (see the diagram in figure 8.1). In particular, we shall obtain results for the smallest classical, monotonic, regular, and normal systems – *E*, *M*, *R*, and *K*.

Completeness is all that needs to be shown in each case; soundness follows easily from theorems 7.5 and 9.1 (see exercise 9.1). We begin with the smallest classical modal logic.

THEOREM 9.8. *E is determined by the class of minimal models.*

Proof. Completeness follows from theorem 9.6 and the existence of canonical minimal models: a sentence valid in the class of minimal models is true in any canonical minimal model for *E*, each of which determines *E*.

To prove the completeness of a classical modal logic with respect to a class of minimal models it suffices to show that some canonical minimal model for the logic is in the class. (Recall the discussion in section 2.7.) The problem in each case is to find a suitable canonical model, since it can happen that not every such model for a system is in the class in question (see exercise 9.19(*a*)). For the classical extensions of *E* using M, C, and N we require theorems 9.9, 9.11, and 9.13 below. The first of these concerns monotonic systems, i.e. classical systems containing M or closed under the rule RM. It states that the supplementation of the smallest canonical minimal model for a monotonic logic is itself a canonical model for the system.

THEOREM 9.9. *If $\mathcal{M} = \langle W, N, P\rangle$ is the smallest canonical minimal model for a monotonic system of modal logic, then its supplementation $\mathcal{M}^+ = \langle W, N^+, P\rangle$ is also a canonical minimal model for the system.*

Proof. Let Σ be a monotonic logic. If $\mathcal{M}$ is the smallest canonical minimal model for Σ, then $N_\alpha = \{|A|_\Sigma : \Box A \in \alpha\}$ for each α in $\mathcal{M}$ (see the remarks following theorem 9.6). By the definition of supplementation, 7.6, it follows that $N^+_\alpha = \{X \subseteq W: |A|_\Sigma \subseteq X$ for some $\Box A \in \alpha\}$. To show that $\mathcal{M}^+$ is a canonical minimal model for Σ it is enough to argue that for every sentence A and every world α in $\mathcal{M}^+$,

$$\Box A \in \alpha \text{ iff } |A|_\Sigma \in N^+_\alpha.$$

The result is trivial from left to right: if $\Box A \in \alpha$, then $|A|_\Sigma \in N_\alpha$, since $\mathcal{M}$ is canonical for Σ. But $N_\alpha \subseteq N_\alpha^+$; so $|A|_\Sigma \in N_\alpha^+$. For the reverse, suppose that $|A|_\Sigma \in N_\alpha^+$, so that for some $\Box B \in \alpha$, $|B|_\Sigma \subseteq |A|_\Sigma$. By theorem 2.22(2) this means that $\vdash_\Sigma B \rightarrow A$. Because Σ is monotonic we can infer by RM that $\vdash_\Sigma \Box B \rightarrow \Box A$, which means that $\Box A$ is Σ-deducible from α (since this set contains $\Box B$). Hence by theorem 2.18(1), $\Box A \in \alpha$, as we wished to prove.

It follows at once that any monotonic modal logic is complete with respect to the class of minimal models closed under supersets, i.e. the class of supplementations. For a sentence in this class is true in the model $\mathcal{M}^+$ and so is a theorem of any monotonic system. In particular, *M*, the smallest monotonic system, is complete with respect to the class of supplementations; since it is also sound with respect to this class (exercise 9.1(*b*)), we have the following theorem.

THEOREM 9.10. *M is determined by the class of supplementations.*

The next theorem provides the basis for the completeness of classical modal logics that contain the schema C. According to it, the smallest canonical minimal model for such a logic is closed under intersections.

THEOREM 9.11. *Let* $\mathcal{M} = \langle W, N, P \rangle$ *be the smallest canonical minimal model for a classical system containing* C. *Then for every* α, X, *and* Y *in* $\mathcal{M}$, *if* $X \in N_\alpha$ *and* $Y \in N_\alpha$, *then* $X \cap Y \in N_\alpha$.

Proof. Let Σ be a classical logic that contains the schema C, and let $\mathcal{M}$ be the smallest canonical minimal model for Σ, i.e. that in which, for every α, $N_\alpha = \{|A|_\Sigma\colon \Box A \in \alpha\}$. If we suppose that X and Y are propositions in N_α, then $X = |B|_\Sigma$ and $Y = |C|_\Sigma$, for sentences $\Box B$ and $\Box C$ in α. Now $\vdash_\Sigma (\Box B \wedge \Box C) \rightarrow \Box(B \wedge C)$, from which it follows that $\Box(B \wedge C)$ is Σ-deducible from α and so is in α. By the canonicity of $\mathcal{M}$, $|B \wedge C|_\Sigma$ belongs to N_α. But by theorem 2.22(7), $|B \wedge C|_\Sigma = |B|_\Sigma \cap |C|_\Sigma$. So $X \cap Y$ is in N_α, which is what we wished to show. Therefore, $\mathcal{M}$ is closed under intersections.

From theorem 9.11 it follows that classical logics in which C is a theorem are complete with respect to the class of minimal models closed under intersections. The smallest system of this sort, *EC*, is indeed determined by this class, since it is sound with respect to it (exercise 9.1). We are also in a position to see that every regular modal logic is complete with respect to the class of quasi-filters. Recall that the regular logics are just the monotonic systems containing C (theorem 8.11), and that a minimal model is said to be a quasi-filter just in case it is supplemented and closed under intersections (see the text before theorem 7.5). For the completeness result it is enough to note that by exercise 7.16(a) and theorem 9.11 the supplementation of the smallest canonical minimal model for a regular logic is closed under intersections. Thus a sentence is valid in the class of quasi-filters only if it is true in this canonical model and so is a theorem of any regular system. Because R (= EMC), the smallest regular logic, is sound with respect to the class of quasi-filters (exercise 9.1(c)), we arrive at the following theorem.

THEOREM 9.12. *R is determined by the class of quasi-filters.*

To complete our account we have to treat classical systems that contain N. The information needed here is that any canonical minimal model for such a logic contains the unit. Formally:

THEOREM 9.13. *Let $\mathcal{M} = \langle W, N, P\rangle$ be a canonical minimal model for a classical system containing* N. *Then $W \in N_\alpha$ for every α in $\mathcal{M}$.*

Proof. Let $\mathcal{M}$ be any canonical minimal model for a classical system Σ containing N. Then, since $\Box\top \in \alpha$ for every α in $\mathcal{M}$, $|\top|_\Sigma \in N_\alpha$ for every α in $\mathcal{M}$. But, as in any canonical model, $W = |\top|_\Sigma$. So $\mathcal{M}$ contains the unit.

Thus any classical logic containing N is complete with respect to the class of minimal models containing the unit, and we have determination results for the four remaining systems, EN, EMN, ECN, and $EMCN$. Let us focus on the last of these, otherwise known as K, the smallest normal system (the others are left as exercises). From theorems 9.9, 9.11, and 9.13 it follows that the supplementation of the smallest canonical minimal model for a normal modal logic is a filter, i.e. a quasi-filter containing the unit (see the text before theorem 7.5). So any normal system, K in particular, is complete with respect to the class of filters. Combining this with exercise 9.1(d), we have the following theorem.

THEOREM 9.14. *K is determined by the class of filters.*

This result can be improved. We pointed out in section 7.3 that any normal modal logic is complete with respect to the class of filters that are augmented, i.e. the class of minimal models $\mathcal{M} = \langle W, N, P\rangle$ such that for every α and X in $\mathcal{M}$,

$$X \in N_\alpha \text{ iff } \bigcap N_\alpha \subseteq X.$$

We can now prove directly that a normal system is complete with respect to the class of augmented models, simply by demonstrating the existence of an augmented canonical minimal model for the logic. This is the point of the next theorem.

THEOREM 9.15. *If $\mathcal{M} = \langle W, N, P\rangle$ is the smallest canonical minimal model for a normal system of modal logic, then its augmentation $\mathcal{M}^! = \langle W, N^!, P\rangle$ is also a canonical minimal model for the system.*

Proof. Let Σ be a normal system, and suppose $\mathcal{M}^!$ is the augmentation of $\mathcal{M}$, the smallest canonical minimal model for Σ. To prove that $\mathcal{M}^!$ is canonical for Σ we must show that for every α in $\mathcal{M}^!$,

$$\Box A \in \alpha \text{ iff } |A|_\Sigma \in N^!_\alpha.$$

By the definition of augmentation, 7.9, to say that $|A|_\Sigma \in N^!_\alpha$ means that $\bigcap N_\alpha \subseteq |A|_\Sigma$, which in turn means that

$$\bigcap\{|A|_\Sigma\colon \Box A \in \alpha\} \subseteq |A|_\Sigma,$$

since $\mathcal{M}$ is a smallest canonical minimal model. In other words, A is a member of every Σ-maximal set of sentences β in $\bigcap\{|A|_\Sigma\colon \Box A \in \alpha\}$. We leave it as an exercise for the reader to prove that for any β in $\mathcal{M}$,

$$\beta \in \bigcap\{|A|_\Sigma\colon \Box A \in \alpha\} \text{ iff } \{A\colon \Box A \in \alpha\} \subseteq \beta.$$

So what we wish to show is that $\Box A \in \alpha$ if and only if A belongs to every Σ-maximal set of sentences β such that $\{A\colon \Box A \in \alpha\} \subseteq \beta$. But this, once again, is theorem 4.30(1).

Thus we have another determination theorem for the smallest normal system.

THEOREM 9.16. *K is determined by the class of augmentations.*

EXERCISES

9.30. State and prove determination theorems for the systems *EC*, *EN*, *EMN*, and *ECN*.

9.31. In connection with the proof of theorem 9.15, show that for any β in $\mathcal{M}$,

$$\beta \in \bigcap\{|A|_\Sigma : \Box A \in \alpha\} \text{ iff } \{A : \Box A \in \alpha\} \subseteq \beta.$$

(This may have been proved already; see exercise 9.25.)

9.32. Using theorem 7.9, deduce theorem 5.12 from theorem 9.16, and vice versa.

9.33. Prove determination theorems for some classical systems containing the schema K, in particular, for the systems *EK* and *ECK*. Show that *EK* and *ECK* are distinct from each other and from the systems in figure 8.1. Finally, argue that *EK*, *ECK*, and the systems in figure 8.1 are all the classical systems that result from adding to *E* various combinations of the schemas M, C, N, R, and K as theorems. (Exercises 7.30 and 8.14 are relevant.)

9.34. Let $\mathcal{M} = \langle W, N, P \rangle$ be the smallest canonical minimal model for a classical system in which T is a theorem. Prove that $\mathcal{M}$ satisfies the condition (t) in section 7.4.

9.35. Let $\mathcal{M} = \langle W, N, P \rangle$ be the largest canonical minimal model for a classical system in which 5 is a theorem. Prove that $\mathcal{M}$ satisfies the condition (v) in section 7.4.

9.36. Using the results in exercises 9.8 and 9.27, prove that the system *M* is determined by the class of minimal models in which modalities are evaluated as in exercise 7.9.

9.37. Collecting results from exercises 9.12 and 9.28, prove determination theorems for the systems in figure 8.1 with respect to classes of models of the kind introduced in exercise 7.10.

9.38. Collecting results from exercises 8.33, 9.13, and 9.29, prove determination theorems for the systems in figure 8.1 with respect to classes of algebraic models.

9.4. The schemas D, T, B, 4, and 5

Let us deal briefly with classical systems containing D, T, B, 4, 5, and the like by proving determination theorems for the systems *ET*, *E5*, and *ET5*. These results are not meant to be exhaustive, but the proofs are illustrative of the problems involved in demonstrating the completeness of classical logics containing the five schemas. Further theorems of this kind are in the exercises.

We wish to consider minimal models $\mathcal{M} = \langle W, N, P\rangle$ in which, for every α and X in $\mathcal{M}$,

(t) if $X \in N_\alpha$, then $\alpha \in X$,

and

(v) if $X \notin N_\alpha$, then $\{\beta \text{ in } \mathcal{M}: X \notin N_\beta\} \in N_\alpha$.

In virtue of theorems 7.11 and 9.1, *ET*, *E5*, and *ET5* are sound with respect to classes of minimal models satisfying (t), (v), and (t) and (v) (see exercise 9.3). For completeness it suffices to show that these logics have canonical models that satisfy these conditions.

We may begin by observing that if $\mathcal{M} = \langle W, N, P\rangle$ is the smallest canonical minimal model for a classical system Σ containing the schema T, then $\mathcal{M}$ satisfies the condition (t). For if $X \in N_\alpha$, then $X = |A|_\Sigma$ for some $\Box A \in \alpha$. So $A \in \alpha$, since T is a theorem of Σ, and this means that $\alpha \in |A|_\Sigma = X$. Moreover, $\mathcal{M}$ continues to satisfy (t) if to any N_α we add any collection of non-proof sets X (in Σ) for which it holds that $\alpha \in X$. We state this formally.

THEOREM 9.17. *Let Σ be a classical system containing* T, *and let $\mathcal{M} = \langle W, N, P\rangle$ be any canonical minimal model for Σ in which for every α,*

$$N_\alpha = \{|A|_\Sigma : \Box A \in \alpha\} \cup \mathcal{X},$$

where $\mathcal{X}$ is any collection of non-proof sets X (in Σ) such that $\alpha \in X$. Then $\mathcal{M}$ satisfies (t).

It follows that every classical *ET*-system is complete with respect to the class of minimal models that satisfy (t). Thus we obtain our first desired result:

THEOREM 9.18. *ET is determined by the class of minimal models satisfying* (t).

Next, for *E5*, we note that the largest canonical minimal model $\mathcal{M} = \langle W, N, P\rangle$ for a classical logic Σ containing 5 satisfies (v). The argument for this is as follows. Suppose that $X \notin N_\alpha$. Then, since $\mathcal{M}$ is the largest canonical minimal model for Σ, X is the proof set of some sentence whose necessitation is not in α; i.e. $X = |A|_\Sigma$ for some $\Box A \notin \alpha$. By Df$\Box$ and theorem 2.18 (5, 9), $\Diamond\neg A \in \alpha$; So $\Box\Diamond\neg A \in \alpha$, since 5 is a theorem of Σ. By the definition of a canonical minimal model, 9.3, $|\Diamond\neg A|_\Sigma \in N_\alpha$. But this is what we wished to prove, since $|\Diamond\neg A|_\Sigma =$

$\{\beta \text{ in } \mathcal{M}: X \notin N_\beta\}$ (exercise 9.17). So $\mathcal{M}$ satisfies (v). Notice that $\mathcal{M}$ still meets condition (v) if from any N_α we subtract any collection of non-proof sets X (in Σ) for which it holds that $\{\beta \text{ in } \mathcal{M}: X \notin N_\beta\} \in N_\alpha$. Thus we can state the following theorem.

THEOREM 9.19. *Let Σ be a classical system containing* 5, *and let $\mathcal{M} = \langle W, N, P\rangle$ be any canonical minimal model for Σ in which for every α,*

$$N_\alpha = \{|A|_\Sigma : \Box A \in \alpha\} \cup \mathcal{X},$$

where $\mathcal{X}$ is any collection of non-proof sets X (in Σ) such that $\{\beta \text{ in } \mathcal{M}: X \notin N_\beta\} \in N_\alpha$. Then $\mathcal{M}$ satisfies (v).

Therefore, every classical *E5*-system is complete with respect to the class of minimal models for which (v) holds. Hence our second desired result:

THEOREM 9.20. *E5 is determined by the class of minimal models satisfying* (v).

Now for *ET5*. Let Σ be a classical system containing T and 5, and let $\mathcal{M} = \langle W, N, P\rangle$ be the smallest canonical minimal model for Σ. Then no N_α contains any non-proof set relative to Σ. By theorem 9.17 $\mathcal{M}$ satisfies (t). To see that the model satisfies (v), note that by theorem 9.19 it is enough to show that $\{\beta \text{ in } \mathcal{M}: X \notin N_\beta\} \in N_\alpha$ whenever X is not a proof set in Σ. But for each such X, $\{\beta \text{ in } \mathcal{M}: X \notin N_\beta\} = W$, and $W = |\top|_\Sigma$ (theorem 2.22(4)). By theorem 8.15 N is a theorem of Σ, which means that $\Box\top$ belongs to every α in $\mathcal{M}$. Hence by the definition of a canonical minimal model, $|\top|_\Sigma \in N_\alpha$ for each α in $\mathcal{M}$, which is what we wished to prove. Therefore, every classical *ET5*-system is complete with respect to the class of minimal models for which both (t) and (v) hold. So we achieve our last desired result:

THEOREM 9.21. *ET5 is determined by the class of minimal models satisfying* (t) *and* (v).

It often happens that in seeking an appropriate canonical minimal model for a classical system neither the smallest nor the largest will serve; the trick then is to find a model 'in between'. There are some examples of this phenomenon in the exercises.

EXERCISES

9.39. Prove determination theorems like 9.18, 9.20, and 9.21 for some classical, monotonic, and regular systems containing as theorems various combinations of the schemas D, T, B, 4, and 5, as well as such schemas as G, P, $\overline{\text{P}}$, $\text{G}^{k,l,m,n}$, $\text{G}_{\text{c}}^{k,l,m,n}$, and U. In particular, prove that the following systems are determined by classes of minimal models satisfying the indicated conditions.

(*a*) *ED*: (d)
(*b*) *EB*: (b)
(*c*) *E4*: (iv)
(*d*) *EG*: (g)
(*e*) *EP*: (p)
(*f*) $E\overline{P}$: ($\overline{\text{p}}$)
(*g*) $EG^{k,l,m,n}$: ($\text{g}^{k,l,m,n}$)
(*h*) $EG_{\text{c}}^{k,l,m,n}$: ($\text{g}_{\text{c}}^{k,l,m,n}$)
(*i*) *EU*: (u)

See section 7.4 and exercises 9.4–9.6 for the conditions.

9.40. Using determination theorems and results in exercise 7.49, prove the decidability of the systems *E*, *M*, *R*, and *K*. (See the discussion in section 2.8.)

9.41. Extend the results of exercise 9.37 by proving determination theorems for systems like those in exercise 9.39 (but include the schemas T and 5) with respect to classes of models of the kind defined in exercise 7.10. (Results in exercises 7.41 and 9.12 are useful.)

9.42. Extend the results of exercise 9.38 by proving determination theorems for systems like those in exercise 9.39 with respect to classes of algebraic models. In particular, prove such theorems for the smallest classical, monotonic, regular, and normal systems containing as theorems various combinations of the schemas D, T, B, 4, 5, G, P, $\text{G}^{k,l,m,n}$, and $\text{G}_{\text{c}}^{k,l,m,n}$ (see the conditions in exercise 8.42).

9.5. Decidability

We focus on the smallest classical, monotonic, regular, and normal logics – *E*, *M*, *R*, and (again) *K*. The reasoning involved in these cases can be adapted for the remaining four classical systems in figure 8.1, as well as others.

THEOREM 9.22. *E, M, R, and K are axiomatizable.*

Proof. Each can be axiomatized by a finite number of schemas together with the reasonable rules RPL and RE.

THEOREM 9.23. *E is determined by the class of finite minimal models.*

Proof. Soundness comes from theorem 9.8. For completeness, suppose A is not a theorem of *E*, so that by theorem 9.8 it is false in some minimal model $\mathcal{M}$. Let $\mathcal{M}^*$ be a filtration of $\mathcal{M}$ through the set of subsentences of A. Then $\mathcal{M}^*$ is a finite minimal model, and by theorem 7.14 A is false in $\mathcal{M}^*$.

THEOREM 9.24. *M is determined by the class of finite supplementations.*

Proof. Soundness follows from theorem 9.10. For completeness, let A be a non-theorem of *M*. By 9.10, again, A fails in some supplemented minimal model $\mathcal{M}$. Let $\mathcal{M}^{*+}$ be the supplementation of a finest Γ-filtration $\mathcal{M}^*$ of $\mathcal{M}$, where Γ is the set of subsentences of A. Then $\mathcal{M}^{*+}$ is a finite supplementation. By part (1) of theorem 7.16 this model is a Γ-filtration of $\mathcal{M}$, and so by theorem 7.14 it falsifies A.

THEOREM 9.25. *R is determined by the class of finite quasi-filters.*

Proof. The argument uses theorems 7.14, 7.16 (part (3)), and 9.12. Exercise.

THEOREM 9.26. *K is determined by the class of finite filters.*

Proof. Theorem 9.14 implies soundness. For completeness, suppose A is not a theorem of *K*. Then by 9.14 A is rejected by some filter $\mathcal{M}$. Let $\mathcal{M}^{*\pm}$ be the quasi-filtering of a finest Γ-filtration $\mathcal{M}^*$ of $\mathcal{M}$, where Γ is any finite set of sentences, closed under subsentences, that contains A and $\Box\top$. Then $\mathcal{M}^{*\pm}$ is a quasi-filter, and by theorem 7.17 it contains the unit. So it is a finite filter. By part (3) of theorem 7.16 the model is a Γ-filtration of $\mathcal{M}$. So by theorem 7.14 it too rejects A.

THEOREM 9.27. *E, M, R, and K have the finite model property.*

Proof. This is a corollary to theorems 9.23–9.26.

THEOREM 9.28. *E, M, R, and K are decidable.*

Proof. This is a corollary to theorems 9.22 and 9.27.

EXERCISES

9.43. Prove theorem 9.25.

9.44. Prove the decidability of the systems in figure 8.1 other than *E*, *M*, *R*, and *K* (see the results in exercise 7.49).

9.45. Prove the decidability of the systems *EK* and *ECK*.

9.46. Prove some decidability theorems for some classical, monotonic, and regular systems containing various combinations of D, T, B, 4, 5, G, P, $G^{k,l,m,n}$ and $G_c^{k,l,m,n}$ (for selected *k*, *l*, *m*, *n*), and others.

9.47. By a result of exercise 9.38, the system *E* is determined by the class of all algebraic models. Let us prove:

> *E* is determined by the class of finite algebraic models.

Soundness is immediate from the earlier determination result. For completeness, suppose that not $\vdash_E A$, so that by the determination result not $\models^{\mathcal{M}} A$ for some algebraic model $\mathcal{M}$. Where Γ is the set of subsentences of A, define $\mathcal{M}^\Gamma$ to meet conditions (1)–(3) in exercise 7.51. Then, by the filtration theorem in that exercise, not $\models^{\mathcal{M}^\Gamma} A$. But $\mathcal{M}^\Gamma$ is a finite algebraic model, so the theorem is proved.

Similarly, we can prove:

> *K* is determined by the class of finite algebraic models satisfying (m*), (c*), and (n*).

(See exercise 7.27 for these conditions.) We have that *K* is determined by the class of all algebraic models that satisfy (m*), (c*), and (n*) (exercise 9.38), so soundness is again trivial. The proof of completeness is like that for *E*, but more complicated, since $*^\Gamma$ must be defined so as to meet condition (2) in exercise 7.51 as well as (m*), (c*), and (n*). Let us indicate how this may be done, and leave the details of the argument for the reader.

For a point *a* in a finite boolean algebra let $M(a)$ be the boolean union of all the elements *b* in the algebra such that $b \leqslant a$. Note that $M(a)$

exists in the algebra since there are only finitely many points b such that $b \leqslant a$. Moreover,

$$M(a \cap b) = M(a) \cap M(b),$$

for every a and b in the algebra.

Now if Γ is the set of subsentences of a sentence A rejected by an algebraic model $\mathcal{M} = \langle \mathcal{B}, *, P \rangle$ satisfying (m*), (c*), and (n*), then any algebraic filtration $\mathcal{M}^{\Gamma} = \langle \mathcal{B}^{\Gamma}, *^{\Gamma}, P^{\Gamma} \rangle$ of $\mathcal{M}$ through Γ is finite. And if $*^{\Gamma}$ is defined by

$$*^{\Gamma}a = M(*a)$$

– for a in $\mathcal{M}^{\Gamma}$ – then conditions (2) and (m*), (c*), and (n*) are satisfied. So by the filtration theorem in exercise 7.51 A is false in $\mathcal{M}^{\Gamma}$, which completes the proof.

Together with facts about their axiomatizability, these finite determination theorems for E and K lead again to decidability results (as the reader should verify).

Using results in exercises 9.38 and 9.42, prove finite determination theorems with respect to classes of algebraic models for the remaining systems in figures 4.1 and 8.1, and others as well.

10

CONDITIONAL LOGIC

Conditionality affords a good example of a concept susceptible of analysis by means of the kinds of models and systems studied in this book. In section 10.1 we present the basic systems of conditional logic and the classes of models that determine them. In section 10.2 we return to the subject of deontic logic and define a minimal logic for conditional obligation. In section 10.3 we offer a definition of the conditional obligation operator in terms of simple obligation and non-deontic conditionality.

As with chapter 6, the purpose of this chapter is to illustrate the use of our semantic and deductive-theoretic techniques in the analysis of philosophically interesting concepts. Again, as in the earlier chapter, the reader will be the judge of the merit of the endeavor and the extent to which it is successful.

10.1. Conditionality

Into the language of propositional logic we introduce sentences of the form $A \Rightarrow B$. The operator $\Rightarrow$ is meant to express a notion of conditionality – a notion in general distinct from that expressed by $\rightarrow$.

In a *standard conditional model* $\mathcal{M} = \langle W, f, P \rangle$ for the language of conditional logic f is a mapping that selects a proposition (set of worlds) $f(\alpha, X)$ for each world α and proposition, or condition, X. Formally, then, f is a function from $W \times \mathcal{P}(W)$ to $\mathcal{P}(W)$.

To evaluate a conditional $A \Rightarrow B$ at a world α in a model $\mathcal{M} = \langle W, f, P \rangle$ we say:

$$\models_{\alpha}^{\mathcal{M}} A \Rightarrow B \text{ iff } f(\alpha, \|A\|^{\mathcal{M}}) \subseteq \|B\|^{\mathcal{M}}.$$

Thus a conditional is true at a world just in case the proposition expressed by the consequent is entailed by the proposition selected in terms of the world and the condition expressed by the antecedent. Equivalently, we may say that $A \Rightarrow B$ is true at a possible world if and only if B is true in every world selected in terms of the given world and the antecedent's condition.

Viewed in the latter way it is clear that we are construing conditionality as a form of relative necessity: we might revise $A \Rightarrow B$ to read [A]B, and so make graphic the necessity of what B expresses relative to the condition given by A.

This analysis of conditionality leads to the result that the logic of $\Rightarrow$ is, so to speak, classical with respect to the antecedent and normal with respect to the consequent. In other words, the following rules hold.

$$\text{RCEA.} \quad \frac{A \leftrightarrow A'}{(A \Rightarrow B) \leftrightarrow (A' \Rightarrow B)}$$

$$\text{RCK.} \quad \frac{(B_1 \wedge \ldots \wedge B_n) \rightarrow B}{((A \Rightarrow B_1) \wedge \ldots \wedge (A \Rightarrow B_n)) \rightarrow (A \Rightarrow B)} \quad (n \geqslant 0)$$

We call a system of conditional logic based on propositional logic and having these rules *normal*. The smallest normal conditional logic we call *CK*.

CK is determined by the class of all standard conditional models.

By way of example of extensions of *CK* let us consider normal conditional systems containing the following schemas.

I. $\quad A \Rightarrow A$

MP. $\quad (A \Rightarrow B) \rightarrow (A \rightarrow B)$

I expresses a law of identity, and MP is a principle of modus ponens for $\Rightarrow$. Neither schema is valid in the class of all standard conditional models, i.e. neither is a theorem of *CK*. Each is plausible for some notions of conditionality, though not for all.

The constraints on standard conditional models required for the validity of I and of MP are perhaps obvious:

(i) $\quad f(\alpha, X) \subseteq X$

(mp) if $\alpha \in X$, then $\alpha \in f(\alpha, X)$

Of course conditional logics weaker than *CK* can be obtained by eliminating RCK in favor of weaker rules of inference. Let us say that a conditional logic closed under RCEA is respectively *classical*, *monotonic*, or *regular* according as it is closed under the following three rules.

$$\text{RCEC.} \quad \frac{B \leftrightarrow B'}{(A \Rightarrow B) \leftrightarrow (A \Rightarrow B')}$$

$$\text{RCM.} \quad \frac{B \to B'}{(A \Rightarrow B) \to (A \Rightarrow B')}$$

$$\text{RCR.} \quad \frac{(B \wedge B') \to C}{((A \Rightarrow B) \wedge (A \Rightarrow B')) \to (A \Rightarrow C)}$$

Not unexpectedly, the smallest classical, monotonic, and regular conditional logics are called *CE*, *CM*, and *CR*.

To model such weaker conditional logics we employ *minimal* conditional models $\mathcal{M} = \langle W, f, P \rangle$ in which $f(\alpha, X)$ is a collection of propositions in $\mathcal{M}$, for each α and X in $\mathcal{M}$, and the truth conditions of a conditional $A \Rightarrow B$ are given as follows.

$$\models_{\alpha}^{\mathcal{M}} A \Rightarrow B \text{ iff } \|B\|^{\mathcal{M}} \in f(\alpha, \|A\|^{\mathcal{M}}).$$

Intuitively, $A \Rightarrow B$ is true at a world just in case the proposition expressed by B is among those picked out as necessary relative to the condition expressed by A at the world.

The system *CE* is determined by the class of all minimal conditional models. *CM* is determined by the class of such models for which the condition

(cm) if $Y \cap Y' \in f(\alpha, X)$, then $Y \in f(\alpha, X)$ and $Y' \in f(\alpha, X)$

holds. *CR* is determined by the class in which both (cm) and

(cc) if $Y \in f(\alpha, X)$ and $Y' \in f(\alpha, X)$, then $Y \cap Y' \in f(\alpha, X)$

hold. And *CK* is determined by the class of minimal models that satisfy (cm), (cc), and the following condition.

(cn) $W \in f(\alpha, X)$

EXERCISES

10.1. Prove that every normal conditional logic has the rules RCEC, RCM, RCR, and

$$\text{RCN.} \quad \frac{B}{A \Rightarrow B},$$

as well as the following theorems.

CN. $A \Rightarrow \top$

CM. $(A \Rightarrow (B \wedge B')) \to ((A \Rightarrow B) \wedge (A \Rightarrow B'))$

CC. $((A \Rightarrow B) \wedge (A \Rightarrow B')) \to (A \Rightarrow (B \wedge B'))$

CR. $(A \Rightarrow (B \wedge B')) \leftrightarrow ((A \Rightarrow B) \wedge (A \Rightarrow B'))$

CK. $(A \Rightarrow (B \rightarrow B')) \rightarrow ((A \Rightarrow B) \rightarrow (A \Rightarrow B'))$

10.2. Use RCEA and various combinations of theorems and rules in the preceding exercise to characterize normal conditional logics. (Compare theorem 4.3 and exercise 4.5.)

10.3. Describe standard conditional models falsifying the schemas I and MP (thus proving that neither is a theorem of *CK*). Then show that these schemas are valid in the classes of standard conditional models satisfying, respectively, (i) and (mp).

10.4. Prove that a normal conditional logic has I as a theorem if and only if it is closed under the rule

RI. $\dfrac{A \rightarrow B}{A \Rightarrow B}$

10.5. Describe a standard conditional model in which the following principle of augmentation is false.

AUG. $(A \Rightarrow B) \rightarrow ((A \wedge A') \Rightarrow B)$

10.6. Call a normal conditional logic SICK if it contains the schema I and a principle of syllogism:

S. $(A \Rightarrow B) \rightarrow ((B \Rightarrow C) \rightarrow (A \Rightarrow C))$

Prove that AUG in the preceding exercise is a theorem of every SICK system.

10.7. Let Σ be a conditional logic closed under RCEA. With reference to the schemas CM and CC in exercise 10.1, prove:

(*a*) Σ is monotonic iff it contains CM and is closed under RCEC.

(*b*) Σ is regular iff it contains CC and is closed under RCM.

(*c*) Σ is regular iff it contains CC and CM and is closed under RCEC.

(Compare theorem 8.11.)

10.8. Consider the class of models $\mathcal{M} = \langle W, R, P \rangle$ in which R is a ternary relation on W and truth conditions for conditionals are given by:

$\models_{\alpha}^{\mathcal{M}} A \Rightarrow B$ iff for every β and γ in $\mathcal{M}$ such that $R(\alpha, \beta, \gamma)$, if $\models_{\beta}^{\mathcal{M}} A$ then $\models_{\gamma}^{\mathcal{M}} B$.

Prove that the conditional logic determined by the class of all such models is normal and closed under the following rule of inference.

$$\frac{A \rightarrow (A_1 \vee \ldots \vee A_n)}{((A_1 \Rightarrow B) \wedge \ldots \wedge (A_n \Rightarrow B)) \rightarrow (A \Rightarrow B)} \quad (n \geqslant 0)$$

Also prove that a normal conditional logic is closed under this rule just in case it contains the following schemas.

$$\bot \Rightarrow A$$

$$((A \vee A') \Rightarrow B) \rightarrow ((A \Rightarrow B) \wedge (A' \Rightarrow B))$$

$$((A \Rightarrow B) \wedge (A' \Rightarrow B)) \rightarrow ((A \vee A') \Rightarrow B)$$

10.9. Prove that *CK* is determined by the class of all standard conditional models.

10.10. Prove determination theorems for normal conditional logics containing I and MP.

10.11. Define a notion of filtration appropriate to standard conditional models. Use it to prove that *CK* is determined by the class of finite standard conditional models. Conclude that *CK* is decidable.

10.12. What condition on standard conditional models makes $\Rightarrow$ have the logic of $\rightarrow$?

10.13. Prove determination theorems for *CE*, *CM*, *CR*, and *CK* relative to classes of minimal conditional models.

10.2. Conditional obligation

At the end of chapter 6 we defined minimal deontic logic, *D*, as the smallest system, based on propositional logic, closed under the rule of inference

$$\text{ROM.} \quad \frac{A \rightarrow B}{\bigcirc A \rightarrow \bigcirc B}$$

and containing the sentence

$$\text{OD.} \quad \neg \bigcirc \bot.$$

We see now that *D* can be described as the smallest monotonic *MP*$\square$-system for the operator $\bigcirc$.

D is determined by the class of supplemented minimal models $\mathcal{M} = \langle W, N, P \rangle$ that satisfy the condition

$$\text{(p)} \quad \emptyset \notin N_\alpha,$$

and in which truth conditions for $\bigcirc$A are given by:

$$\models_{\alpha}^{\mathcal{M}} \bigcirc A \text{ iff } \|A\|^{\mathcal{M}} \in N_{\alpha}$$

– or equivalently by:

$$\models_{\alpha}^{\mathcal{M}} \bigcirc A \text{ iff for some } X \in N_{\alpha},\ X \subseteq \|A\|^{\mathcal{M}}$$

(compare exercise 7.24). In other words, for each α in $\mathcal{M}$, N_{α} is a collection of non-null propositions that are standards of obligation relative to α, and $\bigcirc$A is true at α just in case the proposition expressed by A is entailed by one of these standards.

Note that none of the following schemas is a theorem of *D*.

OC. $(\bigcirc A \wedge \bigcirc B) \rightarrow \bigcirc(A \wedge B)$

ON. $\bigcirc\top$

OD*. $\neg(\bigcirc A \wedge \bigcirc\neg A)$

This means that there are possible worlds that have no standards of obligation, and that where standards do exist they may support conflicting obligations.

At the end of chapter 6 we also remarked the need for a logic of conditional obligation, for a logic of sentences of the form $\bigcirc(B/A)$. If we construct the logic of $\bigcirc(\ /\ )$ by analogy with that for $\bigcirc$, then we should adopt at least the following two rules of inference.

RCOEA. $$\frac{A \leftrightarrow A'}{\bigcirc(B/A) \leftrightarrow \bigcirc(B/A')}$$

RCOM. $$\frac{B \rightarrow B'}{\bigcirc(B/A) \rightarrow \bigcirc(B'/A)}$$

The import of RCOEA is that when A and A$'$ express the same proposition the one-place operators $\bigcirc(\ /A)$ and $\bigcirc(\ /A')$ are equivalent. This reflects our understanding that it is the conditions expressed by A and A$'$ – and not, for example, the sentences themselves – that determine what is obligatory. According to the second rule, RCOM, conditional obligation is closed under implication.

The question now is what to regard as the conditional analogue of OD. The simple schema

COD^{+}. $\neg\bigcirc(\bot/A)$

does not seem right. The significance of this is that nothing impossible is obligatory under any condition. But this rules out even what is impossible as a condition of an impossible obligation; i.e. COD^{+} yields

$\neg\bigcirc(\bot/\bot)$ as a theorem. It seems more reasonable to assume only that nothing impossible is obligatory under any possible condition. That is to say, the correct counterpart to OD is the weaker schema

$$\text{COD.}\quad \Diamond A \rightarrow \neg\bigcirc(\bot/A).$$

Here $\Diamond$ is to represent some suitable concept of possibility. For the sake of simplicity, let us suppose that $\Diamond$ has the logic of a normal *KT5*-system, so that $\Diamond A$ can be taken to mean that A is true at some possible world.

We may call the logic of $\bigcirc(\,/\,)$ defined in terms of propositional logic, RCOEA, RCOM, and COD *minimal conditional deontic logic*, or *CD*. It is the smallest monotonic conditional logic for $\bigcirc(\,/\,)$ containing the theorem COD.

CD is determined by the class of minimal conditional models $\mathcal{M} = \langle W, f, P\rangle$ that satisfy the condition (cm), in the preceding section, and the condition

$$\text{(cd)}\quad \text{if } X \neq \varnothing, \text{ then } \varnothing \notin f(\alpha, X)$$

and in which truth conditions for $\bigcirc(B/A)$ are given by:

$$\models_\alpha^{\mathcal{M}} \bigcirc(B/A) \text{ iff } \|B\|^{\mathcal{M}} \in f(\alpha, \|A\|^{\mathcal{M}})$$

– or equivalently by:

$$\models_\alpha^{\mathcal{M}} \bigcirc(B/A) \text{ iff for some } X \in f(\alpha, \|A\|^{\mathcal{M}}),\ X \subseteq \|B\|^{\mathcal{M}}.$$

(Because we assume $\Box$ and $\Diamond$ to obey the laws of *KT5* (*S5*), there is no need to introduce an alternativeness relation for them into the models; compare theorem 5.15.)

Given this determination result for *CD*, it is easy to see that the system has none of the conditional analogues of OC, ON, and OD*:

$$\text{COC.}\quad (\bigcirc(B/A) \wedge \bigcirc(B'/A)) \rightarrow \bigcirc(B \wedge B'/A)$$

$$\text{CON.}\quad \bigcirc(\top/A)$$

$$\text{COD*.}\quad \Diamond A \rightarrow \neg(\bigcirc(B/A) \wedge \bigcirc(\neg B/A))$$

Moreover, *CD* does not contain a principle of augmentation:

$$\text{OAUG.}\quad \bigcirc(B/A) \rightarrow \bigcirc(B/A \wedge A')$$

This is as it should be, since part of the point of conditional obligation is that obligations can differ under different conditions.

The foregoing account of conditional obligation takes the operator $\bigcirc(\,/\,)$ as primitive, and this leaves unresolved the connection between this concept and that expressed by the simple $\bigcirc$ – although we might

regard ○ as defined in terms of ○(/), for example by ○(/⊤). Another way to explain the connection between the operators is to define ○(/) in terms of ○ and a suitable non-deontic notion of conditionality. We turn to this idea in the next section.

EXERCISES

10.14. Explain why the system *D* is the smallest monotonic *MP*□-system for ○.

10.15. Prove the equivalence of the truth conditions given in section 10.2 for sentences of the form ○A.

10.16. Describe models for the system *D* in which OC, ON, and OD* are false.

10.17. Prove the equivalence of the truth conditions given in section 10.2 for sentences of the form ○(B/A).

10.18. Describe models for the system *CD* in which COC, CON, COD*, and OAUG are false.

10.19. Prove that if COC is a theorem of a monotonic *CD*-system then COD* is too.

10.20. Prove that *D* is determined by the class of supplemented minimal models for ○ that satisfy the condition (p) (compare exercise 9.39).

10.21. Sketch a proof that *CD* is determined by the class of minimal conditional models satisfying (cm) and (cd). Along the way, explain why □ and ◇ do not have to be represented in these models.

10.3. Conditional obligation defined

Let *CKD* be the smallest normal system for the conditional ⇒ in which the schema

CD. $\Diamond A \rightarrow \neg(A \Rightarrow \bot)$

is a theorem. *CKD* is determined by the class of standard conditional models $\mathcal{M} = \langle W, f, P\rangle$ that satisfy the following condition.

(cd) if $X \neq \varnothing$, then $\varnothing \notin f(\alpha, X)$

As we remarked in the preceding section, the system *D* for the operator

$\bigcirc$ is determined by the class of supplemented minimal models $\mathscr{M} = \langle W, N, P \rangle$ that satisfy the following condition.

$$\text{(p)} \quad \varnothing \notin N_\alpha$$

If we combine the systems D and CKD, the resulting logic – $D+CKD$ – is determined by the class of models $\mathscr{M} = \langle W, N, f, P \rangle$ in which N behaves as it does in the models for D and f behaves as it does in the models for CKD. (Again, as in the preceding section, we ignore $\Box$ and $\Diamond$.)

Now let us define the conditional obligation operator $\bigcirc(\ /\ )$ in terms of $\bigcirc$ and $\Rightarrow$, thus:

$$\bigcirc(B/A) = A \Rightarrow \bigcirc B.$$

Given this definition, the logic of $\bigcirc(\ /\ )$ is precisely CD, the minimal conditional deontic logic described in the preceding section.

If the models $\mathscr{M} = \langle W, N, f, P \rangle$ satisfy the further condition

$$\text{(def)} \quad f(\alpha, W) = \{\alpha\},$$

then the following schema becomes a theorem.

$$\text{DEF.} \quad A \leftrightarrow (\top \Rightarrow A)$$

This is a reasonable law for non-deontic conditionality, and it has the virtue of yielding

$$\text{ODEF.} \quad \bigcirc A \leftrightarrow \bigcirc(A/\top)$$

as a special case. We suggested earlier that if $\bigcirc(\ /\ )$ is taken as primitive $\bigcirc$ could plausibly be defined in this way (and if both deontic operators are taken as primitive one might expect ODEF to be a theorem).

EXERCISES

10.22. Describe models of the system $D+CKD$ that falsify COC, CON, COD*, and OAUG.

10.23. Derive ODEF from DEF, and prove that DEF is true in any model for $D+CKD$ that satisfies the condition (def).

10.24. Describe a model of $D+CKD$ in which I $(A \Rightarrow A)$ is true, but the schema

$$\text{OI.} \quad \bigcirc(A/A)$$

is false.

10.25. Prove that

$$\text{OMP.} \quad \bigcirc(\mathrm{B}/\mathrm{A}) \rightarrow (\mathrm{A} \rightarrow \bigcirc \mathrm{B})$$

is true in any model for *D*+*CKD* that satisfies the condition (mp) in section 10.1.

10.26. Let Σ be a system extending *D*+*CKD* that is closed under ROM, RCEA, and RCK. Prove:

(*a*) $\mathrm{A} \rightarrow \bigcirc \mathrm{A}$ is a theorem of Σ if the system contains OI and OMP (see exercises 10.24 and 10.25).

(*b*) Σ is inconsistent if it contains I ($\mathrm{A} \Rightarrow \mathrm{A}$) and COD+ ($\neg\bigcirc(\bot/\mathrm{A})$; see section 10.2).

10.27. Prove that principles of dilemma –

$$\text{DIL.} \quad ((\mathrm{A} \Rightarrow \mathrm{B}) \wedge (\mathrm{A}' \Rightarrow \mathrm{B})) \rightarrow ((\mathrm{A} \vee \mathrm{A}') \Rightarrow \mathrm{B})$$

$$\text{ODIL.} \quad (\bigcirc(\mathrm{B}/\mathrm{A}) \wedge \bigcirc(\mathrm{B}/\mathrm{A}')) \rightarrow \bigcirc(\mathrm{B}/\mathrm{A} \vee \mathrm{A}')$$

– are true in any models for *D*+*CKD* that satisfy the following condition.

$$\text{(dil)} \; f(\alpha, X \cup Y) \subseteq f(\alpha, X) \cup f(\alpha, Y)$$

What is the situation in models for which the converse of (dil) holds?

10.28. Prove that *D*+*CKD* is determined by the class of models $\mathcal{M} = \langle W, N, f, P \rangle$ satisfying the conditions stated in section 10.3.

SELECT BIBLIOGRAPHY

Boolos, George S. *The unprovability of consistency: an essay in modal logic.* Cambridge University Press, 1978.

Boolos, George S. and Jeffrey, Richard C. *Computability and logic.* Cambridge University Press, 1974.

Chellas, Brian F. 'Imperatives'. *Theoria* 37 (1971), 114–29.

– 'Conditional obligation'. In *Logical theory and semantic analysis: essays dedicated to Stig Kanger on his fiftieth birthday* (edited by Sören Stenlund), pp. 23–33. Dordrecht: D. Reidel Publishing Company, 1974.

– 'Basic conditional logic'. *Journal of philosophical logic* 4 (1975), 133–53.

Chellas, Brian F. and McKinney, Audrey. 'The completeness of monotonic modal logics'. *Zeitschrift für mathematische Logik und Grundlagen der Mathematik* 21 (1975), 379–83.

Feys, Robert. *Modal logics* (edited with some complements by Joseph Dopp). Louvain: E. Nauwelaerts, 1965.

Fine, Kit. 'An incomplete logic containing S4'. *Theoria* 60 (1974), 23–29.

Gabbay, Dov M. 'Montague type semantics for non-classical logics'. U.S. Air Force Office of Scientific Research contract no. F 61052-68-C-0036, Scientific Report no. 4, October 1969.

– *Investigations in modal and tense logics with applications to problems in philosophy and linguistics.* Dordrecht: D. Reidel Publishing Company, 1976.

Gerson, Martin. 'The inadequacy of the neighbourhood semantics for modal logic'. *Journal of symbolic logic* 40 (1975), 141–48.

Hansson, Bengt and Gärdenfors, Peter. 'A guide to intensional semantics'. In *Modality, morality and other problems of sense and nonsense: essays dedicated to Sören Halldén*, pp. 151–67. Lund: CWK Gleerup Bokförlag, 1973.

– 'Filtrations and the finite frame property in boolean semantics'. In *Proceedings of the Third Scandinavian Logic Symposium* (edited by Stig Kanger), pp. 32–39. Amsterdam: North-Holland Publishing Company, 1975.

Hilpinen, Risto (ed.). *Deontic logic: introductory and systematic readings.* Dordrecht: D. Reidel Publishing Company, 1971.

Hughes, G. E. and Cresswell, M. J. *An introduction to modal logic.* London: Methuen and Co. Ltd., 1968. (Reprinted with corrections 1972.)

Kaplan, David. Review of 'Semantical analysis of modal logic I. Normal modal propositional calculi', by Saul A. Kripke. *Journal of symbolic logic* 31 (1966), 120–22.

Kripke, Saul A. 'Semantical analysis of modal logic I. Normal modal propositional calculi'. *Zeitschrift für mathematische Logik und Grundlagen der Mathematik* 9 (1963), 67–96.

Lemmon, E. J. (in collaboration with Dana S. Scott). *The 'Lemmon notes': an introduction to modal logic* (edited by Krister Segerberg). *American philosophical quarterly* Monograph Series (edited by Nicholas Rescher), no. 11. Oxford: Basil Blackwell, 1977.

Lewis, Clarence Irving. *A survey of symbolic logic*. Berkeley and Los Angeles: University of California Press, 1918.

Lewis, Clarence Irving and Langford, Cooper Harold. *Symbolic logic* (2nd edn). New York: Dover Publications, Inc., 1959.

Lewis, David K. *Counterfactuals*. Cambridge, Mass.: Harvard University Press, 1973.

Prior, Arthur N. *Time and modality*. Oxford: Clarendon Press, 1957.

– *Formal logic* (2nd edn). Oxford: Clarendon Press, 1962.

– *Past, present and future*. Oxford: Clarendon Press, 1967.

Sahlqvist, Henrik. 'A note on irreflexive, asymmetric and intransitive Kripke models'. Preprint Series, Institute of Mathematics, University of Oslo, no. 11, April 1972.

– 'Completeness and correspondence in the first and second order semantics for modal logic'. Cand. real. thesis, University of Oslo, 1973.

Segerberg, Krister. 'Decidability of S4.1'. *Theoria* 34 (1968), 7–20.

– 'Decidability of four modal logics'. *Theoria* 34 (1968), 21–25.

– *An essay in classical modal logic*. 3 vols. Uppsala: University of Uppsala, 1971.

Stalnaker, Robert C. 'A theory of conditionals'. In *Studies in logical theory* (edited by Nicholas Rescher), pp. 98–112. *American philosophical quarterly* Monograph Series (edited by Nicholas Rescher), no. 2. Oxford: Basil Blackwell, 1968.

Stoll, Robert R. *Set theory and logic*. San Francisco and London: W. H. Freeman and Company, 1963.

Thomason, S. K. 'An incompleteness theorem in modal logic'. *Theoria* 60 (1974), 30–34.

von Wright, Georg Henrik. *An essay in modal logic*. Amsterdam: North-Holland Publishing Company, 1951.

INDEX OF SYMBOLS

INDEX OF SCHEMAS, RULES, AND SYSTEMS

Schemas

Rules

INDEX OF SUBJECTS